Plasma Surface Modification of Polymers:
Relevance to Adhesion

PLASMA SURFACE MODIFICATION OF POLYMERS: RELEVANCE TO ADHESION

Editors: M. Strobel, C.S. Lyons and K.L. Mittal

CRC Press
Taylor & Francis Group
Boca Raton London New York

CRC Press is an imprint of the
Taylor & Francis Group, an **informa** business

First published 1994 by VSP Publishing

Published 2018 by CRC Press
Taylor & Francis Group
6000 Broken Sound Parkway NW, Suite 300
Boca Raton, FL 33487-2742

© 1994 by Taylor & Francis Group, LLC
CRC Press is an imprint of Taylor & Francis Group, an Informa business

First issued in paperback 2019

No claim to original U.S. Government works

ISBN 13: 978-0-367-44946-9 (pbk)
ISBN 13: 978-90-6764-164-7 (hbk)

Visit the Taylor & Francis Web site at
http://www.taylorandfrancis.com

and the CRC Press Web site at
http://www.crcpress.com

CIP-DATA KONINKLIJKE BIBLIOTHEEK, DEN HAAG

Plasma

Plasma surface modification of polymers: relevance to
adhesion / [ed.] M. Strobel, C.S. Lyons, K.L. Mittal. -
Utrecht : VSP
Originally publ. in: Journal of adhesion science and
technology = ISSN 0169-4243; vol. 7, no. 10 (1993),
vol. 8, no. 4 (1994). - With ref.
ISBN 90-6764-164-2 bound
NUGI 813
Subject headings: plasma technology / polymers.

Contents

3. Practical Applications of Plasma-treated Surfaces

Plasma Surface Modification of Polymers, pp. vii–viii
M. Strobel, C. Lyons and K. L. Mittal (Eds)
© VSP 1994

Preface

This book collects into one volume the papers previously published as two special issues of the *Journal of Adhesion Science and Technology* (JAST): Vol. 7, No. 10 (1993) and Vol. 8, No. 4 (1994). The topic of those special issues, and of this book, is the plasma surface modification of polymers. Because numerous individuals expressed an interest in acquiring these two special issues separately, the editors and the publisher decided to produce this book.

This book is a collection of invited papers written by internationally recognized researchers actively working in the field of plasma surface modification. We believe that this book provides a current, comprehensive overview of the plasma treatment of polymers that should be a valuable resource for the scientific community for many years.

The focus of this book is the plasma or glow-discharge, which can be defined as a partially ionized gas normally generated by an electrical discharge at near-ambient temperatures and reduced pressures. It is important to distinguish the topic of this book, plasma surface modification, from the related plasma process, plasma polymerization. Plasma polymerization refers to the deposition of solid polymeric materials from a plasma. Gases or vapors that polymerize in a plasma will alter the surface properties of both organic and inorganic substrates by the deposition of a thin polymer film. By contrast, plasma surface modification reactions do not cause thin-film deposition, and therefore can only modify the surface properties of organic substrates. This modification occurs by altering the chemistry of the outermost few molecular layers of the polymer.

It is interesting to note that although studies of plasma polymerization have tended to dominate the scientific literature, plasma surface modification processes account for most of the commercial use of plasma technology. There are a number of reasons for the growing use of plasma surface modification in industry. As compared with plasma polymerization, plasma surface modification is far easier to implement on an industrial scale and is a more cost-effective process. Perhaps even more importantly, plasma surface modifications are fast, efficient methods for improving the adhesion properties and other surface characteristics of a variety of polymeric materials. Also, plasma treatments can effectively modify polymer surfaces at exposure times of less than one second. This combination of desirable process characteristics is leading to a rapid expansion in the commercial use of plasma surface modification, particularly in the biological, automotive, aerospace, packaging, and electronics industries. We believe that this book is particularly opportune now, during this growth in interest in plasma surface modification.

The focus of this volume is on adhesion phenomens, surface properties, and the surface characterization of plasma-treated materials. In this book, the papers have been rearranged in a more logical sequence relative to the order in which they were published in the special issues of the JAST. This book opens with a critical review of the plasma surface modification of polymers for improved adhesion. In this work, Ted Liston, Ludvik Martinu, and Mike Wertheimer have crafted an insightful and thorough review of plasma treatment. This review should be of great interest not only to novices in plasma technology, but also to experienced researchers and those charged with implementing plasma technology on an industrial scale. The opening paper is a veritable 'critical' review in the sense that it interprets the field in light of several unifying concepts.

We have divided the remainder of the papers in this volume into two sections, one dealing with the characterization of plasma-treated surfaces and the second concerned with various practical applications of plasma-treated surfaces. These papers were selected to provide a broad portrait of the many activities currently underway throughout the world.

We hope that this book will be well received by plasma researchers and adhesion scientists. We thank the many authors who contributed to this volume. Their hard work and fine research has produced an excellent compilation of plasma surface-treatment technology.

Mark Strobel
Christopher S. Lyons
Corporate Research Process Technology Laboratory
3M Company
3M Center, Buiding 208-1-01
St. Paul, MN 55144, USA
and
K. L. Mittal
Editor, Journal of Adhesion Science and Technology

1
Technology Review

Plasma Surface Modification of Polymers, pp. 3–39
M. Strobel, C. Lyons and K. L. Mittal (Eds)
© VSP 1994

Plasma surface modification of polymers for improved adhesion: a critical review

E. M. LISTON,* L. MARTINU and M. R. WERTHEIMER†

Groupe des Couches Minces, and Department of Engineering Physics, École Polytechnique, Box 6079, Station 'A', Montreal, Québec H3C 3A7, Canada

Revised version received 26 May 1993

Abstract—Since the earliest systematic research during the 1960s, the field of materials surface modification by 'cold', low-pressure plasma treatment has undergone an enormous expansion. Much of this expansion has taken place in recent years, particularly in the surface modification of polymeric materials, for which there now exist numerous industrial applications (enhancement of paint adhesion, improved bonding in polymer matrix composites, etc.). In this paper, we provide a critical review of the development and trends in this field; reference is also made to other surface modification techniques, particularly to corona treatment, and comparisons are made wherever appropriate. We begin with a brief overview of adhesion theory, and of the physics and chemistry of 'cold' plasmas. Next, interaction mechanisms between a plasma and a polymer surface are examined; these include physical bombardment by energetic particles and by ultraviolet photons, and chemical reactions at or near the surface. The resulting four main effects, namely cleaning, ablation, crosslinking, and surface chemical modification, occur together in a complex synergy, which depends on many parameters controlled by the operator. In spite of this complexity, for there are still many unanswered questions, it is nevertheless possible to optimize the main set of parameters governing a given process, and then to reliably reproduce the process outcome. Three industrially important systems, for which many research results exist, are then separately examined, namely: (i) polymer–polymer bonding, (ii) polymer–matrix composites, and (iii) metal–polymer bonding. Finally, we present a brief overview of commercial plasma reactors for industrial (non-semiconductor) purposes, and of process considerations for efficient use of such equipment. We foresee that the use of plasma processes will continue to expand, because they have unique capabilities, are economically attractive, and are 'friendly' towards the environment.

Keywords: Low-pressure plasma; polymers; surface modification; adhesion; polymer–matrix composites.

1. INTRODUCTION

1.1. Background

In recent years, we have witnessed a remarkable growth in the use of synthetic organic polymers in technology, both for high-tech and for consumer-product applications. Polymers have been able to replace more traditional engineering materials such as metals, on account of their many desirable physical and chemical charateristics (high strength-to-weight ratio, resistance to corrosion, etc.) and their relativc'y low cost. However, fundamental differences between polymers and other engineering solids have also created numerous important technical challenges, which manufacturing operations must overcome. An

---

*Consultant, 42 Peninsula Road, Belvedere, CA 94920, USA.
†To whom correspondence should be addressed.

important example is the characteristic low surface energy of polymers and their resulting intrinsically poor adhesion [1–4]; the term 'adhesion', as it is used here and in the following text, may be briefly defined as the mechanical resistance to separation of a system of bonded materials [5]. Since adhesion is fundamentally a surface property, often governed by a layer of molecular dimensions, it is possible to modify this near-surface region without affecting the desirable bulk properties of the material.

Over the years, several methods have been developed to modify polymer surfaces for improved adhesion, wettability, printability, dye uptake, etc. These include mechanical treatments, wet-chemical treatments, exposure to flames, corona discharges, and glow discharge plasmas. A basic objective of any such treatment is to remove loosely bonded surface contamination and to provide intimate contact between the two interacting materials on a molecular scale, for molecular energies across an interface decrease drastically with increasing intermolecular distance [1]. The simplest method that one can envisage for improving adhesion is to mechanically roughen a surface, thereby enhancing the total contact area, and mechanical interlocking, which is one of several basic mechanisms that have been proposed to explain adhesion. Theoretical adhesion models have been proposed by various authors to account for a wide range of related experimental observations (for reviews of these theories, see, for example, refs 6 and 7); they are, very briefly:

(i)   the adsorption or chemical reaction theory, which states that bond strength is mainly determined by physi- or chemisorption at the interface;

(ii) the electrostatic theory of Deryagin [8], which is based on contact charging when two dissimilar materials are intimately joined;

(iii) the diffusion theory of Voyutskii [9], which claims that the bond strength of polymers is governed by diffusion across their interface;

(iv) the rheological theory of Bikerman [10], which states that the performance of a bonded system is governed by the mechanical properties of the materials comprising the joint, and by local stresses in the joint; and

(v)   the mechanical interlocking or 'hooking' theory, based on the micro-geometry of the interface, as already mentioned above.

As discussed in some detail by Sharpe [7], each of these 'theories' has certain merits, but also many weaknesses; at present, none of them taken alone can adequately account for any large subset of all of the experimental observations relating to the bond strengths of joined materials. We feel, however, that there is a growing body of evidence that chemical reactions at the interface [mechanism (i) above] can play a key role in many cases, and we illustrate this with many examples in the remainder of this review.

A concept which has been gaining much support among adhesion scientists in recent years is the existence of an 'interphase', loosely defined as a region intermediate to two contacting solids, which is distinct in structure and properties from either of the two contacting phases. Sharpe [7] argues very convincingly that interphases exist in many macro-systems such as adhesive joints, coating/substrate systems, and fibre- or particulate-reinforced composites; that they may control the overall mechanical behaviour of these systems; and that failure to take them into account will likely lead to flawed models. We are in general agreement with these statements, and we present numerous examples of 'interphase' effects in the course of this review.

Returning to surface treatments for adhesion enhancement, mechanical roughening alone has limited effectiveness; wet-chemical treatments with solvents, strong acids or bases, or the sodium/liquid ammonia treatment for fluoropolymers [11], are becoming increasingly unacceptable because of environmental and safety considerations. Furthermore, wet-chemical treatments tend to have inherent problems of uniformity and reproducibility, criticisms which are also often levelled against flame treatment. Modification of polymer surfaces by plasma* treatment, both corona and low-pressure glow discharges, presents many important advantages and overcomes the drawbacks of the other processes mentioned above—reasons why these plasma processes have been gaining wide acceptance over the years in diverse industrial applications. Corona treatment is the longest established and most widely used plasma process [12]; it has the advantage of operating at atmospheric pressure, the reagent gas usually being ambient air. However, this circumstance is also an important limitation in that chemical effects other than surface oxidation, achievable by using reagent gases other than air, tend to be uneconomical or hazardous, or both. These restrictions are largely absent in the case of low-pressure glow discharge treatments, the plasma type that we will address almost exclusively in the remainder of this text. We have recently compared corona and low-pressure plasma surface treatments [13], and have reviewed polymer surface modification by glow discharge plasma [14, 15], as other authors have done [16]. The purpose of the present paper is to provide a critical, up-to-date overview of plasma surface modification science and technology for polymers from the viewpoint of adhesion, wettability, and surface chemistry. Considering the very extensive literature which now exists in this field, we have not attempted to list all relevant publications, but a selection which represents the particular viewpoints advanced in this paper.

1.2. Low-pressure plasma processes

1.2.1. General comments. The industrial use of plasma processing has been spearheaded by the microelectronics industry since the late 1960s (a) for the deposition of thin film materials [17], and (b) for plasma etching of semi-conductors, metals, and polymers such as organic photoresists [18–20]. Several dozen plasma equipment manufacturers in North America, Europe, and the Far East cater to this market, and their combined annual sales exceed US$ 10^9; they have recently been joined by other equipment builders who specialize in 'industrial' (non-semiconductor oriented) plasma applications.

Contrary to processes (a) and (b) above, where material is added to or removed from the surface, respectively, the third type of plasma process—surface modification (of particular interest in the context of this review)—does neither of these in significant amounts. Instead, the composition and structure of a few molecular layers at or near the surface of the material are changed by the plasma.

---

*A plasma may be succinctly defined as a partially ionized gas, with equal number densities of positive and negative charge carriers, in which the charged particles are 'free' and possess collective behaviour.

Largely thanks to this third process category, we are currently witnessing a vigorous expansion of industrial plasma processes into areas other than micro-electronics, namely into the automotive, aerospace, and packaging sectors, to name but a few examples.

In spite of this proliferation of applications, there are still many unresolved questions regarding the most efficient use of plasma processing, largely due to the inherent complexity of the plasma state. In order to ensure the high quality and the reproducibility of a given plasma process, numerous parameters must be controlled with care, such as the pressure and flow rate of the reagent gas or gas mixture, the discharge power density, the surface temperature and electrical potential of the workpiece, etc. Currently, the effects of excitation frequency and of plasma–surface interactions are still only partially understood, and these are the objects of much ongoing research. These are discussed briefly in the remainder of this section, along with low-pressure plasma principles.

1.2.2. The physics and chemistry of low-pressure plasmas. In a low-pressure (≤ 1 Torr $= 133$ Pa), high-frequency (≥ 1 MHz) discharge, the heavy particles (gas molecules and ions) are essentially at ambient temperature (~ 0.025 eV), while the electrons have enough kinetic energy (several eV) to break covalent bonds, and even to cause further ionization (that is, to sustain the discharge). The chemically reactive species thus created can partake in homogeneous (gas-phase) or heterogeneous reactions with a solid surface in contact with the plasma. Since this type of plasma chemistry takes place at near-ambient temperature, it is well suited for processing thermally sensitive materials such as semiconductors and polymers [21, 22].

As mentioned above, the creation of reactive species (radicals, ions, molecular excited states, etc.) in a plasma results primarily from inelastic collisions between 'hot' electrons with energy $u = (mw^2)/2e$ (w, m, and e are the electron velocity, mass, and charge, respectively) and ground-state atoms or molecules. The rate coefficient C_j for excitation of a particular species or state 'j' is given by

$$C_j = \left[\frac{2e}{m}\right]^{1/2} \int_0^\infty \sigma_j(u) F_0(u) u \; du, \tag{1}$$

where $\sigma_j(u)$ is the particular process cross-section and $F_0(u)$ is the electron energy distribution function (EEDF). It is noteworthy that nearly all processes of importance here possess an energy threshold u_j for which

$$\sigma_j(u) \equiv 0, \quad u \leq u_j. \tag{2}$$

The number density $\dot{n}_j$ of species 'j' produced per second in the plasma from ground-state molecules (of number density N cm^{-3}) clearly also depends on the electron density n; that is,

$$\dot{n}_j = C_j N n. \tag{3}$$

The power balance between the applied electromagnetic field (of frequency $f = \omega/2\pi$) and the plasma can be expressed by

$$P_a = \xi n V, \tag{4}$$

where P_a is the power absorbed in the volume V of plasma and ξ is the average power absorbed per electron. The parameter ξ, which is readily measurable, can also be considered as the power required to sustain an electron-ion pair in the plasma. Ferreira and Loureiro [23] have calculated ξ values for the 'simple' case of a low-pressure argon plasma; their model shows that at constant pressure, ξ decreases as f is raised from 'low frequency' ($\leqslant 100$ MHz) to microwave (MW, > 100 MHz) frequencies. In other words, the efficiency of producing electron-ion pairs is greater at MW than at 'low' frequency, for a given power density absorbed in the plasma. Indeed, the yields of other types of chemically reactive particles such as free radicals, which at 20% or more of all species are the majority constituent, are also found to be higher in MW plasmas. As recently reviewed by Moisan *et al.* [24], this is attributed to a significantly higher fraction of energetic electrons in the tail of the EEDF and to a higher n value than for lower-frequency plasmas. It is therefore not surprising that among recent generations of commercial plasma reactors, an increasing number of these systems operate at the 2.45 GHz MW frequency [20, 22] (see also Section 4).

Rather than use either MW or RF (radiofrequency) power to sustain the plasma, we (LM, MRW) often combine the two power sources to generate a so-called 'mixed' (or dual-) frequency plasma [25]. While MW excitation generates a high concentration of active species in the gas phase, as noted above, the role of the RF power is to create a negative DC self-bias voltage, V_B, on the powered, electrically isolated substrate-holder. This causes ions to be accelerated by the potential drop ($V_P - V_B$) across the RF-induced plasma sheath, or 'dark space', to their maximum kinetic energy:

$$E_{i,\max} = e\,|\,V_P - V_B\,|, \tag{5}$$

where V_P, the plasma potential, is generally a few tens of volts positive with respect to ground. In practical situations, at higher gas pressures of ~ 100 mTorr, the ions lose part of their energy through inelastic collisions. The average energy is then typically

$$\bar{E}_i \simeq 0.4 E_{i,\max}. \tag{6}$$

In other words, in dual-frequency (MW–RF) processing, independent control of the RF power allows us to vary the energy of the ions bombarding the substrate surface, with values ranging from a few eV to several hundreds of eV, and with fluxes of up to $\sim 10^{16}$ ions/cm^2 s [25], conditions comparable to the operating parameters of low-energy ion beam systems. The commercial plasma equipment (whether RF or MW) described in Section 4 can, in principle, be readily modified for use in the dual frequency (MW–RF) mode of operation.

2. INTERACTION OF PLASMAS WITH ORGANIC SURFACES

2.1. Types of plasma–polymer interactions

2.1.1. General discussion. In the plasma treatment of polymers, the subject on which we henceforth focus our attention, energetic particles and photons generated in the plasma interact strongly with the polymer surface, usually via free radical chemistry [15, 19, 21, 22]. For lack of space, we choose not to discuss a related field, plasma polymerization, for which the reader is referred to an excellent recent monograph [19].

In plasmas which do not give rise to thin film deposition (see Section 1.2.1), four major effects on surfaces are normally observed. Each is always present to some degree, but one may be favoured over the others, depending on the substrate and the gas chemistry, the reactor design, and the operating parameters. The four major effects are:

(i) surface cleaning, that is, removal of organic contamination from the surfaces;
(ii) ablation, or etching, of material from the surface, which can remove a weak boundary layer and increase the surface area;
(iii) crosslinking or branching of near-surface molecules, which can cohesively strengthen the surface layer; and
(iv) modification of surface-chemical structure, which can occur during plasma treatment itself, and upon re-exposure of the treated part to air, at which time residual free radicals can react with atmospheric oxygen or water vapour.

All these processes, alone or in synergistic combination, affect adhesion, as will be discussed in detail in Section 3.

We now briefly examine each of these four effects, frequently using an oxygen plasma as an example. The reason for this choice is not only that oxygen plasmas lead to strong effects in all four categories, but that they also have much technological importance in semiconductor and 'industrial' (non-semiconductor) processing.

(i) Cleaning. This is one of the major reasons for improved bonding to plasma-treated surfaces. Most other cleaning procedures leave a layer of organic contamination that interferes with adhesion processes; of course, any clean surface rapidly reacquires a layer of contamination when exposed to ambient atmosphere. For example, it is known that as little as $0.1 \ \mu g/cm^2$ (a single molecular layer) of organic contamination on a surface can interfere with bonding if the bonding material cannot dissolve or remove the contaminant from the surface to be bonded [26]. This amount of contamination is the residue from $0.01 \ cm^3/cm^2$ of a liquid containing 10 ppm of non-volatiles. It is difficult and expensive to obtain solvents or water, in industrial quantities, with less than 10 ppm non-volatiles, so that, almost by definition, a surface will remain contaminated after any cleaning process that finishes with a liquid rinse.

In some cases, it has been reported that contamination on the surface to be bonded did not interfere with bonding. It was concluded that the interaction of the substrate and the adhesive was sufficiently strong to cause displacement of the surface contamination by the adhesive, or that some incorporation mechanism existed between the contamination and the adhesive.

Oxygen-containing plasmas are capable of removing organic contamination from inorganic and polymeric surfaces, but it is critically important to plasma-clean a polymer for a sufficiently long time to remove all of the surface contamination. Almost all commercial polymer films, and most moulded parts, contain additives or contaminants such as oligomers, anti-oxidants, mould release agents, solvents, or anti-block agents, which are oily or wax-like. Most of these are deliberately incorporated into the polymer formulation to improve its properties or manufacturability, and they are designed to 'bloom' to the surface and to coat that surface.

Because these materials often have a very similar chemistry to that of the base polymer, they are generally difficult to detect with X-ray photoeletron spectroscopy (XPS) or other analytical techniques. Typically, these additives can be present in layers 1–10 nm thick, even after solvent cleaning; they simply continue to diffuse to the surface.

The surface contamination will react with the plasma in a similar way to the polymer. That is, if the plasma cleaning is not of sufficient duration to completely remove it, the contaminant will become wettable and will have a modified XPS pattern similar to that of the polymer. However, it will still remain a loosely-bonded, albeit plasma-treated, contaminant layer, not a plasma-modified polymer surface. At normal power levels (typically, a few mW/cm^2), it is necessary to clean most polymers for some tens of seconds. A treatment of a few seconds is generally not long enough (unless very high power densities, many tens of mW/cm^2, are used) to remove the contaminants, only to treat them and give the false impression of a wettable and properly treated surface.

(ii) Ablation. Ablation, or plasma etching, is distinguished from cleaning only by the amount of material that is removed. Ablation is important for the cleaning of badly contaminated surfaces, for the removal of weak boundary layers formed during the fabrication of a part, and for the treatment of filled or semi-crystalline materials. Since amorphous polymer is removed many times faster than either its crystalline counterpart or inorganic filler material, a surface topography can be generated, with the amorphous zones appearing as valleys. For example, plasma surface treatment of fluoropolymers for short times improves their wettability without modifying their surface texture, but overtreatment gives a very porous surface [27, 28]. The same is true for polyethylene terephthalate (PET) [29]. This change in surface morphology can improve mechanical interlocking and it can increase the area available for chemical interactions (see Section 1.1). Some ablation of reinforcing fibres tends to improve composite properties, but the fibres must not be significantly reduced in diameter by overtreatment because thinner fibres will be weaker [30–33] (see Section 3.3).

(iii) Crosslinking. CASING (Crosslinking via Activated Species of INert Gases) was one of the earliest-recognized plasma treatment effects on polymer surfaces [34]. As suggested by the acronym, CASING occurs in polymer surfaces exposed to noble gas plasmas (e.g. He or Ar), which are effective at creating free radicals but do not add new chemical functionalities from the gas phase. Ion bombardment or vacuum ultraviolet (VUV) photons* can break C—C or C—H bonds, and the free radicals resulting under these conditions can only react with other surface radicals or with other chains in chain-transfer reactions; therefore, they tend to be very stable [36]. If the polymer chain is flexible, or if the radical can migrate along it, this can give rise to recombination, unsaturation, branching, or crosslinking. The latter may improve the heat resistance and bond strength of the surface by forming a very cohesive skin, the effect so dramatically illustrated by the early CASING experiments [34].

In the past it has been assumed that it was necessary to use an inert gas to obtain crosslinking, but there are some reports on the crosslinking of poly-

---

*Vacuum ultraviolet light is defined as eletromagnetic radiation with wavelengths $\lambda < 175$ nm, the absorption edge of oxygen [35].

propylene (PP) even in oxygen plasmas [37, 38]. In this case, the crosslinked skin is much thinner (~ 300 Å) than on polyethylene, not only because VUV radiation is absorbed in a much thinner layer in PP than in PE [39], but also because O_2 plasma ablates the polymer surface at the same time as the cross-linked layer is being formed in the sub-surface region. This has been discussed in detail in a series of papers by Clark and co-workers (e.g. refs 40 and 41).

(iv) Chemical modification. The most dramatic and widely reported effect of plasma is the deliberate alteration of the surface region with new chemical functionalities capable of interacting with adhesives or other materials deposited on the polymer. Being a core subject of the present review paper, much of the remaining text is devoted to chemical modification and its diverse applications.

2.1.2. Vacuum ultraviolet photochemistry. There have been many studies of the optical emissions from plasmas in the near-ultraviolet and visible region [42]. However, the photon energy at wavelengths longer than the ultraviolet (about 180 nm) region is insufficient to initiate rapid photochemical reactions without some photoinitiator. But the VUV radiation in plasma reactors can break any organic bond and initiate rapid free-radical chemistry. There are very few data in the literature on the emission from real process plasmas in the VUV region, or on the effect of VUV radiation on polymer surfaces. Notable exceptions are the work of Hudis and Prescott [43], and of Clark and co-workers during the 1970s. In an excellent paper [41], Clark and Dilks show the relative effects of VUV and of gas-phase excited species on the depth of reaction in polymers. Their work and other experiments show that the modern plasma reactor is essentially the same light-producing device as those used by researchers in VUV spectroscopy [44], or in some commercial UV lamps [45]. In plasma reactors, the parts are directly immersed in this light source without any intervening windows!

It is important to stress the difference between a 'pure' plasma and a 'real process' plasma: practically all of the published VUV emission data for low-pressure discharges pertain to spectroscopic studies in which extreme care was taken to ensure pure gases. However, plasma reactors, by definition, are used to process materials, and the plasma gas therein will rapidly become contaminated with the volatile byproducts of that processing. For example, when polymers are exposed to oxygen-containing plasmas, 20% or more of the gas phase can be volatile compounds of carbon, hydrogen, and oxygen, all of which can interact with the surface, both chemically and energetically.

The VUV emission from a large number of plasmas has been measured [46], and in all cases the spectra were found to include components of intense, extremely energetic VUV radiation. These emission spectra are usually complex and strongly dependent on experimental parameters such as the gas composition, power, pressure, and the concentration of contaminants. In addition to the complexity arising from the variables just mentioned, the plasma excitation frequency also has a major effect. For example, visible and VUV spectral emissions from MW (2.45 GHz) and RF (13.56 MHz) plasmas show important differences in ion and excited species concentrations [20], which result from their different EEDFs [see equation (1), Section 1.2.2]. In RF and MW oxygen plasmas, our measurements show higher concentrations of O^+ and O^*

(electronically excited atomic oxygen) in the former, but more neutral atomic oxygen in the latter.

Figure 1 illustrates the differences in VUV emission between RF and MW plasmas in nitrogen. This spectrum is a semi-logarithmic plot of the photo-multiplier current vs. the wavelength λ, so apparently small differences in the line height, in fact, represent very significant differences in intensity. Also, it must be stressed that because all of the radiation at $\lambda < 200$ nm is photochemically active, the RF plasma is much more active than its MW equivalent, while the dual-frequency (MW–RF) plasma is more active than MW or RF alone [47].

This VUV radiation will increase the rate of initiation of surface reactions on polymers, while the increased gas-phase concentration of ground-state free radicals (from the MW excitation) will enhance the overall rate of surface reactions. This synergistic effect between MW and RF excitation, along with RF-induced self-bias, can help to explain the advantages that we observe in mixed-frequency plasma processing [13, 25, 48, 49].

As mentioned above, VUV radiation is sufficiently energetic to break any organic covalent bond [40, 50]. These photons are absorbed in a very shallow layer near the polymer surface (or in the surface contamination), typically a few tens of nanometres deep [38, 51–54]. Indeed, differences in the absorption spectra among polymers [51, 54] make it possible, in principle, to tailor the plasma emission spectra to maximize the photochemical effect on different materials. This is illustrated by the data of Egitto and Matienzo [55] shown in Fig. 2; it shows that the receding contact angle of water on polyethylene (PE) is not affected by light with wavelengths longer than about 170 nm (the cut-off of quartz). The absorption spectrum of PE is also shown in Fig. 2. These data indicate that the contact angle is affected by photons with an absorption co-efficient greater than about 10^4 (an absorption depth of about 50 nm). It is obvious that there is a very strong relationship between the two sets of data. Our results show that an oxygen plasma is not very effective for treating poly(tetra-fluoroethylene), PTFE; a possible reason for this is that PTFE does not strongly

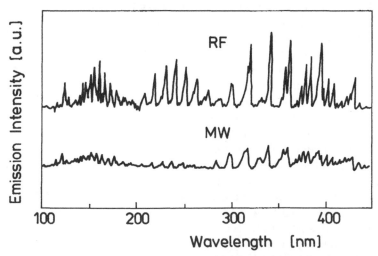

Figure 1. VUV spectra emitted from a MW and an RF discharge in nitrogen. The intensity is shown on a logarithmic scale.

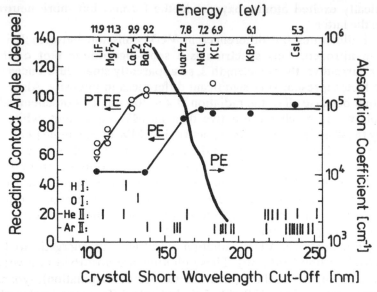

Figure 2. Effect of the short wavelength cut-off on the receding water contact angle on PE and PTFE (after ref. 55), with an overlay of the PE absorption spectrum (after ref. 52).

absorb the 130.5 nm emission from an oxygen plasma. We do not have the absorption spectrum for PTFE, so this cannot be confirmed at the present time, but the similarity in shape between the PTFE and PE data from Egitto and Matienzo suggests that it is probably true. In addition, it is known that hydrogen-containing plasmas (which emit at 121.5 nm) and He plasmas (emitting at 59.0 nm) are efficient at reducing the contact angle on PTFE.

Clearly, then, a very important phenomenon operating in plasma reactors is the VUV photochemistry at surfaces. For the case of polymer surfaces, this is supported by literature references to improved wetting [55], cleaning [56], crosslinking [43, 57], fluorination [58], functionalization [48, 59], and free-radical generation [36]. In all of these examples, the polymer surfaces were separated from the active plasma by a VUV-transparent window, which allowed VUV photons to reach the sample surface, but not the reactive radicals or ions from the plasma. In some cases, the reported results showed that VUV radiation alone can account for as much as 80% of the surface reactions that result when the sample is directly immersed in the plasma.

A recent paper [60] dealt with the treatment of polypropylene in a down-stream plasma system that also used a mercury quartz lamp to illuminate the sample. These authors came to the erroneous conclusion that UV had no influence on surface treatment. This conclusion is unjustified for two reasons: first, the 185 nm radiation from a mercury lamp has a wavelength longer than the absorption edge of PP [39] ($\lambda \approx 170$ nm, see Fig. 2), so it will not be effective for the treatment of this polymer; second, the PP surface was about 40 cm below the surface of the mercury lamp. Oxygen-plasma products (O, O_3) and their reaction products with PP (CO, CO_2, H_2O, etc.) all possess strong VUV absorption cross-sections [35], so that the 40 cm-long optical path through these species results in absorption of most of the UV radiation from the lamp before it can reach the PP surface. This absorption of radiation is one of the main reasons for using downstream strippers in the semiconductor industry [20], that is, machines in which photoresist removal is due to long-lived oxygen species only.

In conclusion, the reaction of polymer surfaces near ambient temperature is very slow in the presence of only ground-state atoms or molecules because these do not possess sufficient energy to cause rapid reaction rates [61, 62]. However, when acting in synergy with elevated surface temperature, bombardment by energetic ions, exposure to VUV photons, or a combination of these, the overall reaction rate will increase by several orders of magnitude. The complex interdependence of these effects is the primary reason why theoretical analysis and modelling of plasma chemical reactions is so difficult.

2.2 Characterization of modified surfaces

2.2.1. General description of techniques. As discussed in Section 2.1, energetic particles and VUV photons generated in the plasma interact strongly with organic polymer surfaces. This interaction gives rise to cleaning, ablation or etching, breakage of bonds, reaction of surface free radicals, crosslinking, and incorporation of chemical groups originating from the plasma or from subsequent exposure to selected gases or vapours. Since all these processes can affect adhesion, it is crucial that one be able to characterize a given plasma treatment in terms of the resulting changes in surface chemical composition, structure, and physical or functional properties.

A variety of surface-specific techniques are available for the characterization of polymers. Among the most powerful and frequently used are X-ray photoelectron spectroscopy (XPS or ESCA) [3, 4, 13, 16, 40, 41, 49, 58], static secondary ion mass spectrometry (SSIMS) [3, 63], Fourier transform infrared spectroscopy (FTIR) [3, 4, 38], and contact angle goniometry [1, 4, 28, 48, 55]. Also very useful are high-resolution electron energy loss spectroscopy (HREELS) [64], scanning electron microscopy (SEM), and techniques based on ion beam probes such as elastic recoil detection (ERD) analysis [65], ion scattering spectroscopy (ISS) [66], and Rutherford backscattering spectroscopy (RBS) [66]. Of course, a wide variety of 'functional' test methods have been devised to evaluate the relative merits of a given surface treatment, e.g. mechanical peel or lap-shear tests, electrical property measurements, and others. These are too numerous to recite here, and the reader is directed to ref. 67. The same applies to plasma diagnostic techniques [42], the most common of which are optical emission and mass spectrometries, and electric probe measurements.

2.2.2. Surface chemical structure. In Sections 1.1 and 2.1.1, we have already expressed our belief that the chemical modification of a polymer surface is one of the most powerful ways of enhancing its bond strength with another surface; the inherently low surface free energy of untreated polymers hinders the wetting by liquid adhesive systems [68, 69], or the interaction with deposited layers of metal or other materials. Typically, a reactive plasma is used to add polar functional groups, which can dramatically increase the surface free energy of the polymer. For example, oxidation is known to enhance metal–polymer adhesion [48, 64], while surface nitrogenation with nitrogen-containing plasmas introduces basic groups that can enhance dyeability with acid dyes [70], printability, or cell affinity in biocompatibility [71]. Some of these are discussed in more detail further below. Of course, plasma can also be used for surface fluorination [72–74] and silylation [75], surface-chemical changes which tend to impart greater hydrophobicity, i.e. reduced wettability and bond strength (this being desirable in some instances). In the remainder of this sub-section, we briefly illustrate and discuss some of the above-mentioned chemical changes.

2.2.2.1. Nitrogenation. In the opening remarks above, we already mentioned how basic groups deriving from surface nitrogenation of polymers can be useful. Two commercial polymers, Kapton® polyimide (PI) and linear, low-density polyethylene (PE) have been exposed to MW and MW–RF plasmas in pure N_2 gas. Figure 3 shows C($1s$) XPS spectra of the untreated, clean polymer surfaces (lower spectra), and of surfaces following MW and MW–RF plasma treatments [49]. Plasma treatment of PE is seen to result in three new spectral features (peaks C2 to C4), besides the original C1 peak at 285 eV associated with carbon bonded to carbon or hydrogen only; the new features arise from the chemical bonding of nitrogen in amine (C2), imine (C3), and amide or nitrile (C4) groups. The relative concentrations of these various functionalities (proportional to the relative peak areas) show a strong, systematic dependence upon V_B, the RF-induced bias voltage during MW–RF plasma treatment. While the N_2 plasma treatment leads to bonded nitrogen primarily in the form of imine or imide groups (C3: C=N, 287.0 eV), amine groups (C2: C—N, 285.8 eV) are in the majority following NH_3 plasma treatment. In the N_2 plasma case, the total bonded-nitrogen concentration can exceed 40 at % on PE [49], while exposure to NH_3 plasma at comparable conditions systematically results in less nitrogen uptake. It has been estimated that up to 20% of the total nitrogen uptake can be photochemically induced by VUV radiation with $\lambda \gtrsim 120$ nm [48]; the VUV radiation from an NH_3 MW plasma was found to be more pronounced than that from an N_2 plasma due to the intense H_α emission at 121.5 nm; this observation is in agreement with other published data [15, 43]. Following the plasma treatment and the subsequent exposure of samples to the atmosphere during their transfer to the XPS instrument, oxygen can also become incorporated. Typically, the highest concentration of oxygen in PE (~8 at %) is found after N_2 plasma treatment, while NH_3 plasma treatment results in a somewhat lower oxygen concentration (~4 to 6 at %) [49].

Referring again to Fig. 3, strong effects may also be noted for the case of Kapton® PI [poly(N,N'-P,P'-oxydiphenylene) pyromellitimide], whose virgin C($1s$) spectrum is evidently much more complex than that of PE: the C1 peak is attributed to carbon in oxydiphenylene, while the C2 peak corresponds to carbon singly bonded to oxygen and nitrogen, plus the carbons in the pyro-mellitic dianhydride ring. The C3 and C4 peaks are associated with the carbonyl carbon in imide linkages and the shake-up satellite associated with aromatic structure, respectively. Following N_2 plasma treatment, the spectra show

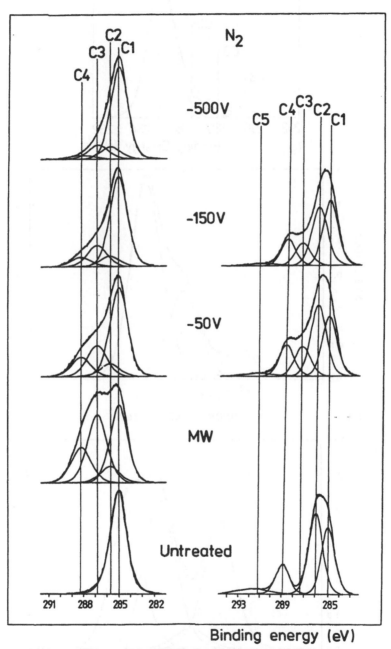

Figure 3. C(1s) XPS spectra for MW and MW–RF N_2 plasma-treated polyethylene (left column) and Kapton® polyimide (right column). Modified after ref. 49.

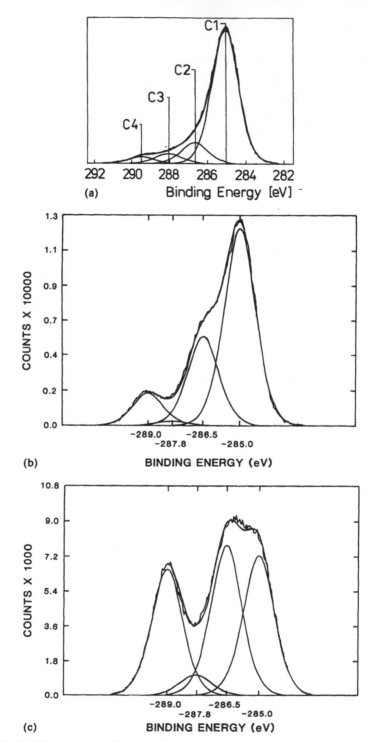

Figure 4. C(1*s*) XPS spectra: (a) PE treated in a low-pressure air MW plasma for 30 s, which results in O/C = 0.30 (after ref. 114). (b) Untreated epoxy, and (c) epoxy degraded by exposure to a corona discharge for 900 h. After ref. 84.

important changes in structure and composition, such as a sharp increase in the amide functionality (C3). Particularly noteworthy, however, is the surface damage induced by ion bombardment. It is manifested by decreases in C4 and C5 intensities, caused by the breakage of C=O bonds in the PI structure, and to the opening of benzene rings, respectively.

The results above, obtained in this laboratory, are presented only by way of illustration, for similar observations have been reported by many others [30, 33, 38, 63, 71, 76, 77].

2.2.2.2. Oxidation. The vast majority of commercial processes involving the corona treatment of polymer surfaces, carried out in ambient atmospheric air, are designed to create polar moieties (carboxyl, ether, carbonyl, hydroxyl, etc.) by reaction with activated oxygen species [12, 78–80]. As will be shown in Section 2.2.3, the presence of such polar groups raises the surface free energy of the polymer, which permits wetting by printing inks and substantially higher bond strengths than those obtained with the untreated surfaces. Low-pressure air or oxygen plasmas may also be used for this purpose, but the need to operate in a partially evacuated chamber adds to the treatment cost. Low-pressure O_2 plasma is, however, employed extensively in microelectronics for the removal (stripping) of polymeric resists, used for microlithography [18, 20], and for the surface treatment of complex shapes, such as car bumpers, where corona cannot be implemented (see also Section 4).

Figure 4a represents the curve-fitted C(1s) XPS spectrum of PE treated in an air MW plasma. This spectrum (compared with that of 'virgin' PE, see Fig. 3) clearly reveals the new peaks resulting from plasma oxidation, namely C2 at 286.5 eV due to C—O groups (hydroxyl, ether, or epoxide), C3 at 288.0 eV due to C=O or O—C—O (carbonyl or double ether), and C4 at 289.4 eV due to O=C—OH or O=C—O—C.

The total surface oxygen concentration can be controlled by varying process parameters such as the plasma treatment time and the power density, but care must be exercised not to 'overtreat' the surface, a problem well known to the corona community. Gross overtreatment can lead to excessive bond breakage and oxidation, manifested by the massive formation of 'low-molecular-weight oxidized molecules' (LMWOM) [28, 38, 80–84], which appear on the surface as solid debris or liquid droplets, readily visible under an optical or a scanning electron microscope [82–84]. An extreme example of this can occur during the faulty operation of high-voltage equipment, when an organic insulation (e.g. epoxy resin) may be exposed to an air corona for tens or hundreds of hours. Figures 4b and 4c show, respectively, the XPS spectra of the untreated epoxy and resulting liquid and solid LMWOM (formic, glycolic, and oxalic acids) [84]. In order to distinguish LMWOM from other low-molecular-weight products in the following text, we refer to the latter as 'LMW species'.

2.2.2.3. Fluorination and silylation. As mentioned above, it is sometimes advantageous to render a surface hydrophobic, that is, non-wettable by water or aqueous solutions. An obvious example is the surface treatment of textiles;

perhaps less well known is the fluorination of the inner surfaces of the plastic gasoline tanks of automobiles, to provide a barrier against gasoline-vapour permeation. For several years this has been done using molecular fluorine, a process that is now becoming outlawed for environmental and safety reasons. Plasma-based fluorination, on the other hand, uses stable, saturated fluorine compounds (perfluorocarbons such as CF_4, C_2F_6, or sulphur hexafluoride, SF_6), and these in much smaller quantities. Surface fluorination can also be achieved by the deposition of a thin layer of a perfluorinated plasma polymer; here, an unsaturated fluorocarbon feedgas such as C_2F_4 or C_4F_8 is best used [85], but excellent water repellancy can also be obtained by the plasma polymerization of organosilicone 'monomers' such as hexamethyldisiloxane [75, 86, 87].

Figure 5 shows the C(1*s*) XPS spectra of PE film following 30 s of treatment in a CF_4 or a C_2F_4 MW plasma (upper and lower spectra, respectively; unpublished results from this laboratory). While the C_2F_4-based plasma polymer is richer in fluorine (F/C = 1.40, as compared with 0.70 for the CF_4 case), the two surfaces displayed amost identical water contact angles ($\theta_a = 122°$, $\theta_r \approx 60°$, see

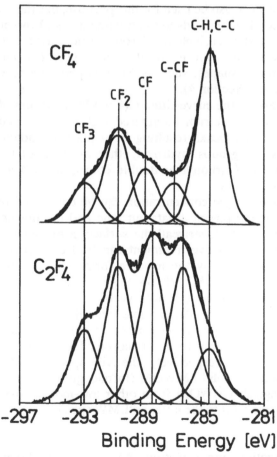

Figure 5. C(1*s*) XPS spectra of PE treated in a low-pressure MW plasma for 30 s: (top) CF_4 plasma; (bottom) C_2F_4 plasma. Peak positions and assignments are in accord with the literature (see, for example, ref. 85).

Section 2.2.3). Very similar results have also been obtained here for the case of kraft paper (lignocellulose, a natural polymer) [74, 88], and numerous other workers have also reported data resembling the ones above [58, 72, 73, 85].

2.2.2.4. *Post-plasma reactions.* In the description of crosslinking by the CASING treatment (Section 2.1.1), we mentioned that free radicals are formed at or below the polymer surface; these radicals may be stable for long durations, and they thereby have an opportunity to react. This opportunity often arises on exposure to atmospheric oxygen or water vapour, which explains the omnipresence of bonded oxygen species in the XPS spectra of He, Ar, or N_2 plasmatreated polymers [38, 49, 77].

Although usually undesirable, post-plasma reactions can also be used deliberately to bond desirable species onto the surface of polymers. In the 1970s, the grafting of acrylic acid to synthetic fabrics, immediately following surface activation in a pure Ar plasma (without intermediate exposure to air), was one of the earliest industrial applications of plasma chemistry [89].

2.2.3. *Wettability.* Because modified wettability is one of the most apparent results of plasma treatment, a common method of characterizing a given treatment is to measure the contact angles of drops of various liquids on the surface. Untreated polymeric surfaces are usually hydrophobic, displaying advancing and receding water contact angles from 60° to 90° or more. Plasmaproduced polar groups increase the surface free energy, γ, of the polymer; the accompanying decrease in contact angle, θ, usually correlates with better bonding of adhesives, and θ has often been used as an estimate of bonding quality [15]. The wettability enhancement differs among various polymers, but it is always significant, even for fluorocarbons [27, 90] and silicones [91].

The value of γ is usually determined from measurements of the static contact angles of a selected series of liquids [92, 93]. However, for rigorous interpretation of contact angle measurements one must take into account the facts that real surfaces possess microroughness, and that the relative concentrations of hydrophobic and hydrophilic molecular groups may vary. The surface heterogeneity gives rise to contact angle hysteresis, that is, a difference between the advancing, θ_a, and receding, θ_r, contact angles. One must therefore take into account deviations from an ideal surface (rigid, smooth, and homogeneous) when evaluating experimental and theoretical data, as discussed in detail in a recent review by Morra *et al.* [94].

The effect of nitrogen concentration, produced by NH_3 plasma treatment, on the contact angle hysteresis of low-density polyethylene (PE) and polyimide (PI, DuPont Kapton® H) is illustrated in Fig. 6. Even on the untreated PE surface, the θ_r value is seen to be lower than θ_a; this is probably due to the presence of surface roughness, or of polar surface contaminants, or of both. With increasing nitrogen concentration, the θ_r values drop rapidly because of the resulting rise in the concentration of hydrophilic groups. However, the hysteresis $(\theta_a - \theta_r)$ remains high until the concentration of hydrophobic groups has become small, at which point θ_a also decreases. Similar behaviour is seen to apply to polyimide, with the difference that this polymer already contains nitrogen and oxygen in its original structure. The measurement of contact angle hysteresis can thus serve as

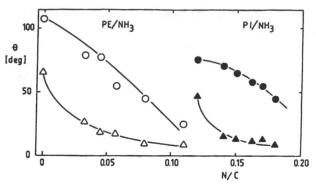

Figure 6. Effect of the XPS atomic ratio N/C on the advancing (O, ●) and receding (△, ▲) water contact angles on PE (O, △) and PI (●, ▲) treated in an NH_3 MW–RF plasma. After ref. 108.

an indicator of the relative concentrations of hydrophilic and hydrophobic groups, and also of the Lewis acid–base characteristics of the surface, a concept of rapidly growing importance in adhesion science [95–97]. The surface characteristics just mentioned are also accessible through other experimental techniques, such as inverse gas chromatography [98, 99].

In practical applications, the stability of the plasma-treated polymer surface is an important issue, but very often a surface rendered wettable by the treatment is found to revert to a less wettable state with time [16, 28, 29, 38, 48, 91, 100–105]. This process, which we call ageing, may result from a combination of four effects [106]: (i) the thermodynamically driven reorientation of polar moieties away from the surface into the sub-surface; (ii) the diffusion of mobile additives or oligomers from the polymer bulk to the surface; (iii) the formation of LWM species in the sub-surface and their subsequent migration to the surface; and, finally, (iv) the reaction of residual free radicals, i.e. chemical change.

The effect of storage time on the wettability of NH_3 plasma-treated PE is illustrated in Fig. 7. The initial contact angle values, $\theta_a = 105°$ and $\theta_r = 70°$, drop to $\theta_a = 30°$ and $\theta_r = 5°$ after plasma exposure. These values increase only slightly during the first few days of storage and remain stable, even at the elevated temperature of 90°C. However, some increase in θ_a and θ_r is noted when the samples are exposed to high humidity, probably resulting from chemical reaction, mechanism (iv) in the above-cited list.

In practice, it is difficult to distinguish the relative contributions of the four effects listed above to the observed ageing (reversion) process. In the case of very clean polymer films, Occhiello *et al.* [102] have used plasma-induced isotopic labelling to demonstrate the importance of entropy-driven molecular relaxation [mechanism (i) above], the rate of ageing apparently depending on the mobility of polar moieties about the polymer chain. The most pronounced ageing process has been observed for polypropylene (PP); poly(ethylene terephthalate) (PET) and polystyrene (PS) showed a lesser tendency to age [107], while PE [48, 107] (see Fig. 7) and PI [48] are quite stable. Attempts have been made to suppress ageing phenomena by stabilizing the surface layer via crosslinking using controlled ion bombardment during MW plasma exposure [48, 49, 108].

As a rule, therefore, ageing rate data cannot be generalized or predicted. The rate of reversion must be measured for every new batch of material, because

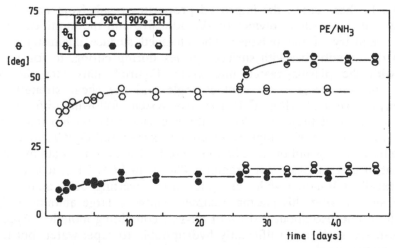

Figure 7. Advancing (θ_a) and receding (θ_r) water contact angles on PE treated at substrate temperatures of 20 and 90°C in MW plasma of NH₃, plotted vs. the storage time in ambient air and at 90% relative humidity. After ref. 48.

small changes in polymer formulation can produce large rate changes. It is insufficient to measure the wetting once; it must be followed with time unless the parts are bonded, printed, or painted immediately after plasma treatment.

3. BONDING ENHANCEMENT THROUGH PLASMA TREATMENT

3.1. General considerations

The various effects of the plasma–surface interactions discussed in the earlier sections contribute in a synergistic manner to the different mechanisms of adhesion. Indeed, bonding enhancement can be regarded as resulting from the following overlap of effects: (a) removal of organic contamination and of weak boundary layers by cleaning and ablation; (b) cohesive strengthening of the polymeric surface by the formation of a thin crosslinked layer that mechanically stabilizes the surface and serves as a barrier against the diffusion of LMW species to the interface; and (c) creation of chemical groups on the stabilized surface that result in acid–base interactions and in covalent linkages believed to yield the strongest bonds.

Experience has shown that the plasma treatment conditions necessary to achieve maximum bond strength must be optimized for any given materials combination. This is accomplished by exercising control over the effects mentioned above, a process in which the chemistry of the substrate material, the energetics at the plasma–polymer interface (ions, photons), and the gas-phase plasma chemistry play the major roles. In the following, we discuss the effects of plasma treatment for different materials combinations, divided into three representative groups: polymer–polymer bonding, polymer–matrix composites, and metal–polymer bonding.

3.2. Polymer–polymer bonding

Plasma treatment can be used with great effect to improve the bond strength to polymers. In these cases, the improved properties result from both increased wettability of the treated polymer by the adhesive and the modification of the surface chemistry of the polymer. The changed surface chemistry facilitates reaction of the adhesive with surface species during curing, to form covalent bonds with the plasma-treated interphase. Figure 8 illustrates the relation between the advancing water contact angle and the shear strength for the polyphenylene sulphide (Ryton® R-4)/epoxy system, following different plasma treatments. Most of these data follow the expected relationship that the bond strength will improve with improved wetting; oxygen and O_2/CF_4 mixtures give both the best wetting and the best bond strength. However, the data in Fig. 8 also show that in certain cases (for example, 4% O_2 in 96% CF_4), it is possible to obtain excellent bonding with very poor water wetting. XPS analysis of the polymer surface after this plasma treatment shows a large amount of fluorine substitution, but also about 5–10% of oxygen-containing moieties. Apparently, the fluorinated surface is sufficiently hydrophobic to repel water, but it is still wettable by the epoxy.

Reference 109 presents what is believed to be the first direct experimental evidence for the formation of covalent bonding between plasma-generated surface functionalities and an epoxy adhesive. The proof of this covalent bonding is extremely important because it would predict greatly improved hot–wet (100% relative humidity, 100°C) stability of the adhesive bonds to plasma-treated polymers. In fact, it has been reported that O_2 plasma treatment of the surface of graphite/PI parts improves the hot–wet stability by a factor of 2, as compared with solvent wiping [110].

Tables 1 and 2 give examples of typical bonding improvement data for a number of polymers. It must be stressed that these data are intended only to show the trend and the magnitude of improvement that can be obtained through

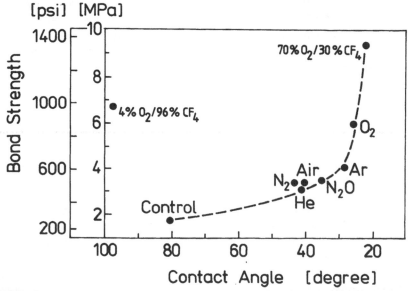

Figure 8. Shear strength vs. static water contact angle for polyphenylene sulphide (Ryton® R-4)/epoxy laminates for different plasma treatments. After ref. 15.

Table 1.
Typical examples of lap-shear bonding improvement after various plasma treatments

	Control		Plasma-treated	
	MN/m^2	psi	MN/m^2	psi
Polyimide (PMR®-15)/graphite	2.90	420	17.93	2600
Polyphenylene sulphide (Ryton® R-4)	2.00	290	9.38	1360
Polyether sulphone (Victrex® 4100G)	0.90	130	21.65	3140
Polyethylene/PTFE (Tefzel®)	—	Very low	22.06	3200
HDPE	2.17	315	21.55	3125
LDPE	2.55	370	10.00	1450
Polypropylene	2.55	370	21.24	3080
Polycarbonate (Lexan®)	2.83	410	6.40	928
Nylon®	5.86	850	27.58	4000
Polystyrene	3.93	570	27.58	4000
Mylar A®	3.65	530	11.45	1660
PVDF (Tedlar®)	1.93	280	8.96	1300
PTFE	0.52	75	5.17	750

Table 2.
Typical examples of peel-strength improvements after various plasma treatments.

	Control		Plasma-treated	
	N/m	lb/in	N/m	lb/in
Silicone (red, Durometer 50)	70.1	0.4	3330	19
RTV silicone (D.C. type E)	—	Very low	403	2.3
Perfluoroalkoxy (PFA)	17.5	0.1	1450	8.3
Fluorinated ethylene propylene (FEP)	17.5	0.1	1820	10.4
Tefzel®	17.5	0.1	2770	15.8
PTFE	17.5	0.1	385	2.2
Polyimide (Kapton®)	700	4	2800	16

plasma treatment. These data only represent typical values, which will differ from one experimenter to the other depending on the polymer formulation, the type and amount of additives, the adhesive, the cure cycle, the time between plasma treatment and bonding, and, of course, the other plasma parameters.

It is also important to stress that there is no 'standard' plasma process. There are normal starting processes for various polymers that are learned through experience, but is is usually necessary to optimize these to obtain the best processing conditions for each application. Under optimum conditions for a given polymer/adhesive system, the bond strength can generally be improved to the point where the bond failure is cohesive in the weakest material, and not in the bond line. For example, a five- to seven-fold bond strength improvement has recently been found for PP to epoxy [111]. Improvements by factors of 3 and 7 were observed for the peel strength of pressure-sensitive adhesive tape to PFA [27] and to PE [81], respectively, and a four-fold improvement in the epoxy bonding of PE or PP to aluminium [6, 112].

Besides polymeric laminated structures using adhesives, efforts are continuing to achieve good polymer–polymer bonding by mere plasma activation. Interfacial phenomena have recently been studied for the cases of PE/PE and PE/PET laminates without adhesives, following treatment in a low-pressure MW air plasma or in an ambient air corona discharge [113, 114]. The adhesion force was found to exhibit a pronounced maximum for a surface concentration of bound oxygen between 11 and 14 at %, independent of the type of treatment (see Fig. 9). Based on high-resolution XPS, such as that illustrated in Fig. 4a, it has been concluded that the maximum adhesion occurs when the XPS peak C2 is highest (i.e. when the concentration of hydroxyl, ether, or epoxide groups is highest) and that of carboxyl (acid) groups is lowest (see Fig. 10). These results

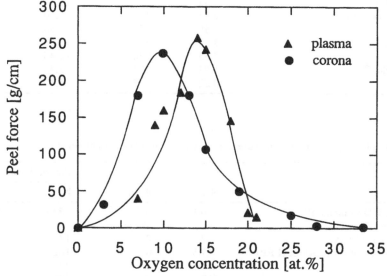

Figure 9. Peel force of PE/PE laminates without adhesive as a function of the oxygen concentration after low-pressure plasma (▲) or corona (●) treatment in air. After ref. 114.

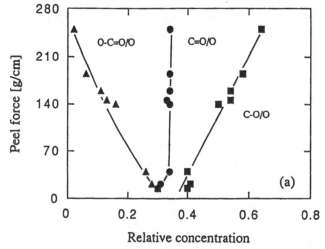

Figure 10. Peel force of PE/PE laminates without adhesive plotted against the relative concentration of oxygen-containing groups following treatment in a low-pressure air plasma. After ref. 114.

suggest that the highest adhesion force appears when the surface is mechanically stabilized by crosslinking and when the effect of a weak boundary layer due to excessive LMWOM is minimal (low carboxyl concentration) [114]. Association of the C4 (carboxyl) group with LMWOM is confirmed by the fact that rinsing the plasma-treated surface with distilled water substantially decreases the C4 peak intensity. It has yet to be established whether the maximum bond strength, observed above, results in part from direct covalent bonding across the interface between the two laminated surfaces.

The acidity of a strongly oxidized surface suggests that another type of chemical bonding mechanism might be of importance here, namely acid–base interactions [95–97]. Adhesion of dye molecules, an example drawn from our own experience, strongly supports this view: non-polar polymers such as PE or polypropylene (PP) cannot be treated with inexpensive water-based acid dyes. If, however, basic (Lewis base) moieties are grafted to the surface, the acid–base interactions between these and the dye (Lewis acid) can give rise to strong (chemical) adhesion at the polymer surface. Surface nitrogenation (amination) using a plasma of N_2, NH_3, or volatile amines has been shown [70] to yield the desired results, which are attractive for both economic and environmental reasons.

3.3. Polymer–matrix composites

In the case of composite materials, the plasma treatment of the filler can be very effective in promoting adhesion because of the large treated surface area. The modification of interfaces not only enhances the bond strength between the components, but it can also improve the electrical and ageing characteristics of the composite by reducing the penetration of water vapour and other con-taminants.

Regarding the mechanical properties, in many fibre-reinforced polymer composites, toughness is derived from energy dissipation that occurs when bonds between the fibres and the matrix fail. After plasma treatment, the peel and flex properties of the composite tend to improve, but the toughness can decrease because the above-described energy loss mechanism is no longer available. Therefore, toughness must be achieved through another mechanism. The easiest way to develop the requisite toughness is to use a toughened (lower modulus) matrix, an approach that will not only add energy dissipation, but also increase the critical flaw size and thus increase the strength of the composite.

The following are typical illustrations of the trends discussed above. Once again, due to large variations among the reinforcing filler and matrix materials, detailed optimization must be performed for each proposed composite system, in order to obtain reliable engineering data.

Much research has been done over the years on the treatment of particulate filler materials, such as $CaCO_3$ powder [115] and mica flakes [116, 117], by exposure to selected plasma gases. The objective was to impart either acidic or basic properties by chemical surface modification, or to deposit strongly adhering plasma–polymer layers, thereby enhancing interfacial compatibility with various polymeric matrix materials. Among the matrices examined were PE, PP, PS, and polyvinyl chloride (PVC). In all cases, mechanical properties (for example, tensile strength) could be significantly enhanced in the plasma-treated sampled, as compared with untreated control samples. Depending on the particular materials/treatment combinations, the improved properties could be explained in terms of acid–base interactions [115] or 'bridging' via compatible polymer/plasma–polymer homologues [116, 117].

Aromatic polyamide or 'aramid' (DuPont Kevlar®-29) fibres or fabric were also modified by plasma treatment or coated with plasma polymer before encapsulation in a triazine resin matrix [118]. Here too, very significant increases in the bond strength (peel strength of two-ply laminates) could be attributed to the plasma treatments; this was not entirely surprising, since untreated aramid fibres are known to have superb mechanical characteristics, but are also known to present bonding difficulties.

In a detailed study by Ismail and Vangeness [119], different graphite composites were compared following oxygen-plasma treatment or high-temperature oxidation. The authors found that the plasma treatment had no substantial effect on the surface area of pitch fibre or carbonized poly-acrylonitrile (PAN), but that it increased the BET surface area (measured by Kr adsorption) of graphitized rayon by almost an order of magnitude. For comparison, the high-temperature treatment of PAN (without plasma) increased the BET area by three orders of magnitude. Even though the BET area of the graphite fibres increased only slightly following both types of treatment, the O_2 adsorption on the plasma-modified surface rose by 430%. This difference has been attributed to chemical adsorption of the O_2 by active sites left on the fibre surfaces. The authors recommended that plasma treatment be used on pitch-carbon fibres if primarily chemical bonding is sought, while high-temperature oxidation is preferred for primarily physical bonding.

Jang and Das [120] reported the results of tests on a high modulus graphite/epoxy composite following both O_2 and N_2 plasma treatments. They found that after a 1 min N_2 plasma treatment of the graphite, the composite shear strength increased by 55%, the flex strength decreased by 7%, and the toughness decreased by 24%. However, the toughness could be restored by the addition of less than 5% chlorotributylnitride (CTBN) rubber to the matrix. Significant improvements have also been observed for high modulus graphite in high modulus epoxy (Thornel T-300 in DER 332/T403) [33]. Increases of the T-peel force by 41%, 32%, and 22% were observed following plasma treatment in air, Ar, and NH_3, respectively, while a decrease by 18% after O_2 plasma exposure was felt to indicate that the treatment conditions had not been fully optimized.

Interesting results have also been reported for polyaramide/epoxy composites [33, 121]. Allred et al. [121] studied commercial Kevlar® cloth; SEM examination of the surface showed that the fibres were very smooth before and after NH_3 plasma treatment, which they believed healed small surface flaws. After plasma treatment, they observed a change in the failure mode of the composite from matrix–fibre debonding to cohesive failure in the fibre (splitting, fibrillation, separation of the skin and core of the fibre, and crack propagation through the epoxy matrix), implying covalent bonding between the epoxy and the fibres. After 1 min of NH_3 plasma treatment, a 114% increase in T-peel, a 31% increase in interlaminar strength, and a 5% decrease in toughness were observed.

Chemical derivatization analysis of the plasma-treated surface using a dye showed that there was about one NH_2 group per polymer-repeat unit after 1 min of N_2 plasma treatment. Oxygen contamination of the NH_3 plasma interfered with the incorporation of NH_2 into the surface, probably because of the competition for the active sites by oxygen species. No change in the NH_2 concentration on the surface was observed, even after 18 months of storage in air.

The plasma treatment also reduced the water absorption of the composite system by a factor of 3, and it converted from a capillary (wicking along the fibre–matrix interface) to a Fickian (diffusive) absorption through the body of the matrix [121]. The rate of water absorption in the composite was found to be less than that in a block of neat epoxy of the same size; water uptake was found to be equivalent to the volume fraction of epoxy in the composite. This change in the absorption mechanism would suggest a great improvement in the hydrothermal stability of plasma-treated composites.

The results of a series of T-peel tests on three different types of polyaramide and two different polyethylene yarns, each treated for 1 min in four different plasma gases [33], are summarized in Table 3. Kevlar® 29 has a lower modulus and a greater elongation than Kevlar® 49; when type 29 is plasma-treated and bonded to the matrix, it has a much greater peel strength than the best results for type 49, namely 5.25 N/mm vs. 2.70 N/mm (30 lb/in vs. 15.4 lb/in). This is another indication that a lower-modulus system is needed in composites that possess good bonds between the matrix and the fibres. The first material, 49/7100, showed no ageing or loss in performance even after 300 days of storage in air. The T-peel performance of polyethylene yarns illustrated in Table 3 shows large variations in peel strength depending on which plasma gas is used, illustrating, once again, that the plasma process must be optimized for each material used. These data pertain to a medical-grade material, while the standard-grade material had so much surface contamination that plasma treatment made no difference to the performance of the composite.

The data shown above illustrate the great improvements that can be achieved in most composites after plasma treatment of the reinforcing material, without affecting the other properties of the composite.

Table 3.
Per cent improvement of the T-peel force for two types of high-modulus fibres (after ref. 33)

Fibre type	Denier	Plasma			
		O_2	Air	Ar	NH_3
Aramid (DuPont Kevlar® 49)	7100	318	218	204	169
Aramid (DuPont Kevlar® 49)	380	129	154	169	83
Aramid (DuPont Kevlar® 29)	1500	2383	–	–	2486
Polyethylene yarn (Spectra® 900)	1200	360	380	220	150
Polyethylene yarn (Spectra® 1000)	660	153	100	87	45

3.4. Metal–polymer bonding

There is strong evidence in favour of the chemical reaction mechanism (see Section 1.1), the one which should, in principle, lead to the strongest bonds, to explain strong metal–polymer adhesion. Burkstrand [122] was among the first to show that evaporated metals can react with oxygen-containing polymer surfaces, which can lead to metal–oxygen–carbon (M—O—C) type linkages.

Plasmas can be used with success to provide the necessary surface functionalities that can form strong bonds. For example, *in situ* XPS studies have revealed the presence of Ag—O—C and Ag—N—C linkages [123] after exposing PE to oxygen and nitrogen plasmas, respectively, as illustrated by the XPS spectra in Fig. 11. For this particular system, the metal–PE adhesion was found to improve according to the following sequence of plasma gases used: $Ar < O_2 < N_2$. Similar effects were observed in another *in situ* study of Mg on PP [124]. The highest sticking probability for evaporated Mg atoms was found on a PP surface following exposure to an N_2 plasma or, for comparison, following argon ion bombardment at a dose of 5×10^{15} ions/cm^2.

In practical situations, the plasma conditions needed for treatment of a given polymer surface must be optimized for every metal/polymer combination. Departure from optimum treatment conditions can lead to various effects, often involving the presence of a weak boundary layer. For example, if the surface is insufficiently treated, the ubiquitous contamination layer (lubricant, mould release, anti-blocking agents) is incompletely removed or additives from the polymer's bulk can diffuse to the interface and contribute to the formation of a weak boundary layer. Therefore, even if the surface is found to be wettable and to contain the desired oxygen or nitrogen moieties, adhesion does not improve

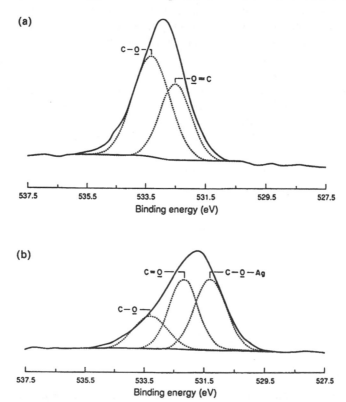

Figure 11. Line-shape analysis of the O(1s) spectrum of oxygen-plasma-treated PE: (a) before Ag deposition; (b) after deposition of nominally 5×10^{14} atoms/cm^2 of silver. After ref. 123.

because one is attempting to bond to a plasma-treated layer of contamination. Also, the surface can be grossly overtreated, which leads to the formation of excessive LMWOM [80, 83, 84], as already discussed in Section 2.2.2.2.

An example of the mechanical failure of a weak boundary layer is the case of polytetrafluoroethylene (PTFE); this material can readily be plasma-treated for cleaning and surface modification, to give good wetting and to provide oxygen-containing or nitrogen-containing groups. The adhesion, however, is still found to be poor, because the sub-surface structure of PTFE remains weak and not easily stabilized by crosslinking.

It follows that an 'optimum' plasma treatment should yield two simultaneous effects: surface crosslinking and the formation of the requisite chemical functionalities. Such a composite interphase is illustrated schematically in Fig. 12. The crosslinking process not only leads to mechanical strengthening of the interface, but the crosslinked interface sublayer simultaneously provides a barrier against diffusion of LMW species from the polymer bulk. In practice, the control of crosslinking can be achieved by using plasma gases which strongly emit VUV radiation, the efficiency of which roughly follows the sequence $He > Ne > H_2 > Ar \approx O_2 \approx N_2$ [125]. Therefore, plasma surface treatments using mixtures such as He/O_2 or He/N_2 should, in principle, lead to the most pronounced adhesion improvements.

It follows that the thickness and properties of the composite interphase in Fig. 12 depend on the substrate material, on the penetration depth of the crosslinking agents (photons, ions), and on the plasma characteristics. Therefore, as already stressed on several occasions above, it is important to optimize the quality of this interphase for every materials combination.

In some cases, improved adhesion cannot be achieved merely by plasma

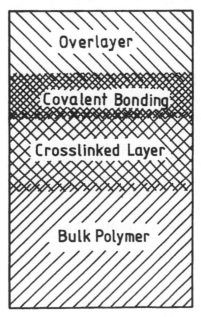

Figure 12. A schematic diagram of the interphase structure between the adhering layer and a polymer base.

surface modification, and another intermediate layer is needed; such a layer increases the thickness of the interphase and may also play additional roles. Metal-containing plasma polymers have been suggested as a means of improving the adhesion of structures such as Au–PTFE [126], where chemical bonding is practically excluded. In this case, metal clusters are incorporated into the intermediate layer by simultaneously sputtering or evaporating metal during the plasma polymerization of a fluorocarbon or hydrocarbon monomer [127]. The metal concentration can be gradually increased from the polymer side towards the metal side, so that adhesion improvement is achieved by complete mechanical interlocking. The composite metal–polymer structure can also be used to tailor the optical properties for use in decorative applications [127].

Low-pressure plasma treatment has been suggested as a means of improving the adhesion of copper to polyimide or to fluorocarbons, as used in multilayer printed circuit boards. In a simple case, polyimide must be pretreated by oxygen plasma to increase the adhesion of a Cr intermediate layer, onto which Cu is then deposited by electroless plating [128]. When PTFE is used, on account of its good thermal and dielectric properties, the surface can be activated by a hydrogen plasma to achieve wettability, following which copper formate is applied by spin coating [129]. The copper formate is then converted to metallic copper by chemical reduction, again in a hydrogen plasma.

Metallic intermediate layers of Pd, Pt, Au, or Cu, obtained by the plasma decomposition of organometallic compounds, have also been investigated for improved adhesion of electroplated Cu to PTFE [130].

Intermediate layers appear to be necessary for good adhesion in corrosive and other chemically active environments. Their role is to suppress the diffusion of destructive agents towards the interface. In this capacity, plasma polymerized hydrocarbon (CH_4) films have been found to improve the bonding of Pt electrical contacts to Parylene® or other polymers that are used as implantable, biocompatible sensors [131].

4. PLASMA EQUIPMENT FOR INDUSTRIAL PROCESSING

Plasma technology has been used in the laboratory for at least 50 years but it has not been a practical industrial or commercial technology until the last 15 years. As already mentioned above (Section 1.2.1), commercial equipment development was originally driven by the requirements of the semiconductor market, but more recently equipment designed specifically for the industrial market has also become available. The 'industrial market' is defined here as any application that does not involve semiconductor wafers.

Plasma equipment usually consists of six modules or functions: vacuum system, power supply, matching network, power monitor, reactor chamber, and controller.

Low-pressure plasma systems for surface modification generally operate in the pressure range from 10 to 500 Pa (about 0.1–3.5 Torr), with a continuous gas flow into the reactor. Therefore, the vacuum system must be able to maintain this pressure/flow regime; this moderate vacuum level does not require sophisticated pumps, so that two-stage mechanical pumps are generally satisfactory. The pump package is usually sized to allow pumpdown to the operating pressure in less than 1 min and to maintain an inlet gas flow of 50 sccm for the smallest systems to several 1000 sccm for systems of several cubic metres in volume. Pump maintenance, and the perfluorinated pump fluid that is required if an oxygen plasma is to be used, tends to be the largest part of the total plasma-system operating and maintenance costs.

A major advantage of plasma surface treatment as compared with most other treatment processes is the lack of harmful byproducts from the treatment process. There are no toxic or hazardous liquids or gases that must be disposed of. Usually, the main process byproducts are CO, CO_2, and water vapour, none of which is present in toxic quantities. We are unaware of any users of plasma for polymer surface treatment who need scrubbers on their pump exhausts; in other words, plasma surface treatment is very benign towards the environment.

The plasma excitation power of industrial systems generally ranges from 50 to 5000 W, again depending on the size of the reactor. Plasma reactors have been built utilizing a wide range of excitation frequencies, from DC to microwave; DC plasmas are not advantageous, primarily because of the need for a current-limiting resistor to stabilize the plasma and prevent arcing, and because of the very large bias effects that are present. Most plasma reactors, therefore, use AC electrical power supplies, operating at audio-, radio-, or microwave frequency [21, 22]. More specifically, international regulatory agencies have allocated certain 'ISM' (industrial, scientific, medical) frequencies for use in applications other than telecommunications, and it is these frequencies which equipment builders favour for obvious technical and economic reasons. Commercial plasma systems, therefore, usually operate in the low-frequency (LF, 50–450 kHz), radiofrequency (RF, 13.56 or 27.12 MHz), or microwave (MW, 915 MHz or 2.45 GHz) ISM frequencies. There are numerous vendors who offer highly efficient and cost-effective power supplies, matching networks, and other accessory hardware in these frequency ranges.

LF plasmas (50–450 kHz) are sometimes used because the generators are somewhat less expensive than those operating at other frequencies, and because they do not require precise impedance matching. However, our studies have shown that the reaction rates tend to be significantly slower than those at RF or MW frequencies, and that there is more of a tendency to arc at LF.

RF plasmas (13.56 MHz) are easily generated with equipment that is stable and reliable, and which has been commercially available from several vendors for many years. It is necessary to use a matching network to match the impedance of the plasma to that at the generator output (usually 50 Ω, resistive), a task that can be accomplished either manually or with automatic, servo-driven devices. At 13.56 MHz or higher, the plasma is very stable and reactive because the quench time of the plasma species is longer than the half-cycle period of the applied field.

MW plasma (2.45 GHz) is often more reactive than RF, as explained in Section 1.2.2, and the MW generator may be less expensive. However, if the cost of all of the peripheral equipment, such as wave guides, power meters, dummy loads, stub tuners, and applicators is included, the total system cost tends to be similar to that of an RF system. Nevertheless, in recent years there has been a gradual shift from RF to MW, a trend that we believe will continue in the foreseeable future.

A recurring problem in characterizing plasma processes, or in transferring a plasma process from one type or size of reactor to another, is the specification of the 'power density' in the plasma. This can be defined in at least three ways, namely (1) power per unit area of electrode; (2) power per unit volume of primary plasma; or (3) power per unit volume of the entire reactor. As each of these is deficient in some way, it is usually difficult to predict the best process parameters for transferring a process form one reactor to another. Instead, one needs to start with a reasonable set of parameters and then optimize the process experimentally in the new reactor.

There are several types of plasma reactor chambers: dielectric (e.g. fused quartz) or metal, and batch or continuous. Quartz chambers are important in the semiconductor industry because of the requirements for extreme cleanliness and particle-free operation, but they may also have advantages in some industrial applications. In the present context, plasma treatment for improved adhesion, extreme cleanliness is rarely required, and the danger of breakage favours the use of metal. Aluminium is the metal of choice for constructing plasma reactors, because it has excellent thermal and electrical conductivity, and chemical resistance. Aluminium is not readily attacked by any plasma gas except the heavy halogens (Cl_2, Br_2, or I_2). Aluminium has been fabricated into cylindrical vessels (known as 'barrel' reactors), and into rectangular reactor chambers, with shelf or cage electrodes. It is also possible to fabricate specially shaped electrodes for particular applications. The only size limitation on metal reactors is the strength of large vacuum vessels; the largest currently available commercial reactors have dimensions in excess of 2 m [132, 133]. These are used to process plastic automobile parts (e.g. bumpers or dash assemblies) in plasma generated either in an electrodeless MW plasma from a cluster of MW sources or in large, specially shaped RF electrodes.

Most commercially available plasma systems are designed for batch operation, which involves loading a batch of parts, evacuation, plasma processing, purging to atmospheric pressure, and removal of the parts. While this has been satisfactory for many applications, the use of polymers in ever more sophisticated applications, such as composite structures, increasingly calls for the continuous processing of filaments, yarn, film, and fabric. Therefore, plasma systems have been built for continuous processing, either using 'air-to-air' or 'cassette-to-cassette' ('batch-continuous') configurations.

In air-to-air systems, there are sequentially pumped chambers on either side of the reactor chamber, which are connected by some form of material feed-through system. This feed-through system makes it possible to continuously bring material from atmospheric pressure to the low reactor pressure, and then back to ambient atmosphere; the major engineering problem in air-to-air systems is the design of the feed-through.

Cassette-to-cassette systems have been built in which the source and takeup spools are both under vacuum. This configuration minimizes the abrasion of fragile materials, such as graphite yarn, when they are drawn through a feed-through, and it facilitates materials treatment in plasma gases with very little oxygen contamination. One commercially available RF (13.56 MHz) plasma machine (Fig. 13) has a double-sided electrode system that is 0.67 m (26 in.) square, so that the film or yarn traverses 1.34 m of plasma on each pass through

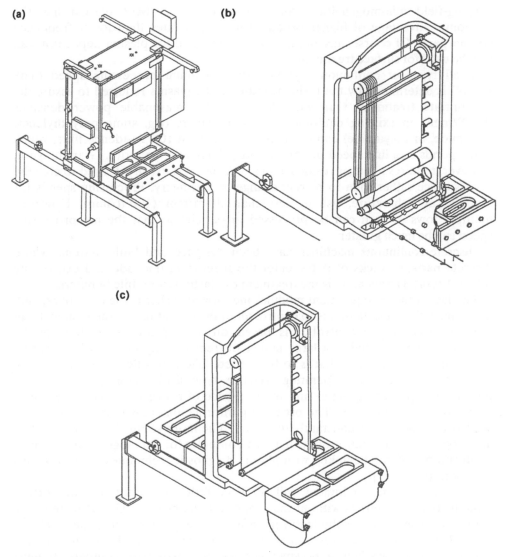

Figure 13. Drawings of industrial-scale continuous plasma treaters: (a) overview; (b) yarn-treatment configuration; (c) continuous cassette-to-cassette treatment configuration. (Courtesy of GaSonics/ International Plasma Corp.)

the plasma zone [134]. The design of the upper and lower rollers makes it possible to run yarn through the plasma several times, thus assuring a sufficiently long residence time in the plasma, even at high line speed. This machine is available in either an air-to-air (Figs 13a and 13b) or a cassette-to-cassette configuration (Fig. 13c).

Two large 'batch-continuous' machines have been built for plasma deposition onto flexible web substrates, one for LF (50 kHz) operation [135], the other operating at MW (2.45 GHz) frequency [136]. The latter, based on the LMP® (Large volume Microwave Plasma) principle [137, 138], uses pairs of counter-activated, linear slow-wave microwave applicators which mutually compensate fringing-field inhomogeneities. As mentioned above, MW plasma has the economic advantage of higher product throughput rates than lower frequency counterparts. Of course, any plasma system designed for plasma deposition can also be used for surface treatment.

In all types of continuous systems, the maximum processing speed ('line speed') is determined by the residence time in the plasma required to assure the appropriate treatment. Our data show that for reasonable power densities (0.1 W/cm^3) in existing equipment, clean polymers (e.g. some polyethylenes, polyimide, or polyamide) can be treated in less than 10 s residence time, which corresponds to a line speed of 150 m/min (500 ft/min) or more. Graphite yarn, however, requires a longer residence time because it is usually desired to etch the fibre surface rather than to merely change its chemistry. Processing speeds for graphite are, therefore, usually less than 30 m/min (100 ft/min). However, multiple tows of yarn can be processed in parallel so that the net processing speed can be much greater.

Discrete-continuous machines have been designed and built, systems where discrete parts, or racks of parts, enter the reactor on one side and exit on the other. A typical application is the treatment of plastic automobile bumpers.

The last of the six equipment modules mentioned earlier is the controller, and, as the name implies, its role is to control all of the operations of the system. In all modern systems, the controller is based on a microprocessor that can be programmed to remember a large numer of 'recipes', any one of which can be called up, depending on what is to be treated. The controller determines all of the machine parameters (operating pressure, type of plasma gas, gas flow rate, power level, processing time) and the sequence of processing steps, if a multi-step process is being used. The processing information is displayed on a screen, so that the operator is constantly informed. The more sophisticated controllers are capable of communication with a host computer for data logging and process verification, and can be linked with plasma diagnostic techniques such as optical emission spectroscopy.

The best plasma equipment will do very little of benefit for the surface modification of materials without the proper plasma process. Unfortunately, this is the area where there is the least scientific knowledge, and where the most 'art' is involved. As we have shown in Section 2, the interactions between a plasma and a surface, especially a polymer surface, are extremely complex. It must always be remembered that a 'real' plasma contains not only the feed gas, but also all of the volatile products of the interaction between the starting plasma and the surface; the composition and concentration of these products are very rarely known, especially in the boundary layer (the few millimetres closest to the polymer surface) where most of the reactions occur that affect the surface. It is extremely difficult to perform analyses in this region because any sampling technique will affect the composition and concentration of the unstable transient species in the boundary layer. Analysis of the plasma outside the boundary layer is of relatively little value because of the myriad reactions that the boundary-layer species undergo during their diffusion into the bulk of the plasma.

The development of the proper process is nearly always the result of a series of optimization tests in which the plasma parameters are varied, and these changes are then correlated with the results obtained by testing the properties of the treated parts. The literature, for example this paper, and the experience of others working in this field, especially the manufacturers of plasma equipment, will give a good starting point for process development; however, final refinement and implementation of any given plasma process must be accompanied by verification, testing, and optimization.

5. CONCLUSIONS

We have shown in this review that 'cold', low-pressure plasma treatment can give rise to profound changes in the surface and interfacial properties of materials, particularly of polymers. Corona treatment, a similar process in several respects, has been in industrial use for several decades. However, unlike corona, in which the reactive gas is usually air and which is generally restricted to simple surface geometries such as flexible webs, plasma treatment is extremely versatile in its capabilities.

Following brief reviews of adhesion theory and of the physics and chemistry of 'cold' plasmas, we examined the all-important mechanisms by which a plasma interacts with a polymer surface. These include physical bombardment by energetic particles and by ultraviolet photons, particularly vacuum ultraviolet (VUV, $\lambda \leqslant 175$ nm), which are energetic enough to break most organic bonds, and cause chemical reactions at or near the surface. The resulting four main effects—cleaning, ablation or etching, crosslinking, and surface chemical modification—occur together in a complex synergy, which depends on many parameters controlled by the operator.

Plasmas can readily be applied to objects of all possible geometries, ranging from webs or films to large solid objects with complex shapes, and to small discrete parts in large quantities (even including fine particles such as powders, fibres, or flakes). Industrial plasma systems are available for treating webs, either in a continuous air–vacuum–air or in a batch-continuous (cassette-to-cassette) mode of operation, while other reactor designs are readily adapted for treating discrete parts with the characteristics listed above.

The most important advantage of a low-pressure plasma, however, is the fact that it offers virtually limitless possibilities for tailoring the surface structure and chemistry of a given polymer. Of course, like corona, this plasma treatment can comprise a controlled surface oxidation with air or oxygen designed to create polar (e.g. carboxyl, hydroxyl, etc) moieties. Polar groups greatly increase the surface energy of the polymer, thereby enhancing its wettability by liquids and adhesives; this in turn, improves bond strength, printability, and dye uptake, to name but a few attributes. Nitrogen-containing functionalities can be created by exposing the surface to a plasma of nitrogen, or nitrogen-containing gases (e.g. ammonia), with additional advantages. In contrast, a surface can also be rendered more hydrophobic (*less* wettable) by plasma fluorination or silylation with a variety of reagent gases or vapours, all offering particular characteristics and advantages. It is noteworthy that only very small quantities of reagents are consumed, and that these can almost always be chosen from among inexpensive, non-toxic compounds. In some processes, hazardous byproducts may be created (e.g. HCN), but only in sufficiently small quantities that can be readily collected from the pump exhaust and then neutralized. These are routine procedures, as has been proved by the semiconductor industry, for many years the leading user of plasma processing.

The very richness of choices offered by plasma processing can also be construed as an important drawback: each new application requires that the process conditions be clearly identified and optimized. Because there are numerous parameters (power, pressure, gas composition and flow rate, treatment duration), optimization can be tedious. Furthermore, owing to the inherent complexities of the plasma state and, even more, of its interactions with solid surfaces, there are still many unanswered questions and much ongoing research. We have pointed out some such unknowns in the preceding text, but have also emphasized that they pose no fundamental obstacles to process optimization by semi-empirical procedures. Once the key process variables have been identified and optimized, modern control instrumentation can ensure that the process outcome is reliably reproduced. In conclusion, we are convinced that industrial plasma processing will continue its vigorous expansion because it has unique capabilities, because it is economically viable, and because it is 'friendly' towards the environment.

Acknowledgements

This work has been supported, in part, by grants from the Natural Sciences and Engineering Research Council of Canada (NSERC), the Fonds 'Formation des chercheurs et aide à la recherche' (FCAR) of Québec, and the Institute for Chemical Science and Technology (ICST). Useful discussions and collaborations with the following persons are gratefully acknowledged: Mrs J. E. Klemberg-Sapieha, and Drs D. W. Dwight, A. Holländer, E. Sacher, S. Sapieha, and H. R. Thomas.

REFERENCES

1. S. Wu, *Polymer Interface and Adhesion*. Marcel Dekker, New York (1982).
2. L. H. Lee (Ed.), *Fundamentals of Adhesion*. Plenum Press, New York (1991).
3. D. T. Clark and W. J. Feast (Eds), *Polymer Surfaces*. John Wiley, New York (1978).
4. G. Akovali (Ed.), *The Interfacial Interactions in Polymeric Composites*, Proc. NATO-ASI, Series E: Applied Sciences, Vol. 230. Kluwer, Dordrecht (1993).
5. L. H. Sharpe and H. Schonhorn, *Adv. Chem. Ser.* **43**, 189–201 (1964).
6. K. L. Mittal, *J. Vac. Sci. Technol.* **13**, 19–25 (1976).
7. L. H. Sharpe, in: ref. 4, Chapter 1, pp. 1–20.
8. B. V. Deryagin and V. P. Smilga, *Proc. 3rd Int. Congr. on Surface Activity*, Köln, Vol. 2, p. 349. Universitätsdruckerei, Mainz (1960).
9. S. S. Voyutskii, *Polymer Reviews 4*. Interscience, New York (1963).
10. J. J. Bikerman, *The Science of Adhesive Joints*, 2nd edn. Academic Press, New York (1968).
11. L. M. Siperko and R. R. Thomas, *J. Adhesion Sci. Technol.* **3**, 157–173 (1989).
12. M. Goldman, A. Goldman and R. S. Sigmond, *Pure Appl. Chem.* **57**, 1353–1362 (1985).
13. J. E. Klemberg-Sapieha, L. Martinu, S. Sapieha and M. R. Wertheimer, in: ref. 4, pp. 201–222.
14. E. M. Liston, in: ref. 4, pp. 223–268.
15. E. M. Liston, *J. Adhesion* **30**, 199–218 (1989).
16. O. M. Küttel and S. Novak, in: *Plasma Processing of Materials*, J. Pouch and S. A. Alterowitz (Eds). Trans Tech Publications, Aedermannsdorf, Switzerland (in press).
17. J. Mort and F. Jansen (Eds), *Plasma Deposited Thin Films*. CRC Press, Boca Raton, FL (1986).
18. D. M. Manos and D. L. Flamm (Eds), *Plasma Etching—An Introduction*. Academic Press, Boston (1986).

19. R. d'Agostino (Ed.), *Plasma Deposition, Treatment and Etching of Polymers.* Academic Press, Boston (1990).
20. J. I. McOmber, J. T. Davies, J. C. Howden and E. M. Liston, *Proc. 9th Plasma Symp.,* The Electrochemical Society. PV 12-18, 104-114 (1992).
21. J. R. Hollahan and A. T. Bell (Eds), *Techniques and Applications of Plasma Chemistry.* John Wiley, New York (1974).
22. H. V. Boenig, *Plasma Science and Technology.* Cornell University Press, Ithaca, NY and London (1982); H. V. Boenig, *Fundamentals of Plasma Chemistry and Technology.* Technomic Publishing, Lancaster, PA (1988).
23. C. M. Ferreira and J. Loureiro, *J. Phys. D: Appl. Phys.* **17**, 1175-1188 (1984).
24. M. Moisan, C. Barbeau, R. Claude, C. M. Ferreira, J. Margot, J. Paraszczak, A. B. Sá, G. Sauvé and M. R. Wertheimer, *J. Vac. Sci. Technol.* **B9**, 8-25 (1991).
25. O. M. Küttel, J. E. Klemberg-Sapieha, L. Martinu and M. R. Wertheimer, *Thin Solid Films* **193/194**, 155-163 (1990).
26. D. M. Brewis and D. Briggs, *Polymer* **22**, 7-16 (1981).
27. T. Kasemura, S. Ozawa and K. Hattori, *J. Adhesion* **33**, 33-44 (1990).
28. M. Morra, E. Occhiello and F. Garbassi, *Surface Interface Anal.* **16**, 412-417 (1990).
29. Y.-L. Hsieh, D. A. Timm and M. Wu, *J. Appl. Polym. Sci.* **38**, 1719-1737 (1989).
30. C. Jones and E. Sammann, *Carbon* **28**, 509-514 (1990).
31. C. Jones and E. Sammann, *Carbon* **28**, 515-519 (1990).
32. G. S. Nadiger and N. V. Bhat, *J. Appl. Polym. Sci.* **30**, 4127-4136 (1985).
33. M. D. Smith, *Surface Modification of High Strength Reinforcing Fibers by Plasma Treatment,* KCP-613-4307. Allied-Signal Aerospace Company, Kansas City, KS (March 1990).
34. R. H. Hansen and H. Schonhorn, *J. Polym. Sci., Polym. Lett. Ed.* **B4**, 203-209 (1966).
35. K. Watanabe, M. Zelikoff and E. C. Y. Inn, *Absorption Coefficients of Several Atmospheric Gases,* AFCRL TR-53-23. Cambridge, MA (June 1953).
36. H. Yasuda, *J. Macromol. Sci. Chem.* **10**, 383-420 (1976).
37. H. Schonhorn, F. W. Ryan and R. H. Hansen, *J. Adhesion* **2**, 93-99 (1970).
38. F. Poncin-Epaillard, B. Chevet and J.-C. Brosse, *Makromol. Chem.* **192**, 1589-1599 (1991).
39. R. H. Partridge, *J. Chem. Phys.* **49**, 3656-3668 (1968).
40. D. T. Clark and A. Dilks, *J. Polym. Sci, Polym. Chem. Ed.* **17**, 957-976 (1979).
41. D. T. Clark and A. Dilks, *J. Polym. Sci., Polym. Chem. Ed.* **15**, 2321-2345 (1977).
42. O. Auciello and D. L. Flamm (Eds), *Plasma Diagnostics.* Academic Press, Boston (1989).
43. M. Hudis and L. E. Prescott, *Polym. Lett.* **10**, 179-183 (1972).
44. J. A. R. Samson, *Techniques of Vacuum Ultraviolet Spectroscopy.* John Wiley, New York (1967).
45. M. G. Ury, Fusion Systems Corp. personal communication.
46. E. M. Liston, *Proc. IUPAC Int. Symp. on Plasma Chem. (ISPC-9),* L7-L12 (Sept. 1989).
47. E. M. Liston, unpublished results.
48. J. E. Klemberg-Sapieha, L. Martinu, O. M. Küttel and M. R. Wertheimer, in: *Metallized Plastics 2: Fundamental and Applied Aspects,* K. L. Mittal (Ed.), pp. 315-329. Plenum Press, New York (1991).
49. J. E. Klemberg-Sapieha, O. M. Küttel, L. Martinu and M. R. Wertheimer, *J. Vac. Sci. Technol.* **A9**, 2975-2981 (1991).
50. E. M. Liston, *Proc. IUPAC Int. Symp. on Plasma Chem. (ISPC-7),* 513-517 (1985).
51. R. H. Partridge, *J. Chem. Phys.* **47**, 4223-4227 (1967).
52. L. R. Painter, E. T. Arakawa, M. W. Williams and J. C. Ashley, *Radiat. Res.* **83**, 1-18 (1980).
53. E. T. Arakawa, M. W. Williams, J. C. Ashley and L. R. Painter, *J. Appl. Phys.* **52**, 3579-3582 (1981).
54. R. H. Partridge, *J. Chem. Phys.* **45**, 1685-1690 (1966).
55. F. D. Egitto and L. J. Matienzo, *Polym. Degrad. Stab.* **30**, 293-308 (1990).
56. N. J. DeLollis, *The Use of RF Activated Gas Treatment To Improve Bondability,* SC-RR-71 0920. Sandia Laboratories, Albuquerque, NM (Jan. 1972).
57. M. Hudis, *J. Appl. Polym. Sci.* **16**, 2397-2415 (1972).
58. R. E. Cohen, R. F. Baddour and G. A. Corbin, *Proc. IUPAC Int. Symp. on Plasma Chem. (ISPC-6),* 537-541 (1983).
59. A. G. Shard and J. P. S. Badyal, *Polym. Commun.* **32**, 217-219 (1991).

60. F. Normand, J. Marec, P. Leprince and A. Granier, *Mater. Sci. Eng.* **A139**, 103–109 (1991).
61. B. G. Ranby and J. F. Rabek, *Photodegradation, Photo-Oxidation and Photostabilization of Polymers.* Wiley-Interscience, New York (1975).
62. M. A. Golub and T. Wydeven, *Polym. Degrad. Stab.* **22**, 325–338 (1988).
63. J. Lub, F. C. B. M. van Vroonhoven, E. Bruninx and A. Benninghoven, *Polymer* **30**, 40–44 (1989).
64. J. J. Pireaux, C. Gregoire, M. Vermeersch, P. A. Thiry, M. Rei Vilar and R. Caudano, in: *Metallization of Polymers*, E. Sacher, J. J. Pireaux and S. Kowalczyk (Eds), pp. 47–59. ACS Symp. Ser. No. 440. American Chemical Society, Washington, DC (1990).
65. S. C. Gujrathi, in: ref. 64, pp. 88–109.
66. Y. De Puydt, P. Bertrand and P. Lutgen, *Surface Interface Anal.* **12**, 486–490 (1988).
67. K. L. Mittal (Ed), *Adhesion Aspects of Polymeric Coatings.* Plenum Press, New York (1983).
68. T. J. Hook, J. A. Gardella and L. Salvati, *J. Mater. Res.* **2**, 132–142 (1987).
69. D. S. Everhart and C. N. Reilley, *Anal. Chem.* **53**, 665–676 (1981).
70. L. Cop, J. Jordaan, H. P. Schreiber and M. W. Wertheimer, US Patent 4,744,860 (1987).
71. J. R. Hollahan, B. B. Stafford, R. D. Falb and S. T. Payne, *J. Appl. Polym. Sci.* **13**, 807–816 (1969).
72. G. A. Corbin, R. E. Cohen and R. F. Baddour, *Polymer* **23**, 1546–1548 (1982).
73. M. Anand, R. E. Cohen and R. F. Baddour, *Polymer* **22**, 361–371 (1981).
74. S. Sapieha, M. Verreault, J. E. Klemberg-Sapieha, E. Sacher and M. R. Wertheimer, *Appl. Surface Sci.* **44**, 165–169 (1990).
75. E. Sacher, H. P. Schreiber and M. R. Wertheimer, US Patent 4,557,946 (1985).
76. D. S. Everhart and C. N. Reilley, *Surface Interface Anal.* **3**, 126–133 (1981).
77. R. Foerch, N. S. McIntyre, R. N. S. Sodhi and D. H. Hunter, *J. Appl. Polym. Sci.* **40**, 1903–1915 (1990).
78. T. F. McLaughlin, Jr., *Information Bulletin*, Du Pont de Nemours & Co. (1962).
79. G. W. Traver, US Patent 3,018,189 (1962).
80. M. Strobel, C. Dunatov, J. M. Strobel, C. S. Lyons, S. J. Perron and M. C. Morgen, *J. Adhesion Sci. Technol.* **3**, 321–335 (1989).
81. J. W. Chin and J. P. Wightman, *J. Adhesion* (in press).
82. C. Y. Kim, J. Evans and D. A. I. Goring, *J. Appl. Polym. Sci.* **15**, 1365–1375 (1971).
83. M. Gamez-Garcia, R. Bartnikas and M. R. Wertheimer, *IEEE Trans. Electr. Insul.* **22**, 199–205 (1987).
84. C. Hudon, R. Bartnikas and M. R. Wertheimer, *Proc. IEEE CEIDP, IEEE Doc. 91CH 3055-1*, 237–243 (1991).
85. R. d'Agostino, F. Cramarossa, F. Fracassi and F. Illuzzi, in: ref. 19, pp. 95–162.
86. A. M. Wrobel and M. R. Wertheimer, in: ref. 19, pp. 163–268.
87. E. Sacher, J. E. Klemberg-Sapieha, H. P. Schreiber and M. R. Wertheimer, *J. Appl. Polym. Sci., Appl. Polym. Symp.* **38**, 163–172 (1984).
88. S. Sapieha, A. M. Wrobel and M. R. Wertheimer, *Plasma Chem. Plasma Process.* **8**, 331–346 (1988).
89. A. Bradley and J. D. Fales, *Chem. Tech.* **1**, 232–237 (1971).
90. N. Inagaki, S. Tasaka and H. Kawai, *J. Adhesion Sci. Technol.* **3**, 637–649 (1989).
91. M. Morra, E. Occhiello, R. Marola, F. Garbassi, P. Humphrey and D. Johnson, *J. Colloid Interface Sci.* **137**, 11–23 (1990).
92. D. H. Kaelble, P. J. Dynes and E. H. Cirlin, *J. Adhesion* **6**, 23–48 (1974).
93. E. Sacher, in: *Surface Characterization of Biomaterials*, B. D. Ratner (Ed.), pp. 53–64. Elsevier, Amsterdam (1988).
94. M. Morra, E. Occhiello and F. Garbassi, *Adv. Colloid Interface Sci.* **32**, 79–116 (1990).
95. F. M. Fowkes, *J. Adhesion Sci. Technol.* **1**, 7–27 (1987).
96. M. B. Kaczinski and D. W. Dwight, *J. Adhesion Sci. Technol.* **7**, 165–177 (1993).
97. F. M. Fowkes, M. B. Kaczinski and D. W. Dwight, *Langmuir* **7**, 2464–2477 (1991).
98. H. P. Schreiber, in: ref. 4, pp. 21–60.
99. D. R. Lloyd, T. C. Ward and H. P. Schreiber (Eds), *Inverse Gas Chromatography*, ACS Symp. Ser. No. 391. American Chemical Society, Washington, DC (1989).
100. E. Occhiello, M. Morra, G. Morini, F. Garbassi and P. Humphrey, *J. Appl. Polym. Sci.* **42**, 551–559 (1991).

101. T. Yasuda, M. Miyama and H. Yasuda, *Langmuir* **8**, 1425–1430 (1992).
102. E. Occhiello, M. Morra, F. Garbassi, D. Johnson and P. Humphrey, *Appl. Surface Sci.* **47**, 235–242 (1991).
103. F. Garbassi, M. Morra, E. Occhiello, L. Barino and R. Scordamaglia, *Surface Interface Anal.* **14**, 585–589 (1989).
104. M. Morra, E. Occhiello, L. Gila and F. Garbassi, *J. Adhesion* **33**, 77–88 (1990).
105. M. Morra, E. Occhiello and F. Garbassi, *J. Colloid Interface Sci.* **132**, 504–508 (1989).
106. H. L. Spell and C. P. Christenson, *Tappi* **62**, 77–81 (1979).
107. M. Morra, E. Occhiello and F. Garbassi, in: ref. 48, pp. 363–371.
108. M. H. Bernier, J. E. Klemberg-Sapieha, L. Martinu and M. R. Wertheimer, in: ref. 64, pp. 147–160.
109. H. F. Webster and J. P. Wightman, *J. Adhesion Sci. Technol.* **5**, 93–106 (1991).
110. J. D. Moyer and J. P. Wightman, *Surface Interface Anal.* **17**, 457–464 (1991).
111. E. Occhiello, M. Morra, G. Morini, F. Garbassi and D. Johnson, *J. Appl. Polym. Sci.* **42**, 2045–2052 (1991).
112. S. Nowak, H.-P. Haerri, K. Schlapbach and J. Vogt, *Surface Interface Anal.* **16**, 418–423 (1990).
113. L. Martinu, J. E. Klemberg-Sapieha, H. P. Schreiber and M. R. Wertheimer, *Vide, Suppl.* **258**, 13–20 (1991).
114. S. Sapieha, J. Cerny, J. E. Klemberg-Sapieha and L. Martinu, *J. Adhesion* (in press).
115. H. P. Schreiber, M. R. Wertheimer and M. Lambla, *J. Appl. Polym. Sci.* **27**, 2269–2282 (1982).
116. A. Bialski, R. St. J. Manley, M. R. Wertheimer and H. P. Schreiber, *J. Macromol. Sci. Chem.* **A10**, 609–618 (1976).
117. H. P. Schreiber, Y. B. Tewari and M. R. Wertheimer, *J. Appl. Polym. Sci.* **20**, 2663–2673 (1976).
118. M. R. Wertheimer and H. P. Schreiber, *J. Appl. Polym. Sci.* **26**, 2087–2096 (1981); Canadian Patent 1,122,566 (1982).
119. I. K. Ismail and M. D. Vangeness, *Carbon* **26**, 749–751 (1988).
120. B. Z. Jang and H. Das, *Interfaces in Polymer, Ceramic, and Metal Matrix Composites*, pp. 319–333. Elsevier, New York (1988).
121. R. E. Allred, *Proc. 29th Natl. SAMPE Symp. Exhib.* pp. 947–957 (1984); R. E. Allred, E. W. Merrill and D. K. Roylance, *Polym. Sci. Technol.* **27**, 333–375 (1985).
122. J. M. Burkstrand, *J. Vac. Sci. Technol.* **15**, 223–226 (1978).
123. L. J. Gerenser, *J. Vac. Sci. Technol.* **A6**, 2897–2903 (1988).
124. S. Nowak, R. Mauron, G. Dietler and L. Schlapbach, in: ref. 48, pp. 233–244.
125. G. Liebel and R. Bischoff, *Kunststoffe-German Plastics* **4**, 4 (1987).
126. L. Martinu, V. Pische and R. d'Agostino, in: ref. 64, pp. 170–178.
127. H. Biederman and L. Martinu, in: ref. 19, pp. 269–320.
128. K. J. Blackwell, P. C. Chen, A. R. Knoll and J. Y. Kim, *Proc. 35th Annu. Tech. Conf.*, pp. 279–283. Society of Vacuum Coaters (1992).
129. R. Padiyath, M. David and S. V. Babu, in: ref. 48, pp. 113–120.
130. H. Meyer, M. Schulz, H. Suhr, C. Haag, K. Horn and A. M. Bradshaw, in: ref. 48, pp. 121–130.
131. B. D. Ratner, A. Chilkoti and G. P. Lopez, in: ref. 19, pp. 463–516.
132. W. Landman, Model 7500S, GaSonics/International Plasma Corporation, 2730 Junction Avenue, San Jose, CA 95134-1909, USA, product literature.
133. G. Liebel, Model 4002-B, Technics Plasma GmbH, Dieselstrasse 22a, D8011 Kirchheim bei Munchen, Germany, product literature.
134. W. Landman, Model 8150, GaSonics/International Plasma Corporation, 2730 Junction Avenue, San Jose, CA 995134-1909, USA, product literature.
135. J. T. Felts, Model 'Flex-1', AIRCO Coatings Technology, P.O. Box 4105, Concord, CA 94524, USA, product literature.
136. J. Kieser and M. Neusch, *Thin Solid Films* **118**, 203–210 (1984).
137. R. G. Bosisio, M. R. Wertheimer and C. F. Weissfloch, *J. Phys. E.: Sci. Instrum.* **6**, 628–630 (1973); US Patent 3,814,983 (1974).
138. M. R. Wertheimer, German Patent 41325567 (1992).

2

Characterization
of Plasma-treated Surfaces

Plasma Surface Modification of Polymers, pp. 43–64
M. Strobel, C. Lyons and K. L. Mittal (Eds)
© VSP 1994

XPS studies of *in situ* plasma-modified polymer surfaces

L. J. GERENSER

*Analytical Technology Division, Research Laboratories, Eastman Kodak Company, Rochester,
NY 14650-2132, USA*

Revised version received 25 March 1993

Abstract—X-ray photoelectron spectroscopy (XPS) has been used to study the chemical effects of
both inert (argon) and reactive (oxygen, nitrogen, and mixed gas) plasma treatments done *in situ* on a
variety of polymer surfaces. Inert gas plasma treatments introduce no new detectable chemical
species onto the polymer surface but can induce degradation and rearrangement of the polymer
surface. However, plasma treatments with reactive gases create new chemical species which
drastically alter the chemical reactivity of the polymer surface. These studies have also shown that the
surface population of chemical species formed after plasma treatment is dependent on both the
chemical structure of the polymer and the plasma gas. The effects of direct and radiative energy-
transfer processes in a plasma have also been studied. Polymers containing certain functional groups
were found to be more susceptible to damage via radiative energy transfer. Ageing studies of plasma-
modified polymer surfaces exposed to the atmosphere have shown that the ageing process consists of
two distinct phases. The initial phase, which occurs rapidly, involves adsorption of atmospheric
contaminants and, in some cases, specific chemical reactions. The second phase, which occurs slowly,
is due to surface reorganization.

Keywords: Adhesion; degradation; *in situ*; plasma; polymer; stability; surface modification; XPS.

1. INTRODUCTION

The modification of the chemical structure, reactivity, and bonding characteristics of polymer surfaces has considerable technological importance in the areas
of metallization, composite fabrication, and biomedical compatibility. Plasma
(glow discharge) treatment is one type of surface modification that is commonly
used. Plasma treatment can be used to modify the surface of a polymer to improve
adhesion or wettability, to provide a diffusion barrier layer, or to minimize
degradation of a polymer surface during metallization. A unique feature of plasma
modification is that the surface structure of the polymer can be selectively
modified for a specific application while the bulk properties of the polymer are
unaffected. However, it must be noted that the complexity of the plasma itself
makes it difficult to unravel the mechanisms responsible for the surface modification. For example, the roles of direct and radiative energy transfer in plasma-
induced surface modification are still unresolved. The general consensus is that
reactions at the immediate surface are due to a combination of both direct and
radiative energy transfer while the subsurface reactions are dominated by
radiative energy transfer through the UV component of the electromagnetic
spectrum.

X-ray photoelectron spectroscopy (XPS) is an ideal technique for studying
plasma-induced chemistry at polymer surfaces because of its surface sensitivity
(~ 1–10 nm), semi-quantitative nature, and the ability to obtain chemical bonding

information. Through the use of angular-dependent XPS, a composition depth profile of the plasma-modified region can be determined.

Numerous XPS studies have been done to gain insights into the chemistry of plasma-modified polymer surfaces [1–12]. A variety of gases were used in these studies, including argon, oxygen, nitrogen, ammonia, water, and combinations of the above. Primarily RF (radio-frequency) plasmas were employed. In most of these studies, the plasma treatments were done in a separate chamber and transfer of the modified polymer to the spectrometer in air was required. It has been shown that the exposure of these highly reactive modified surfaces to ambient conditions can produce ambiguous results due to adsorption of contaminants or chemical reaction with the ambient [8]. Therefore, some researchers have resorted to *in situ* plasma treatments, alleviating the problems associated with exposure to the atmosphere [5, 7, 8].

In this paper, the results of the plasma modification of several polymer surfaces, including polyethylene (PE), polystyrene (PS), poly(ethylene tereph-thalate) (PET), bis-phenol-A-polycarbonate (BPAPC), and poly(methyl meth-acrylate) (PMMA) will be discussed. The gases employed include argon, oxygen, nitrogen, and argon/oxygen mixtures. All plasma treatments were done *in situ* in the spectrometer preparation chamber.

2. EXPERIMENTAL

The XPS spectra were obtained on a Hewlett Packard 5950A photoelectron spectrometer with a monochromatic Al $K\alpha$ X-ray source (1486.6 eV). The use of a monochromatic source is especially important for polymers to minimize sample radiation damage and to provide the spectral resolution necessary to determine plasma-induced species. All polymers were analyzed at ambient temperature and exhibited no evidence of damage during measurements. Typically, the full width at half-maximum (FWHM) for the individual components of the C $1s$ peak in a clean polymer sample varied from 0.8 to 0.9 eV. All spectra were referenced to the C $1s$ peak for neutral carbon in the polymer, which was assigned a value of 284.6 eV. Where necessary, line-shape analyses were done using a least-squares deconvolution routine employing line shapes with 90% Gaussian/10% Lorentzian character. All spectra were taken at an electron takeoff angle (ETOA) of 38° unless noted otherwise. The ETOA is the angle between the sample surface and the electron lens. Angle-resolved measurements were made using a Surface Science Laboratory Model 259 angular-rotation probe and altering the magnifi-cation of the four-element electron lens from -5.0 to -2.3 [13]. This procedure provides improved resolution at high ETOAs and also reduces the electron acceptance solid angle of the electron lens from a 3.5° to a 2.8° half-angle cone.

The plasma treatments were done in the preparation chamber of the spectrometer (base pressure 5×10^{-9} Torr) by applying a potential to a 6 mm diameter Al rod mounted on a high-voltage feedthrough using a 60 Hz, 5 kV power supply. The sample rod was floated with the chamber walls at ground potential. The sample-to-electrode distance was approximately 50 mm with the sample residing in the positive column of the plasma during treatment. The sample was not in a direct line-of-sight of the powered electrode. The typical operating conditions were 10 W of primary power under flowing gas at 50 mTorr pressure for 5–90 s. The gases used were Matheson Research Grade argon, oxygen, nitrogen, and argon/oxygen mixtures.

The PS was additive-free, narrow molecular-weight distribution (110 k M_w, 1.04 M_w/M_n) obtained from Scientific Polymer Products. It was dissolved in cyclohexane and spin-coated onto a silicon wafer to a thickness of ~ 20 Nm.

The PMMA was additive-free, narrow molecular-weight distribution (107 k M_w, 1.10 M_w/M_n) obtained from Scientific Polymer Products. It was dissolved in tetrahydrofuran and spin-coated onto a silicon wafer to a thickness of ~ 20 nm.

The BPAPC was non-UV-stabilized Lexan 101–112 grade polycarbonate (General Electric Company) (49 k M_w, 2.13 M_w/M_n). It was purified by dissolution in methylene chloride followed by filtration and precipitation with acetone. After purification, the BPAPC was dissolved in methylene chloride and spin-coated onto a silicon wafer to a thickness of ~ 20 nm.

The PE (hydrogenated polybutadiene) was additive-free, narrow molecular-weight distribution (108 k M_w, 1.32 M_w/M_n) obtained from Scientific Polymer Products. PE films were prepared by compression molding above the melting point in a Tetrahedron press between sheets of polyimide (Upilex 125S).

The PET was 0.18 mm thick, biaxially oriented Kodak ESTAR. Since the PET was not prepared in the laboratory and its surface cleanliness varied, the PET samples were washed consecutively in a series of solvents (heptane, dichloromethane, ethanol, and ethyl acetate) and dried in a dry nitrogen atmosphere in a glove bag attached to the entrance port of the spectrometer. Immediately after drying, the PET samples were inserted into the preparation chamber of the spectrometer, where they were evacuated to ~ 5×10^{-9} Torr.

Prior to plasma treatment, all polymers were annealed in the spectrometer preparation chamber (~ 1 h) to drive off any residual solvent, water, or adsorbed gases. The annealing temperature corresponded to a value slightly higher than the polymer T_g, except for PE and PET which were annealed at ~ 100°C. After annealing, the polymer surfaces were characterized with XPS. Both the core and the valence levels were analyzed at low (10°) and high (80°) ETOAs to verify the surface cleanliness and stoichiometry.

3. RESULTS AND DISCUSSION

3.1. Surface chemical structure

PE is an example of a relatively simple polymer to study with XPS because it consists of an aliphatic chain containing only one type of carbon atom. Because of its simplicity, plasma-induced effects are easier to discern compared with oxygen- or nitrogen-containing polymers. PE has been studied in detail previously [8] and will be referred to in comparison with other polymers.

PS is another example of a structurally simple polymer to study with XPS. In addition, PS contains a pendant phenyl ring which provides a distinctive energy-loss feature in XPS due to a $\pi-\pi^*$ shakeup transition. Changes in this energy-loss feature can provide information regarding the extent of ring-opening induced by plasma treatment. PS can be compared with PE to determine whether the aromatic system has an effect on the rate of incorporation and the population of plasma-induced species.

Survey scans for clean untreated, argon-, oxygen-, and nitrogen-plasma-modified PS surfaces are shown in Fig. 1. The spectra for the untreated and argon-plasma-modified PS contain a single peak centered at 284.6 eV due to electrons originating from the C 1s level. The spectrum for the oxygen-plasma-modified PS consists of two intense peaks, the C 1s at 284.6 eV and one centered at 532.0 eV due to electrons originating from the O 1s level. A weak peak centered at 27.0 eV due to electrons originating from the O 2s level and the oxygen KLL Auger peaks at 977.0 and 998.0 eV are also observed in this spectrum. The spectrum for the nitrogen-plasma-modified PS also consists of two intense peaks, the C 1s at 284.6 eV and one centered at 399.0 eV due to electrons originating from the N 1s level. No other peaks are observed in the spectrum for the nitrogen-plasma-modified PS. All four spectra also contain weak bands (0–20 eV) due to the valence electrons.

The survey scans for PS are very similar to those observed for PE except for the sloping background beginning at ~ 100 eV for the untreated, argon-plasma- and oxygen-plasma-modified PS. The sloping background is due to inelastically scattered electrons originating from the silicon substrate which have sufficient mean free paths to escape from the thin polymer overlayer (~ 20 nm). The slope

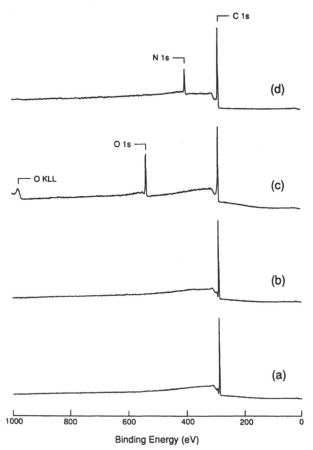

Figure 1. Survey scans for (a) untreated, (b) 15 s argon-plasma-, (c) 15 s oxygen-plasma-, and (d) 15 s nitrogen-plasma-modified PS.

of this background can be used to estimate the thickness of the overlayer when no peak from the substrate is detectable. The slope increases slightly after argon-plasma modification and more significantly after oxygen-plasma modification. The most interesting aspect is the absence of a sloping background for the nitrogen-plasma-modified PS. These results suggest some etching or ablation with argon and oxygen but none with nitrogen. In fact, the spectrum for the nitrogen-plasma modification suggests either a thicker overlayer, a denser overlayer, or possibly the filling-in of pinholes or thin areas of the spin-coated PS. These effects will be discussed in more detail in Section 3.2.

Basd on the survey scans, it is obvious that argon-plasma treatment does not introduce new chemical species into the PS surface, whereas oxygen- and nitrogen-plasma treatments incorporate only oxygen and nitrogen, respectively, into the PS surface. No other species were detected on the modified PS surfaces. These results are very similar to those found for plasma-modified PE [8] except for the amount of incorporated oxygen and nitrogen. A greater amount of oxygen (18% compared with 15%) and nitrogen (15% compared with 8%) is incorporated into the PS surface under identical treatment conditions and ETOAs. The difference in the oxygen incorporation is small ($\sim 20\%$) but significant; however, the difference in nitrogen incorporation is a factor of 2. These results suggest that the surface structure of PS (π system, surface crystallinity, etc.) has an effect on the rate of incorporation of plasma-induced species. Similar effects have been observed for the plasma treatment of PE and PS by other researchers [14, 15].

Survey scans are only suitable for providing elemental information about the polymer surface. In order to elucidate the plasma-induced chemical changes, high-resolution scans of the core and valence levels are necessary. High-resolution scans of the C 1s region for PS before and after various plasma treatments are shown in Fig. 2. The C 1s peak for the untreated PS consists of a relatively narrow primary peak (0.90 eV FWHM), centered at 284.6 eV and the $\pi-\pi^*$ shakeup peak at 290.6 eV. The area ratio of the $\pi-\pi^*$ shakeup peak to the primary peak is 6.3% ± 0.1. Although the survey scan for the argon-plasma-modified PS does not indicate the incorporation of new species, the high-resolution scan of the C 1s region indicates broadening of the primary peak (1.2 eV FWHM) and a loss of intensity of the $\pi-\pi^*$ shakeup peak. The area ratio of the $\pi-\pi^*$ shakeup peak to the primary peak decreases to 2.5% ± 0.1. These results suggest sample damage during plasma treatment involving bond breaking of the phenyl ring and possibly chain scission of the PS backbone. In contrast, the C 1s spectra for the oxygen- and nitrogen-plasma-modified PS contain definite features at high binding energy. The high binding energy structure is indicative of the formation of carbon–oxygen and carbon–nitrogen bonds in the oxygen- and nitrogen-plasma-modified PS, respectively. Both spectra also exhibit a loss in intensity of the $\pi-\pi^*$ shakeup peak similar to that of the argon-plasma-modified PS.

A more detailed understanding of these new species can be obtained using a line-shape analysis deconvolution routine as shown in Fig. 3. The line-shape analysis of the C 1s spectrum for the oxygen-plasma-modified PS indicates the formation of three distinct carbon–oxygen species (C—O at 286.1 eV, C=O at 287.5 eV, and O—C=O at 288.9 eV), with C—O the most prevalent. However,

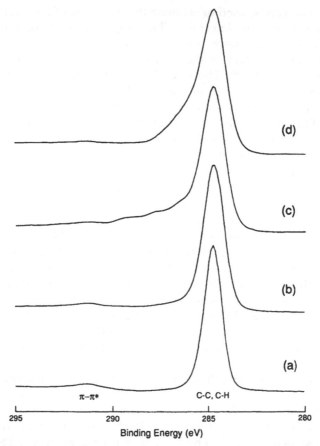

Figure 2. C 1s spectra for (a) untreated, (b), 15 s argon-plasma-, (c) 15 s oxygen-plasma-, and (d) 15 s nitrogen-plasma-modified PS.

the envelope occupied by each of the three carbon–oxygen peaks can contain more than one unique carbon–oxygen species [8]. For example, C—O can be a hydroxyl, an ether, or an epoxy group. Based on the integrated area under the O 1s peak and the total integrated area under the various carbon–oxygen species in the C 1s spectrum, an estimate can be made of the relative amounts of hydroxyl, ether, and/or epoxy groups. For these ratios to be internally consistent, less than 50% of the peak due to the C—O species must be composed of two carbon atoms bonded to one oxygen atom. This is in contrast to oxygen-plasma-modified PE, where ether and/or epoxy groups constitute the major portion of the C—O peak [8].

The O 1s spectrum for oxygen-plasma-modified PS is relatively broad and non-descript, with a slight skew to higher binding energy. The broadness of the peak definitely suggests more than one unique oxygen environment. Although attempts were made to deconvolute the spectrum into separate components, more than one unique fit could be made depending on the number and position of the component peaks used. Therefore, the total integrated area under the O 1s peak was only used as an internal check on the consistency of the component analysis of the C 1s spectrum.

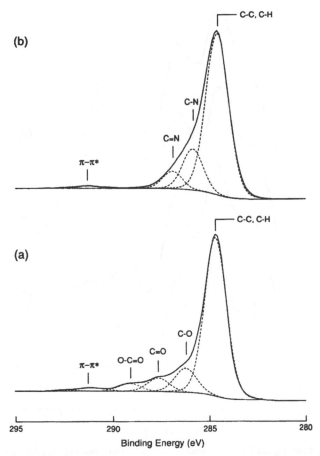

Figure 3. Line-shape analysis of the high resolution C 1s spectra for (a) 15 s oxygen-plasma and (b) 15 s nitrogen-plasma-modified PS.

A line-shape analysis of the C 1s spectrum for the nitrogen-plasma-modified PS provides a distinctly different fit compared with the oxygen-plasma-modified PS. For this spectrum, only two peaks can be fitted to the high binding energy region. These peaks have been assigned to amine (285.8 eV) and imine (287.0 eV) carbon species and are present in a 3:1 ratio. The N 1s spectrum is broad but asymmetric with a definite skew to high binding energy, indicating the presence of more than one nitrogen species. A line-shape analysis of the N 1s spectrum provides two peaks at 399.0 eV (amine) and 400.3 eV (imine), also in a 3:1 ratio. In order for the integrated areas under the C 1s and N 1s peaks to be internally consistent, the majority of the amine and imine species must contain a terminal nitrogen (primary amine or imine). These results are similar to those found for nitrogen-plasma-modified PE [8].

Analysis of the valence band region can provide additional insights into plasma-induced chemical changes. The valence band spectra for plasma-modified PS are shown in Fig. 4. All plasma treatments produce considerable band-broadening, suggesting bond breaking in the polymer. In addition, the oxygen- and nitrogen-plasma-modified PS spectra exhibit significant intensity changes.

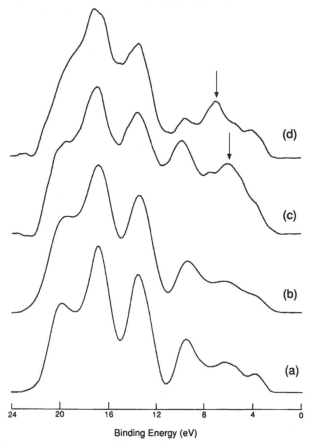

Binding Energy (eV)

Figure 4. Valence band region for (a) untreated, (b) 15 s argon-plasma-, (c) 15 s oxygen-plasma-, and (d) 15 s nitrogen-plasma-modified PS. The arrows indicate the primary plasma-induced features.

The most noteworthy effect is the large increase in the density of states near the top of the valence band (2–10 eV) for both the oxygen- and the nitrogen-plasma-modified PS. These bands probably contain significant O $2p$ and N $2p$ character and are associated with the incorporated oxygen and nitrogen, respectively. The intensities of these bands are similar to those observed for high surface energy polymers and can be directly responsible for improved wetting and adhesion. We have demonstrated the importance of these bands in interfacial chemistry occurring during the initial stages of the nucleation of evaporated silver [16–18]. Molecular orbital calculations have shown that the O $2p$ and N $2p$ orbitals, which provide the majority of the intensity for these bands, overlap with silver $4d$ orbitals to form chemical bonds with evaporated silver [19]. This bond formation is responsible for the improved adhesion observed for evaporated silver on oxygen- and nitrogen-plasma-modified polymer surfaces.

XPS interpretation of plasma-induced changes on oxygen- or nitrogen-containing polymers is more difficult due to the more complex chemical structure of the polymer. A combination of line-shape deconvolution and spectral subtraction techniques is often required to elucidate the plasma-induced species for these polymers. An example is shown in Fig. 5 for oxygen-plasma-modified

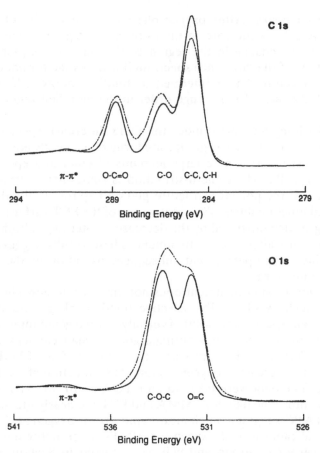

Figure 5. Overlay of C 1s and O 1s spectra for untreated (solid line) and 15 s oxygen-plasma-modified (dashed line) PET.

PET. The C 1s spectrum for PET contains four distinct peaks: the carbon atoms in the phenyl ring at 284.6 eV, the methylene carbon atoms singly bonded to oxygen at 286.1 eV, the ester carbon atoms at 288.6 eV, and the $\pi-\pi^*$ shakeup peak at 290.6 eV. The area ratio of the three primary peaks is 3:1:1, as expected based on PET stoichiometry. The $\pi-\pi^*$ shakeup peak is ~4.5% of the total integrated area of the primary peaks. After oxygen-plasma modification, the area ratio of the three primary peaks changes to 2:1:1. The high binding energy peak centroids are also shifted slightly to higher energy (~0.2 to 0.3 eV) compared with the untreated PET, and the valley between them at ~287.6 eV begins to fill in. These differences are more apparent in the C 1s difference spectrum. The C 1s difference spectrum suggests that a nearly equal distribution of O—C=O, C=O, and C—O species is produced and that possibly a small amount of carbonate groups is also formed.

Information from the O 1s spectrum is more difficult to interpret. The O 1s spectrum for the clean PET surface contains three distinct peaks due to the carbonyl oxygen atoms at 531.8 eV, the ester oxygen atoms at 533.5 eV, and the $\pi-\pi^*$ shakeup feature at 538.5 eV. The main peaks are present in a 1:1 ratio, as expected for clean PET. The $\pi-\pi^*$ shakeup peak is ~2.5% of the total integrated area of the main peaks. After oxygen-plasma treatment, the O 1s spectrum appears to have a large increase in intensity in the region of the ester oxygen (533.5 eV) with no change in the region of the carbonyl oxygen (531.8 eV). However, analysis of the difference spectrum illustrates the formation of a broad band with a peak centroid at an intermediate binding energy (532.8 eV). This is consistent with the lack of long-range order in the modified region, resulting in one broad peak.

Based on the integrated area under the O 1s difference spectrum, primarily carboxyl, carbonyl, and hydroxyl groups are formed in a nearly equal distribution. This is in contrast to PE, where large amounts of ether and epoxy groups are formed [8], and to PS, where the distribution is somewhat intermediate. These results suggest that the polymer structure plays a key role in the distribution of plasma-induced functionalities. The reactive site of the PET surface appears to be the phenyl ring, as demonstrated by the decrease in intensity of both the C 1s and O 1s shakeup peaks after plasma treatment. These results suggest that plasma treatments induce ring-opening and subsequent oxidation at the PET surface similar to that found for PS.

Although argon-plasma treatment does not incorporate new chemical species into polymers such as PE or PS, selective bond breakage and desorption of various short-chain species can occur. Typically, for oxygen-containing polymers, the oxygen/carbon ratio decreases during argon-plasma treatment, probably due to the loss of CO or CO_2. Examples are shown in Fig. 6 for PMMA, PET, and BPAPC, where the decrease in oxygen is plotted as a function of the argon-plasma treatment time. The data were taken at an ETOA of 10° to maximize the near-surface contribution. The data for PMMA and PET are nearly identical; however, the data for BPAPC suggest a much more rapid loss of oxygen at short treatment times. The initial rapid loss of oxygen for BPAPC is probably due to the labile nature of the carbonate group and will be discussed in Section 3.3. All three polymer surfaces reach an equilibrium value of 30–35% oxygen loss. Thus, argon

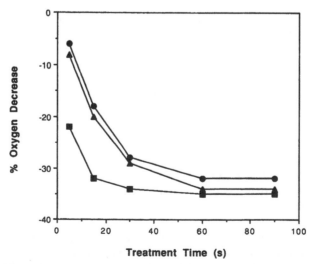

Figure 6. Decrease in surface oxygen as a function of the argon-plasma treatment time for PMMA (●), PET (▲), and BPAPC (■). The data were taken at an ETOA of 10°.

is not the gas of choice for producing improved adhesion or wettability. In fact, our previous studies have shown that argon-plasma treatment has no positive effect on the adhesion of evaporated metals [16–19].

Unlike plasma treatments with pure argon, plasma treatments done with small amounts of oxygen mixed with argon introduce reactive functional groups into polymer surfaces. Mixtures of 1, 5 and 10% oxygen in argon were used to modify various polymer surfaces (Table 1). Significant amounts of oxygen were introduced in all cases, although the amount of incorporated oxygen is less than that observed for pure oxygen. In fact, the amount of incorporated oxygen approximately tracks the percentage of oxygen in the gas mixture with the argon/ 10% oxygen approaching the value for pure oxygen. These results are consistent with the improved adhesion observed for evaporated metals on polymers treated with argon/oxygen mixtures as compared with pure argon [20]. These results illustrate the effect that small amounts of oxygen can have on argon-plasma-modified polymer surfaces and should be considered when improved wettability or adhesion is claimed for argon-plasma treatment.

Table 1.
Atomic percent oxygen determined by XPS at an ETOA of 38° for several polymer surfaces modified for 15 s with argon and argon/oxygen gas plasmas

Plasma gas	PE	PS	PET
Untreated	0	0	28
Argon	0	0	26
Argon/1% oxygen	2	3	30
Argon/5% oxygen	5	7	32
Argon/10% oxygen	11	13	35
Oxygen	15	18	38

3.2. Depth of modification

Angular-dependent XPS measurements can be used to nondestructively depth-profile the plasma-modified region as shown in Fig. 7 for nitrogen-plasma-modified PET. The concentrations of nitrogen and oxygen (inherent in the PET) are plotted as a function of the ETOA. The near-surface contributions are enhanced at smaller ETOAs. To a first approximation, the ETOA corresponds to the analysis depth in angstroms for 950–1100 eV electrons in a polymer.

The data in Fig. 7 all suggest a concentration gradient of plasma-induced nitrogen species in the PET surface. The depth and shape of this concentration gradient change as a function of the plasma treatment time. Based on the angular-dependent measurements, the thickness of the modified overlayer varies between 1 and 5 nm depending on the treatment time. The angular-dependent data illustrate the importance of specifying the ETOA when reporting the concentration of plasma-induced species.

The minimum treatment level for optimum surface modification can be predicted from these measurements. Qualitatively, these plots demonstrate the reason for the optimum adhesion strength observed for evaporated silver between

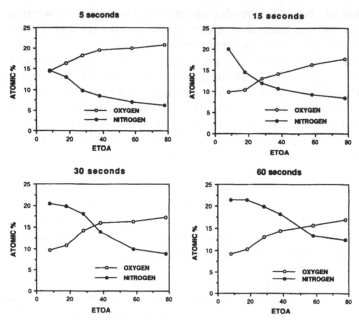

Figure 7. Angular-dependent XPS data for nitrogen-plasma-modified PET as a function of the plasma treatment time.

15 and 30 s of plasma treatment time [19, 20]. At 5 s, the uppermost surface (10° ETOA) has not yet reached saturation in terms of nitrogen incorporation. However, for treatment times of 15 s and longer, the uppermost surface region (10° ETOA) reaches a saturation level and further nitrogen incorporation occurs at greater depths (30–80° ETOA) into the polymer. Similar effects were observed for oxygen-plasma-modified PE [20]. Because only the first few atomic layers of the polymer are responsible for bonding with an overlayer, incorporation of nitrogen into the subsurface should not have a positive effect on adhesion. On the contrary, longer treatment times may induce more damage to the polymer subsurface, producing a weak boundary layer and subsequent loss of adhesion. Also, to minimize incompatibility between the modified and unmodified PET, a gradual transition from the nitrogen-rich region to the unmodified PET is desirable. Of the treatment times studied, the shape of the angular-dependent curves for the 15 s treatment time comes closest to meeting these requirements.

The angular-dependent XPS measurements raise the question of surface roughness and its effect on the data. The mild plasma treatments used in the studies reported here produced no roughening observable by scanning electron microscopy up to a magnification of ×20 000. Studies using atomic force microscopy (AFM) on PET suggest that the surfaces are extremely smooth (root mean square surface roughness ~ 1 nm) even with plasma treatment times of 60 s, although there are some long-range changes in surface morphology. As discussed by Fadley [21], the important criteria in determining the effect of surface roughness on angular-dependent data is the aspect ratio a/λ, where a is the peak-to-valley height and λ is the distance between peaks and valleys. The aspect ratios determined from AFM measurements before and after plasma treatment suggest that shadowing effects would be minimal for ETOAs as low as 10°.

Based on ellipsometry measurements of the thickness before and after various plasma treatments of spin-coated PS, PMMA, and BPAPC, etch rates $\leqslant 6$ nm/min were estimated for the plasma treatment conditions discussed here. The ellipsometry measurements assume that the optical constants of the spin-coated polymer do not change after plasma modification. This assumption may not be valid because the modified region (1–5 nm) can be a significant portion of the spin-coated polymer (20 nm). These etch rates were confirmed from XPS measurements of the spin-coated polymers. XPS could not detect any signal from the silicon substrate at an ETOA of 80° for treatment times up to 30 s, suggesting that the polymer layer is $\geqslant 8$–10 nm. At treatment times $\geqslant 60$ s, evidence of the silicon substrate was detected by XPS in all cases except the nitrogen-plasma-modified PS. Using the appropriate value for the electron inelastic mean free path for Si $2p$ electrons through an organic overlayer, the thickness of the remaining polymer layer was estimated. Based on the XPS overlayer-thickness measurements, etch rates of 0–2 nm/min for nitrogen-, 3–4 nm/min for argon-, and 5–6 nm/min for oxygen-plasma treatment were estimated. The etch rates indicate some variation depending on the polymer, but the data are not conclusive. These low etch rates for the spin-coated polymers explain the inability of AFM to detect surface roughening on the plasma-modified PET.

3.3. Degradation

BPAPC is an example of a polymer that undergoes severe bond breakage, desorption of short-chain species, and rearrangement when exposed to an argon plasma. In contrast to PMMA and PET, where the surface structure remains nearly intact, argon-plasma treatment of BPAPC produces a surface significantly depleted in carbonate groups (Figs 8 and 9). A 15 s argon-plasma treatment of PET (Fig. 8) produces line-broadening in the C $1s$ and O $1s$ levels and some loss of oxygen; however, the characteristic features due to the ester groups are still apparent in both spectra. These characteristic features remain apparent even at a treatment time of 60 s. By comparison, after 15 s argon-plasma treatment (Fig. 9), the O $1s$ fingerprint characteristic of a carbonate (2:1 intensity ratio of C—O to C=O) is no longer present. A best-fit line-shape analysis provides two peaks in a 1:1 ratio, with both peak positions shifted to a lower binding energy as compared with the untreated surface. These results are consistent with the C $1s$ level, where a large decrease in intensity of the peak due to the carbonate species and some loss of intensity of the π–π^* shakeup feature are observed. These changes are accompanied by the formation of a broad band extending from the primary peak to the carbonate peak, suggesting the formation of hydroxyl, carbonyl, carboxyl, and ester functionalities. These results illustrate the labile nature of the carbonate group in BPAPC.

Since the carbonate group appears to be very susceptible to damage in an argon plasma, the following experiment was done to separate the effects of direct and radiative energy transfer. BPAPC was exposed to UV radiation in vacuum via a mercury vapor lamp (8.4 W/cm^2 at 360 nm) through a quartz window attached to the spectrometer preparation chamber. Although the spectral emission lines for mercury are different from those for argon, the work of Egitto and Matienzo [22] suggest that the modification of polymers by UV radiation requires photons with

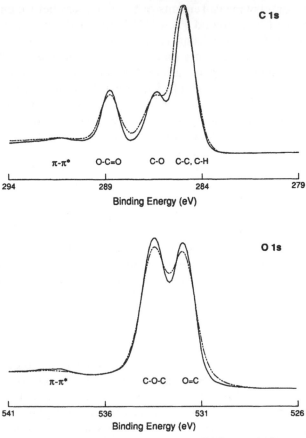

Figure 8. Overlay of C 1s and O 1s spectra for untreated (solid line) and 15 s argon-plasma-modified (dashed line) PET.

energies equal to or greater than the first ionization potential of the polymer. The most intense emission lines for Hg I (254, 297, and 360 nm) are all greater in energy than the first ionization potential of BPAPC (~ 3.3 eV as determined from the XPS valence band spectrum) and thus should induce UV modification.

The effects of UV exposure are shown in Figs 10 and 11 for two different exposure times. For a 15 min exposure (Fig. 10), the oxygen/carbon ratio remains unchanged at the stoichiometric value of 0.19; however, significant changes are observed in both C 1s and the O 1s spectra. In the C 1s spectrum, the peak due to carbonate species at 290.4 eV decreases in intensity and a new peak is observed at 288.9 eV. This peak position is consistent with the formation of ester groups. The changes in the O 1s spectrum also suggest loss of carbonate species and the formation of ester groups. The peak at 534.0 eV (C—O) decreases in intensity and the peak at 532.2 eV (C=O) increases in intensity. Also, the centroid of both peaks appears to shift to lower binding energy. In fact, this spectrum consists of a convolution of carbonate, ester, and hydroxyl species. This is consistent with the formation of ester groups (where the typically observed peak positions are 533.4 ± 0.2 eV for C—O and 531.8 ± 0.2 eV for C=O with a 1:1 intensity ratio) and hydroxyl groups (532.8 ± 0.2 eV).

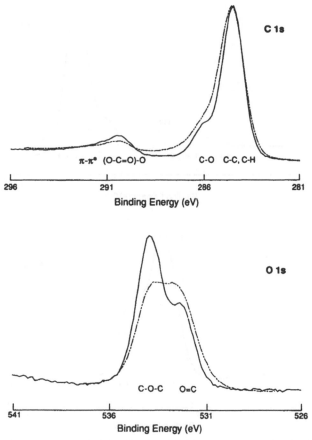

Figure 9. Overlay of C 1*s* and O 1*s* spectra for untreated (solid line) and 15 s argon-plasma-modified (dashed line) BPAPC.

These results suggest that the BPAPC surface undergoes UV-induced degradation similar to that observed for a photo-Fries pathway [23]. The XPS data suggest that neither photooxidation nor reduction due to the loss of CO or CO_2 occurs for short exposures in vacuum since the O:C ratio remains constant. In fact, the data suggest that the reaction produces only an ester group and a ring hydroxyl because no evidence for isolated carbonyls is observed and the intensity of the C—O peak remains unchanged.

For a 60 min UV exposure (Fig. 11), the XPS results suggest that the degradation is not limited to the formation of ester and hydroxyl groups, as illustrated by the C 1*s* spectrum where the formation of carbonyl species (287.6 eV) is apparent, and the increased intensity of the C—O peak (286.1 eV) suggests the formation of considerable hydroxyl species. The O 1*s* spectrum also exhibits more severe changes compared with the short UV exposure, suggesting the formation of various C—O species. The O:C ratio decreases slightly to 0.17, indicating some loss of CO or CO_2. These results clearly show that radiative energy transfer is an important factor to consider in understanding the mechanism responsible for plasma modification.

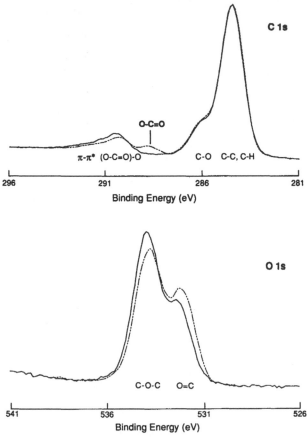

Figure 10. Overlay of C 1s and O 1s spectra for untreated BPAPC (solid line) and BPAPC subjected to 15 min UV exposure in vacuum (dashed line).

3.4. Stability (ageing effects)

The long-term stability of a modified polymer surface is important if the surface is not stored in a controlled environment or coated immediately after treatment. The stability of a modified surface depends on several factors including the chemical structure of the unmodified polymer, the treatment level, the plasma gas, and the storage environment. It has been observed by many researchers that modified polymer surfaces are susceptible to ageing effects [8, 11, 24–28].

The effect of treatment level on surface ageing for oxygen-plasma-modified PE is shown in Fig. 12. Based on angular-dependent measurements, the outer 1–2 nm reach a saturation level of incorporated oxygen at a treatment time of 15 s. Longer treatment times drive the modified region deeper into the PE subsurface and may produce more bond breaking and chain scission. A treatment time of 15 s was found to be optimum and 60 s is considered to be an overtreatment based on the adhesion of evaporated metals on a variety of polymer surfaces [20]. All three treated surfaces exhibit a significant decrease in incorporated oxygen on initial exposure to air. This effect is fairly rapid and occurs within the first few hours of exposure. This initial effect is probably due to adsorption of atmospheric

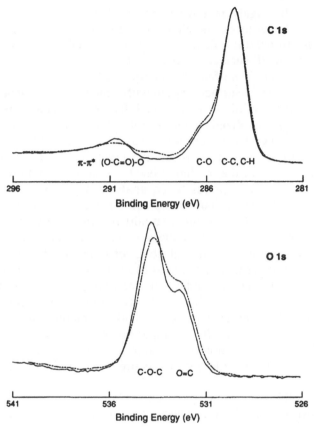

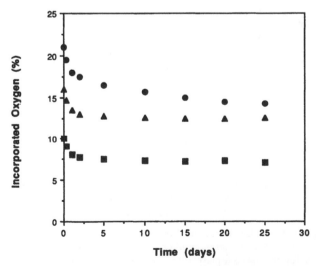

Figure 11. Overlay of C 1s and O 1s spectra for untreated BPAPC (solid line) and BPAPC subjected to 60 min UV exposure in vacuum (dashed line).

Figure 12. Ageing data for a PE surface modified in an oxygen plasma for 5 s (■), 15 s (▲), and 60 s (●). The modified PE surfaces were aged in a clean hood at 25°C and 70% relative humidity (RH). All data were taken at an ETOA of 38°.

contaminants on the highly reactive surfaces. After this initial effect, the two surfaces treated at lower levels remain stable with time. However, the overtreated surface continues to exhibit a decrease in incorporated oxygen with time. After 25 days of ageing, the PE surface treated for 60 s is almost identical in the amount of surface oxygen to the PE surface treated for 15 s.

This slower decrease in surface oxygen with time is due to surface reorganization. Surface reorganization is determined by two mechanisms, diffusion of low-molecular-weight oxidized material into the bulk and macromolecular motions which reorient polar groups away from the surface. Molecular rearrangements are confined to the outermost atomic layers of the polymer surface and should occur regardless of the treatment level. Because the data were taken at an ETOA of 38°, which corresponds to an analysis depth of ~4 nm, molecular rearrangements would be difficult to detect. In order to study such molecuar rearrangements, the modified surfaces should be stored in vacuum to minimize adsorption of contaminants and analysis should be done at low ETOAs to maximize surface sensitivity. Although long-term ageing studies have not been done in vacuum, short-term studies (~24 h) indicate no detectable changes with XPS. The short-term ageing studies in vacuum support the adsorption of atmospheric contaminants as the reason for the initial rapid decrease in surface oxygen. Diffusion of low-molecular-weight oxidized material into the bulk would not be predicted for PE due to the incompatibility of the highly oxidized material with the bulk PE. However, molecular rearrangements alone cannot account for the large effect observed with long-term ageing of PE treated at a high level with an oxygen plasma.

The effect of different plasma gases on the stability of plasma-modified PET surfaces is shown in Fig. 13. Both plasma treatments were done at the optimum

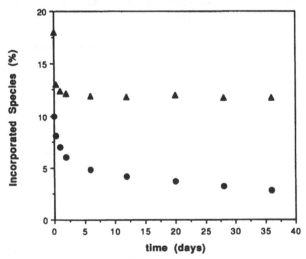

Figure 13. Ageing data for a PET surface modified in a nitrogen (▲) and an oxygen (●) plasma for 15 s. The data for the nitrogen-plasma-modified PET are simply the amount of incorporated nitrogen. The data for the oxygen-plasma-modified PET were determined from the difference between the stoichiometric oxygen value for a clean PET surface and the value measured by XPS for the modified surface. The modified PET surfaces were aged in a clean hood at 25°C and 70% RH. All data were taken at an ETOA of 38°.

treatment time of 15 s to minimize bond breaking and chain scission. Both treated surfaces exhibit the similar initial rapid decrease in incorporated surface species (nitrogen or oxygen) that was observed for PE. However, after the initial rapid change, the nitrogen-plasma-modified surface remains stable with time, but the oxygen-plasma-modified surface continues to lose surface oxygen with time. The fact that the oxygen-plasma-modified PET surface undergoes surface reorganization even at short treatment times may be due to several factors. More bond breaking and chain scission may occur for PET as compared with PE, even at short treatment times, thus producing more low-molecular-weight oxidized material which can diffuse more readily into the bulk. The amount of crosslinking in the near-surface region may be different for the two polymers and may have an effect on the mobility or the rate of diffusion of low-molecular-weight oxidized material into the bulk. The solubility or compatibility of the low-molecular-weight oxidized material in the bulk polymer may also control the rate of diffusion. The ageing results for oxygen-plasma treatment of PET are similar to those reported for the air corona treatment of PET [24, 26, 28].

Except for the initial rapid decrease in incorporated nitrogen, the nitrogen-plasma-modified PET surface is stable with time. The initial rapid decrease in incorporated nitrogen is much greater than that observed for incorporated oxygen in oxygen-plasma-modified PE or PET. Previous studies on nitrogen-plasma-modified PE suggest that plasma-induced imine groups are readily hydrolyzed to the parent carbonyl on exposure to air [8]. The same mechanism may occur for nitrogen-plasma-modified PET. However, it is difficult to follow this reaction sequence on PET since uptake of small amounts of oxygen will not be readily apparent due to the significant amount of oxygen (inherent in the PET) still present in the modified surface.

The effect could be followed quite easily with nitrogen-plasma-modified PS. Although no oxygen was detected on the nitrogen-plasma-modified PS surface initially, after exposure to air approximately 4–5% oxygen was detected and a similar amount of nitrogen was lost. These effects are accompanied by the changes in the N $1s$ spectrum shown in Fig. 14. After air exposure, the integrated area under the peak due to imine species decreases by $\sim 60\%$. The amount of imine species lost is approximately equal to the oxygen uptake. No detectable changes are observed in the C $1s$ spectrum. These results are consistent with the hydrolysis of imines to the parent carbonyl via the following reactions:

$$R-CH=NH + H_2O \rightarrow R-CH=O + NH_3 \tag{1}$$

$$R-CH=NR' + H_2O \rightarrow R-CH=O + R'NH_2. \tag{2}$$

Reaction (1) would result in a loss of nitrogen while reaction (2) would not if R' were part of a polymer chain or a high molecular weight fragment. However, if R' is a low molecular weight fragment that can volatilize in the vacuum of the spectrometer, then nitrogen can also be lost via reaction (2). The XPS data suggest that reaction (1) is predominant since the majority of the imine species produced by plasma treatment are terminal imines. These effects are not easily detectable in the C $1s$ spectrum since the peak positions for C=N (287.0 eV) and C=O (287.6 eV) are separated by only 0.6 eV and the C=N peak is a small component of the overall C $1s$ spectrum.

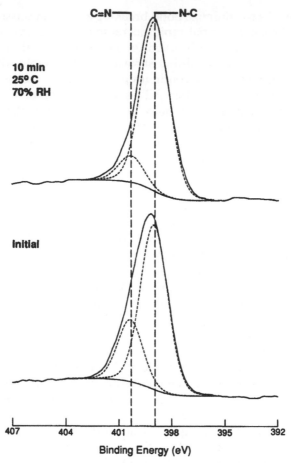

Figure 14. N 1s spectra for a 15 s nitrogen-plasma-modified PS surface before and after 10 min air exposure (25°C and 70% RH).

4. CONCLUSION

XPS has been used to study the chemical effects of both inert (argon) and reactive (oxygen, nitrogen, and mixed gas) plasma treatments done *in situ* on a variety of polymer surfaces. Inert gas plasma treatments introduce no new detectable chemical species onto the polymer surface but can induce degradation and rearrangement of the polymer surface. For all oxygen-containing polymers studied, the oxygen/carbon ratio decreases during argon-plasma treatment, probably due to the loss of CO or CO_2. The effect is very rapid for BPAPC, due to the labile nature of the carbonate group. BPAPC was also found to be extremely susceptible to damage via radiative energy transfer. The process appears to be similar to that observed in a photo-Fries pathway.

Plasma treatments with reactive gases create new chemical species which drastically alter the chemical reactivity of the polymer surface. Small amounts of oxygen (1–10%) in argon were also effective in incorporating oxygen into polymer surfaces. These studies have shown that the surface population of chemical species formed after plasma treatment is dependent on both the chemical structure of the polymer and the plasma gas. The rate of incorporation of plasma-induced species was also found to be dependent on the polymer structure. All the modified surfaces were found to have a concentration gradient within the XPS sampling depth. The shape and depth of this concentration gradient varied depending on the treatment time.

Ambient ageing studies of plasma-modified polymer surfaces have shown that the ageing process consists of two distinct phases. The initial phase, which occurs within the first few hours of exposure, involves the adsorption of atmospheric contaminants and in the case of nitrogen-plasma treatment, the hydrolysis of plasma-produced imines to their parent carbonyl. The second phase, which occurs slowly, is due to surface reorganization. Surface reorganization can be minimized by the choice of plasma gas and by treating at low levels where bond breaking and chain scission are not excessive, thus minimizing the production of low-molecular-weight oxidized material which can diffuse into the bulk. Nitrogen-plasma-modified polymer surfaces were found to be less susceptible to surface reorganization then oxygen-plasma-modified polymer surfaces.

Acknowledgements

I am grateful for the contributions of the following colleagues at Eastman Kodak Company: K. E. Goppert-Berarducci for the preparation of spin-coated polymers and experimental assistance; V. A. De Palma for AFM measurements; R. C. Bowen for SEM measurements; and D. A. Glocker, J. M. Pochan, J. F. Elman and P. M. Thompson for helpful discussions.

REFERENCES

1. H. Yasuda, H. C. Marsh, S. Brandt and C. N. Reilly, *J. Polym. Sci.* **15**, 991–1019 (1977).
2. J. M. Burkstrand, *J. Vac. Sci. Technol.* **15**, 223–226 (1978).
3. A. Dilks, *J. Polym. Sci.* **19**, 1319–1327 (1981).
4. D. T. Clark and R. Wilson, *J. Polym. Sci.* **21**, 837–853. (1983).
5. A. Dilks and A. VanLaeken, in: *Physicochemical Aspects of Polymer Surfaces*, K. L. Mittal (Ed.), Vol. 2, pp. 749–772. Plenum Press, New York (1983).
6. R. G. Nuzzo and G. Smolinsky, *Macromolecules* **17**, 1013–1019 (1984).
7. J. F. Evans, J. H. Gibson, J. F. Moulder, J. S. Hammond and H. Goretzki, *Fresenius' Z. Anal. Chem.* **319**, 841–849 (1984).
8. L. J. Gerenser, *J. Adhesion Sci. Technol.* **1**, 303–318 (1987).
9. T. G. Vargo, J. A. Gardella, Jr. and L. Salvati, Jr., *J. Polym. Sci. A., Polym. Chem.* **27**, 1267–1286 (1989).
10. J. E. Klemberg-Sapieha, O. M. Küttel, L. Martinu and M. R. Wertheimer, *J. Vac. Sci. Technol. A.* **9**, 2975–2981 (1991).
11. R. Foerch, J. Izawa and G. Spears, *J. Adhesion Sci. Technol.* **5**, 549 (1991).
12. R. Foerch and D. Johnson, *Surface Interface Anal.* **17**, 847–854 (1991).
13. R. J. Baird and C. S. Fadley, *J. Electron Spectrosc. Relat. Phenom.* **11**, 39–65 (1977).
14. D. T. Clark, A. Dilks and D. Shuttleworth, in: *Polymer Surfaces*, D. T. Clark and W. J. Feast (Eds), Ch. 9. John Wiley, Chichester (1978).
15. R. Foerch, N. S. McIntyre and R. N. S. Sodhi, *J. Appl. Polym. Sci.* **40**, 1903–1915 (1990).
16. L. J. Gerenser, *J. Vac. Sci. Technol.* **6**, 2897–2903 (1988).
17. L. J. Gerenser, in: *Metallization of Polymers*, E. Sacher, J. J. Pireaux and S. P. Kowalczyk (Eds), Ch. 32. American Chemical Society, Washington, DC (1990).
18. L. J. Gerenser, *J. Vac. Sci. Technol.* **8**, 3682–3691 (1990).
19. L. J. Gerenser and K. E. Goppert-Berarducci, in: *Metallized Plastics 3: Fundamental and Applied Aspects*, K. L. Mittal (Ed.), pp. 163–178. Plenum Press, New York (1992).

20. R. W. Burger and L. J. Gerenser, in: ref. 19, pp. 179–193.
21. C. S. Fadley, *Prog. Solid State Chem.* **11**, 265–303 (1976).
22. F. D. Egitto and L. J. Matienzo, *Polym. Degrad. Stab.* **30**, 293–308 (1990).
23. A. Factor, W. V. Ligon and R. J. May, *Macromolecules* **20**, 2461–2468 (1987).
24. D. Briggs, D. G. Rance, C. R. Kendall and A. R. Blythe, *Polymer* **21**, 895–900 (1980).
25. J. M. Pochan, L. J. Gerenser and J. F. Elman, *Polymer* **27**, 1058–1062 (1986).
26. L. J. Gerenser, J. F. Elman, M. G. Mason and J. M. Pochan, *Polymer* **26**, 1162–1166 (1985).
27. E. Occhiello, M. Morra, G. Morini, F. Garbassi and P. Humphrey, *J. Appl. Polym. Sci.* **42**, 551–559 (1991).
28. M. Strobel, C. S. Lyons, J. M. Strobel and R. S. Kapaun, *J. Adhesion Sci. Technol.* **6**, 429–443 (1992).

Plasma Surface Modification of Polymers, pp. 65–97
M. Strobel, C. Lyons and K. L. Mittal (Eds)
© VSP 1994

Multitechnique study of hexatriacontane surfaces modified by argon and oxygen RF plasmas: effect of treatment time and functionalization, and comparison with HDPE

F. CLOUET,[1,*] M. K. SHI,[1] R. PRAT,[1] Y. HOLL,[1] P. MARIE,[1,*] D. LÉONARD,[2] Y. DE PUYDT,[2,†] P. BERTRAND,[2] J.-L. DEWEZ[3] and A. DOREN[3]

[1] *Groupe des Matériaux Organiques, Institut de Physique et Chimie des Matériaux de Strasbourg, ICS, 6 Rue Boussingault, F-67083 Strasbourg Cedex, France*
[2] *PCPM Laboratory, UCL, 1 Place Croix du Sud, B-1348 Louvain-la-Neuve, Belgium*
[3] *CIFA Laboratory, UCL, 1 Place Croix du Sud, B-1348 Louvain-la-Neuve, Belgium*

Revised version received 18 October 1993

Abstract—Hexatriacontane ($C_{36}H_{74}$) has been used as a model molecule for the study of the surface modifications of high-density polyethylene (HDPE) in argon and oxygen radio-frequency (RF) plasmas. The combination of static secondary ion mass spectrometry (SIMS), ion scattering spectroscopy (ISS), X-ray photoelectron spectroscopy (XPS), and contact angle measurements has constituted a powerful method for the investigation of the surface modifications induced by the plasma treatments. The surface degradation and functionalization are shown to depend on both the nature of the treated material and the nature of the plasma atmosphere. The SSIMS results obtained on plasma-modified hexatriacontane and HDPE are compared in order to identify the nature of the functionalities present at the plasma-treated surfaces. Finally, plasma treatment ^{18}O atmosphere was performed on HDPE, $C_{36}H_{74}$, and polystyrene (PS). In that case, the isotopic specificity of both ISS and SIMS allowed the determination of the relative concentrations of ^{16}O and ^{18}O in relation to the probed depth and plasma atmosphere.

Keywords: Plasma treatments; polymer surfaces; hexatriacontane; high-density polyethylene; polystyrene; ISS; static SIMS; XPS; contact angle measurement; functionalization; isotopic labelling.

1. INTRODUCTION

In order to understand the influence of the physical and chemical structure of polymers on their behaviour under cold plasmas, a study of well-defined model surfaces has been undertaken [1]. In that study, hexatriacontane ($C_{36}H_{74}$) and octadecyloctadecanoate [$CH_3-(CH_2)_{16}CO_2(CH_2)_{17}-CH_3$] were chosen as models of high-density polyethylene [$(-CH_2-CH_2)_n$] and polycaprolactone [$(-(CH_2)_5CO_2-)_n$], respectively. The degradation of model surfaces in argon and oxygen RF (13.56 MHz) plasmas studied by analysing the volatile fragments arising from the surfaces by mass spectrometry showed that the breaking of the $-CO-O-$ bond takes place

---

*To whom correspondence should be addressed.

†Present address: Ecole des Mines de Paris/CEMEF, BP 207, F-06904 Sophia Antipolis Cedex, France.

more rapidly than that of $-CH_2-CH_2-$. The present paper focuses on the study of the functionalization of hexatriacontane as compared with that of its corresponding polymer HDPE.

The complexity of polymer surface analysis requires the use of complementary techniques which are able to probe different depths in the material. X-ray photoelectron spectroscopy (XPS), known as a useful analytical tool for the analysis of plasma-treated polymer surfaces [2–6], allows a 50 Å sampling depth, while static secondary ion mass spectrometry (SSIMS) investigates about 10 Å. The use of SSIMS as a complement to XPS leads to specific information [7–12] originating from molecular specificity. Although quadrupole mass spectrometers provide quite limited information, it is possible to obtain meaningful static SIMS results choosing the appropriate analysis conditions [13–15] or by using isotope labelling [9]. Ion scattering spectroscopy (ISS), characterized by a sensitivity limited to the uppermost monolayer, gives the elemental composition of this layer. Incomplete knowledge of the polymer surface structure in terms of both molecular (elastic shadowing) and electronic (inelastic shadowing) structures limits the quantification of the ISS data obtained from polymer samples. However, some ISS experiments can produce useful data, typically as a complement to other techniques [15–17].

Here, the spectroscopic results are compared with contact angle measurements and 'in situ' mass spectrometry of the plasma gas phase in order to investigate the surface modifications of hexatriacontane and HDPE samples treated in Ar and $^{16}O_2$ RF plasmas. These are discussed in terms of surface degradation and functionalization. Moreover, the isotopic sensitivity of both ISS and static SIMS is used in order to answer the still open question about the 'post-plasma' deactivation (by oxygen from air) of treated polymer surfaces. For this purpose, the ISS and static SIMS results obtained on samples treated in an $^{18}O_2$ plasma are compared with the results obtained on samples treated under the same conditions in an $^{16}O_2$ plasma.

2. EXPERIMENTAL

2.1. Materials

HDPE ($\overline{M_n} = 3.5 \times 10^5$ g/mol) was purchased from Aldrich (purity > 98% and crystallinity $\geqslant$ 80%). HDPE disks (diameter = 5 cm, thickness = 0.5 mm) were modulded under vacuum to avoid oxidation during the moulding and were stored under argon prior to plasma treatment. One cm^2 polystyrene ($\overline{M_n} = 5 \times 10^4$ g/mol) plates were taken from Petri dishes used for biomedical purposes.

Hexatriacontane [$CH_3-(CH_2)_{34}-CH_3$] was purchased from Aldrich (purity > 98%) and recrystallized in distilled hexane to obtain close to 100% crystallinity. $C_{36}H_{74}$ films were prepared either by melting an aliquot of $C_{36}H_{74}$ on aluminium plates placed on a heating pad followed by spreading out with a Conway barrel to form a 3 μm thick film, or by casting a solution of $C_{36}H_{74}$ in thrice-distilled benzene onto aluminium plates followed by evaporation of the benzene in an Ar partial vacuum. The films were stored under Ar to avoid any contamination.

$^{18}O_2$ (isotopic purity 97.4%) was purchased from Isotec Inc. (Matheson Company, USA) and $^{16}O_2$ (natural isotopic abundance 99.76%) was purchased from Air Liquide Company (France).

Table 1.

Initial gas composition in the plasma reactor (the relative values are calculated directly from mass spectrum intensity)

	Component				
	$^{16}O_2$	H_2	H_2O	N_2	Ar
Ar plasma	0.2%	0.2%	0.7%	0.9%	98%
$^{16}O_2$ plasma	97%	0.2%	1%	1.5%	0.3%

2.2. Treatments

The plasma reactor is a stainless-steel cylindrical vessels, 15 cm in height and 23 cm in diameter. The lid forms the electrode, which is capacitively coupled to a 13.56 MHz ENI generator. The bottom, a 15 cm diameter disk, forms the grounded electrode onto which the samples are placed. The plasma reactor is coupled to a quadrupole mass spectrometer VG instrument (SXP 300) which is used to analyse the gas composition before and during the treatment. Before treatment, the reactor was evacuated (reactor base pressure equal to 1.3 Pa) and the gas flow rate (40 sccm) and pressure (40 Pa) were adjusted until the desired gas composition was obtained (Table 1). In the case of the $^{18}O_2$ treatment, traces of $^{16}O_2$ are present ($< 5\%$). The discharge was then excited (power 60 W). After treatment, all the samples analysed in this study were kept for 10 min in the plasma reactor under an $^{16}O_2$ flux and then stored in an air atmosphere before the analysis.

The degradation rate was evaluated by weighing a 50 cm^2 sample before and after treatment with a Mettler AE 160 balance. The calculated error in this gravimetric technique was around 5%.

2.3. Surface characterization

2.3.1. Contact angle measurements. The advancing contact angles were measured with an automated sessile-drop apparatus which uses a video digital-image processing technique [18]. A drop of the test liquid was formed at the tip of a capillary tube of an SMI Micro/Pettor positive displacement syringe and deposited onto the surface to be studied. The syringe releasing the liquid was slowly and smoothly raised during drop expansion by means of a micromanipulator in order to guarantee simple and symetrical streaming conditions. The drop and surface were placed in an environmental control chamber (temperature 23 °C, relative humidity 50%).

The optical system, which consisted of a Nachet M4 macro-zoom microscope coupled to a Panasonic WV-CD50 CCD video camera [500 (H) × 582 (V)], was mounted on an optical bench seated on a vibration-isolation system. The optics were focused by optimizing the video image of a reticule containing a finely ruled grid that was placed at the drop location. The reticule also provided direct calibration of both the vertical and the horizontal magnification factors of the instrument. The drop was back-lit by a white-light source (Dolan-Jenner Fiber-Lite System 181) through a heavily frosted diffuser.

The video image from the camera was sent through a Panasonic WV80 digital timer to a video tape (Hitachi) and a Matrox video frame grabber board resident within

Table 2.

Advancing contact angles measured with water and diiodomethane on $C_{36}H_{74}$ and HDPE surfaces (standard deviation $\pm 2°$)

		Θ_a with H_2O	Θ_a with CH_2I_2
$C_{36}H_{74}$	Untreated	109	74
	O_2 plasma 20 s	82	46
	Ar plasma 20 s	66	25
HDPE	Untreated	105	87
	O_2 plasma 20 s	48	49
	Ar plasma 20 s	56	45

a microcomputer that performed the image digitization. The digitized image was analysed with Visilog software by using a morphological gradient method, adaptative thresholding, and mathematical morphology functions (skeleton and thinning) to extract the arbitrary coordinate points of the drop profile.

The experimental profile was analysed with a robust shape-comparison algorithm developed by Neumann and co-workers [19] and called the axisymmetric drop shape analysis-profile (ADSA-P). Details of the ADSA-P protocol and data processing have been given by Cheng *et al.* [20]. The input parameters are the gravitational acceleration, the density of the liquid, and the drop profile cooridanates. A calculated Laplacian curve is compared with the experimental curve until satisfactory convergence is achieved. The resulting outputs are the contact angle (θ), liquid surface tension (γ_{LV}), the drop volume, the drop-solid contact radius, and the drop-vapour surface area.

Contact angle values obtained with highly pure water and diiodomethane on the $C_{36}H_{74}$ and HDPE surfaces treated in oxygen and argon plasmas are given in Table 2. To follow the evolution of the surface energy during the plasma treatment, we consider the increase in the values of the surface energies calculated from the contact angles of the two tests liquids (an average of five measurements). The following harmonic-mean equation proposed by Wu, and recommended for polymers, was used [21]:

$$\gamma_{LV}(1 + \cos\theta) = \frac{4\gamma_S^d\gamma_L^d}{\gamma_S^d + \gamma_L^d} + \frac{4\gamma_S^p\gamma_L^p}{\gamma_S^p + \gamma_L^p},$$

where γ_L^d, γ_L^p, γ_S^d, and γ_S^p are the dispersive (d) and polar (p) components of the test liquid (L) and the solid (S) surfaces energies, and γ_{LV} is the total test-liquid surface energy. The components of the solid surface energy can be calculated (error limits are 5%) from the surface-energy components of the test liquids (known) and the contact angles of the test liquids (to be measured).

2.3.2. XPS analyses. The XPS analyses were performed 18 h after the plasma treatment with a VG ESCALAB MKII (XPS-S) spectrometer or a few days after the treatment with a Surface Science Instruments (SSI) spectrometer (XPS-L). The VG apparatus uses AlK_α excitation radiation from a non-monochromatized X-ray source operated at 11 kV and 5 mA. The take-off angle of the photoelectrons with respect to the surface was 90° and the analyser pass energy was 50 eV. Because the charging effect was weak (2–4 eV) and almost constant from one sample to another, no flood gun was used. Under these conditions, the escape depth of the electrons arising from the C_{1s} level is about 20 Å and the analysis depth is about 60 Å.

The SSX 100 spectrometer from SSI is equipped with an aluminium anode (10 kV and 11.5 mA) and a quartz monochromator (emission radiation $Al K_\alpha$). The analyser pass energy was 50 eV. Charge compensation was performed using a flood gun (low energy, 6 eV). The angle between the sample surface and the analyser entrance was 35°; the analysed depth was about 50 Å.

The atomic concentration ratios were calculated with sensitivity factors using Scofield cross-sections and taking into account the instrumental parameters (analyser transmission). Thanks to reproducibility studies, the relative error in atomic ratios was 6% and 0.3–1.5% for the VG and SSI instruments, respectively.

2.3.3. SSIMS and ISS analyses. Static SIMS and ISS analyses were performed a few days after the plasma treatment at Louvain-la-Neuve in the same UHV chamber, where a base pressure of 1.3×10^{-7} Pa was maintained. For the ISS analyses, a 2 keV $^3He^+$ ion beam was rastered on a 4.2 mm^2 area. The use of ^3He instead of ^4He allows a reasonable sensitivity for carbon with a total ion dose for one ISS spectrum acquisition of about 1.5×10^{14} ions/cm^2. The charge compensation was performed with an electron gun (VSW-EG2) working at the following conditions: filament current about 2.5 A and electron acceleration voltage about 300 eV. The ISS spectra were analysed using a standard procedure described elsewhere [16, 17].

For the SSIMS analyses, a 4 keV Xe^+ ion beam was rastered on a 33 mm^2 area. Settings of the quadrupole spectrometer were adjusted by optimizing the absolute intensity of a molecular peak characteristic of the analysed molecule or polymer sample: typically around 100 amu in the positive mode (95 for $C_{36}H_{74}$, 105 for HDPE, and 115 for PS) and at 48 amu in the negative mode for all the samples. The typical ion doses used for one SSIMS spectrum acquisition, including the spectrometer setting procedure, were less than 10^{13} ions/cm^2. This value is quite high with respect to the now commonly accepted 'SSIMS' condition; however, it is the result of the necessary compromise between the ion-induced degradation of the polymer samples and the low counting efficiency of the quadrupole spectrometer that we used (low angular acceptance and low transmission). Charge compensation was performed using electrons from a flood gun (positive SSIMS) or a heated tungsten filament (negative SSIMS). The filament current and the electron acceleration voltage were adjusted for each experiment. The raw SSIMS spectra were treated as follows: they were transformed into histograms in which the intensity for each mass unit was normalized with respect to the overall secondary-ion intensity. Using these experimental conditions and data treatment procedures, the precision in the relative intensities measured on identical samples is typically less than 20% for a given peak. The same relative error holds for the sums of the normalized intensities, which are discussed below.

3. RESULTS AND DISCUSSION

3.1. Influence of the plasma treatment time

A previous study concerning the influence of plasma parameters such as power, pressure, and flow rate on the surface degradation and functionalization of $C_{36}H_{74}$ and $[CH_3-(CH_2)_{16}CO_2(CH_2)_{17}-CH_3]$ has been published elsewhere [1, 22]. It appears that the degradation rate is higher in an oxygen plasma than in an argon plasma and higher for an oxygenated surface than for a hydrocarbon one. The functionalization rate is quite dependent on the plasma-induced degradation. The greater the degradation, the less the functionalization. In order to verify this behaviour for polymers, the degradation rate of HDPE was measured. The degradation rate (R_D) is reported in Table 3. The R_D values for $C_{36}H_{74}$ and HDPE are very similar. Although the crystallinity and the molecular weight differences between HDPE and $C_{36}H_{74}$ could influence the degradation rate, both parameters taken together do not modify the overall weight loss.

Figure 1 shows the evolution of the surface energy of hexatriacontane and HDPE surfaces treated in oxygen and argon plasmas (at 60 W, 40 sccm, and 40 Pa) as a function of the plasma treatment time. It appears that after 5–10 s of treatment, the

Table 3.
Degradation rates (in $\mu g\,cm^{-2}\,min^{-1}$) of $C_{36}H_{74}$ and HDPE in oxygen and argon plasmas

Plasma	$C_{36}H_{74}$	HDPE
O_2	10	9.9
Ar	1.7	1.6

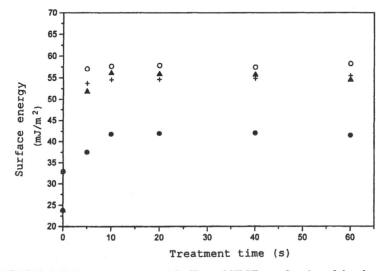

Figure 1. Evolution of the surface energy of $C_{36}H_{74}$ anf HDPE as a function of the plasma treatment time. (•) $C_{36}H_{74}$ in an oxygen plasma; (▲) $C_{36}H_{74}$ in an argon plasma; (○) HDPE in an oxygen plasma; (+) HDPE in an argon plasma.

surface energy remains constant. After the treatment, the surface energies are in the following decreasing order:

$$\text{HDPE, } O_2 \geqslant \text{HDPE, Ar} \sim C_{36}H_{74}, \text{ Ar} > C_{36}H_{74}, O_2.$$

While the argon plasma treatments of $C_{36}H_{74}$ and HDPE and the oxygen plasma treatment of HDPE induce an equivalent surface energy increase, the surface energy of $C_{36}H_{74}$ is increased significantly less by an oxugen plasma.

The contact angles provide information on the uppermost layer of the surface. Aging could lead to surface rearrangement [6, 8], but the measurements were performed a few minutes after treatment, so we assume that this rearrangement is not present here.

The O/C ratio measured by XPS (Fig. 2) reaches a maximum after 5 s of treatment in an oxygen plasma, while it increases until 20 s of treatment in an argon plasma. The O/C ratios are in the same decreasing order as the surface energies. From these results, which agree quite well with those of Clark and Dilks [23], it appears that in an oxygen plasma oxygen incorporation occurs very rapidly, but that the oxidized functions are degraded, as shown by the detection of CO, CO_2, and H_2O in the gas phase by mass spectrometry [1]. An equilibrium between oxygen incorporation and degradation is reached at treatment times of about 10 s.

According to the literature, in a pure argon plasma the ion bombardment on the sample surface induces the breaking of C—C and C—H bonds with subsequent double-bond formation or crosslinking and trapping of radicals [24, 25]. The formation of oxygenated functional groups is then attributed to the reaction of long-lived radicals with the oxygen from air, after plasma treatment has been completed. In

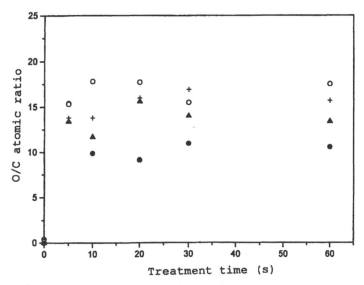

Figure 2. Evolution of the O/C atomic ratio detected by XPS of $C_{36}H_{74}$ and HDPE as a function of the plasma treatment time. (•) $C_{36}H_{74}$ in an oxygen plasma; (▲) $C_{36}H_{74}$ in an argon plasma; (o) HDPE in an oxygen plasma; (+) HDPE in an argon plasma.

our case, oxygen traces in the plasma chamber (0.2% O_2 and 0.7% H_2O) may also participate in the oxygen incorporation observed by XPS, although mass spectrometry has shown that H_2 is the main product of the degradation, which indicates that the contribution of atomic oxygen in the reaction is low [1]. For argon plasma treatments, the maximum oxygen incorporation is observed after about 20 s of treatment.

The SSIMS analysis results published elsewhere [11] have shown that the oxygen incorporation revealed through the occurrence of peaks associated with C_nOH_{2n+1} ions agrees with the contact angle measurements and the XPS results.

3.2. ISS, SSIMS, and XPS analyses of treated $C_{36}H_{74}$ and HDPE surfaces

After studying the evolution of the surface energy and the XPS peaks as a function of time, plasma treatment times were chosen to achieve a maximum level of oxygen incorporation: 20 s for the Ar plasma treatment and 10 s for the O_2 plasma treatment.

The ISS spectra of untreated and plasma-treated HDPE and $C_{36}H_{74}$ samples are presented in Fig. 3. The corresponding atomic per cent oxygen is reported in Table 4. These values give only qualitative information about the oxidation of the uppermost monolayer. Indeed, although these values have been corrected for the scattering cross-sections and the instrumental factors (electron multiplier efficiency and analyser transmission), it is more difficult to take into account elastic and inelastic shadowing and, thus, to provide complete quantitative data [17, 26]. This is a real limitation for quantification of the ISS data.

In the case of HDPE, the oxygen incorporation appears clearly in the ISS results. As with the surface energy, there is no significant difference between a plasma treatment performed in argon or in oxygen. Therefore, the higher degradation rate in an oxygen plasma as compared with an argon plasma does not play any role in the oxygen concentration detected in the uppermost layer of the HDPE.

In the case of hexatriacontane, the ISS results do not give any evidence of surface oxidation after the plasma treatment, either for argon or for oxygen. This result appears in contradiction with the surface energy measurements (Fig. 1). But it should be pointed out that the ISS analysis was performed a few days after the plasma treatment. In the case of $C_{36}H_{74}$, the low-molecular-weight oxidized compounds formed because of chain breaking and functionalization can desorb or migrate towards the

Table 4.
ISS and XPS atomic per cent oxygen measured on $C_{36}H_{74}$ and HDPE. XPS-S and XPS-L results were obtained 18 h and 3 days after treatment, respectively

Technique:	ISS	ISS	XPS-S	XPS-S	XPS-L	XPS-L
Surface:	$C_{36}H_{74}$	HDPE	$C_{36}H_{74}$	HDPE	$C_{36}H_{74}$	HDPE
Treatment						
Untreated	< 0.1	5	0	0.4	0.7	1.7
Ar, 20 s	< 0.1	23	15	16	13	14
O_2, 10 s	< 0.1	24	10	18	8	14

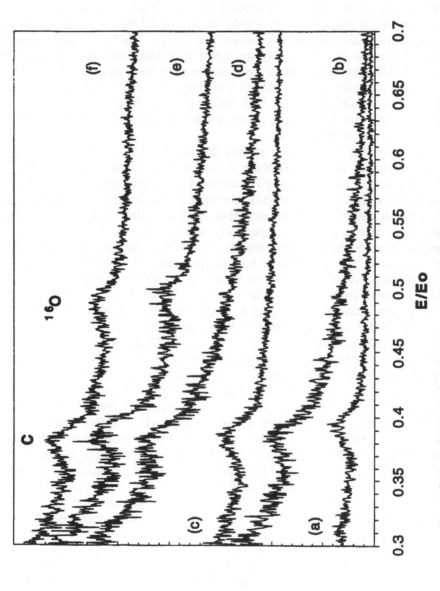

Figure 3. ISS spectra (ion dose: 3×10^{14} ions $^3He^+/cm^2$) of (a) untreated hexatriacontane ($C_{36}H_{74}$); (b) treated $C_{36}H_{74}$, O_2 plasma for 10 s; (c) treated $C_{36}H_{74}$, Ar plasma for 20 s; (d) untreated HDPE; (e) treated HDPE, O_2 plasma for 10 s; (f) treated HDPE, Ar plasma for 20 s (a.u.: arbitrary unit).

bulk more easily than the high-molecular-weight materials formed on the HDPE surface. This effect is well known and has often been shown by a decrease of surface energy during the aging of plasma-treated polymers [6, 9].

From the XPS results (Table 4) it appears that the extent of surface oxidation of $C_{36}H_{74}$ in an argon plasma is very similar to that of HDPE, while in an oxygen plasma HDPE is significantly more oxidized than $C_{36}H_{74}$. The XPS values measured 18 h after treatment are always slightly higher than those measured 3 days after treatment. Besides the difference in the spectrometers used, this may also indicate that some low-molecular-weight oxidized fragments are volatile (in particular, in the case of $C_{36}H_{74}$) or migrate into the bulk deeper than the 50 Å probed by XPS. By comparing the ISS and XPS results, it seems that in $C_{36}H_{74}$ the oxygen is localized deeper after surface rearrangement.

The high-resolution C_{1s} and O_{1s} spectra do not allow the precise identification of the new chemical incorporated at the surface of the plasma-treated samples. In order to characterize more precisely the different functional groups present on the polymer surfaces, studies of the SSIMS spectra of 1-triacontanol, triacontanoic acid, octadecyloctadecanoate, and 10-nonadecanone, used as models of an alcohol, an acid, an ester, and a ketone function, respectively, are in progress. Preliminary results indicate that these model molecules lead to characteristic but similar fragmentation patterns. Moreover, we do not know how the proximity of different functions will influence the fragmentation of the 'new' molecules formed by the plasma treatment. Therefore, the analysis of treated surfaces, done by comparing the spectra of model functional groups with the actual spectra, should be performed very carefully.

Finally, the SSIMS spectra obtained in the positive mode on untreated and treated (10 s in oxygen plasma and 20 s in argon plasma) HDPE and $C_{36}H_{74}$ are presented in Figs 4 and 5, respectively. The spectra of untreated hexatriacontane and HDPE are characterized by all of the carbon-containing ion clusters from C1 (12, 13, 14, and 15 amu from the ^+C, ^+CH, $^+CH_2$, and $^+CH_3$ ions, respectively) to C10 (131 and 133 amu from the $^+CH_{10}H_{11}$ and $^+C_{10}H_{13}$ ions, respectively). A comment should be made about the untreated surfaces: the $C_{36}H_{74}$ surface appears to be cleaner than the HDPE. On the HDPE, slight traces of sodium (peak at 23 amu), polydimethylsiloxane (PDMS; peaks at 73, 133, and 147 amu [27]), and an antioxidizing agent such as Irganox 1010 [27] (peaks at 203 and 219 amu, not shown in Fig. 4) can be observed, in agreement with the ISS and XPS results in which traces of Si and oxygen were detected. These contaminants are seen on both the untreated and the treated HDPE samples. Note in particular that the peaks characteristic of the additives and the sodium are more intense for treated samples. This may indicate that the contaminants are present in higher concentration in the layers deeper than the uppermost layer and are revealed by the etching of the surface by the plasma. Unfortunately, the presence of such contaminants complicates the interpretation of the SSIMS results.

After treatment, signs of oxidation are revealed through new peaks that appear and cannot be confused with hydrocarbon peaks: 31, 45, and 59 amu. They are clearly present in the treated HDPE spectra. In the Ar-plasma-treated $C_{36}H_{74}$ spectrum, only peaks at 31 and 45 amu are discernible. Nevertheless, the intensities of these new peaks are very low, so that we need a more efficient method to account for the functionalization effect. Therefore the sum of the intensities of ions such as $^+C_nOH_{2n+1}$, $^+C_nO_2H_{2n-1}$, and $^+C_nO_2H_{2n+1}$ was calculated.

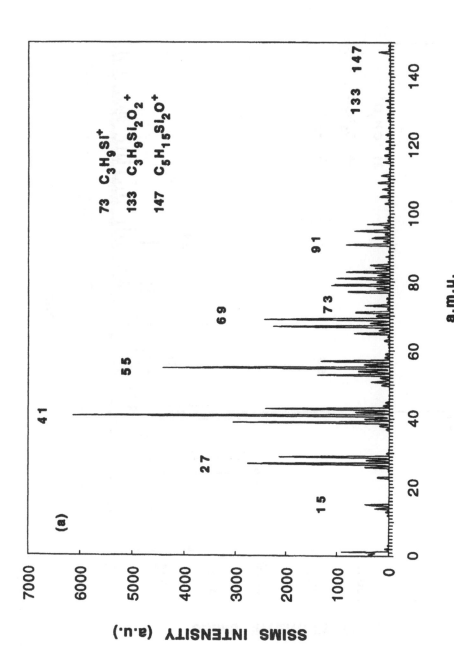

Figure 4a. Positive SSIMS spectra of HDPE: untreated.

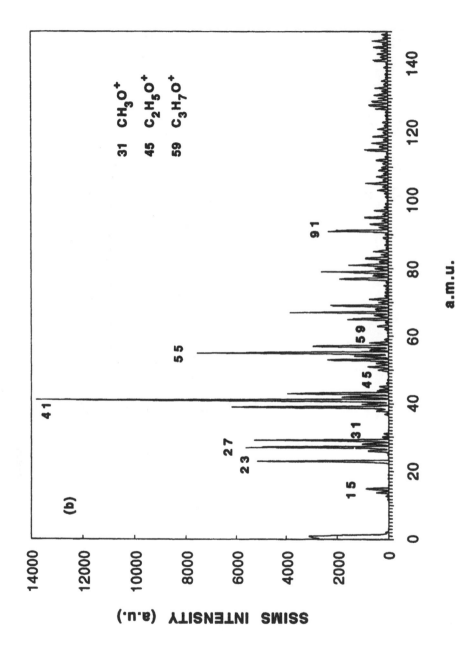

Figure 4b. Positive SSIMS spectra of HDPE: O_2 plasma for 10 s.

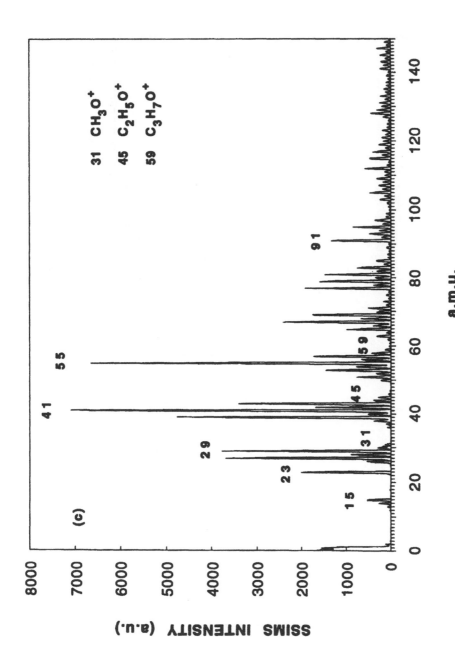

Figure 4c. Positive SSIMS spectra of HDPE: Ar plasma for 20 s.

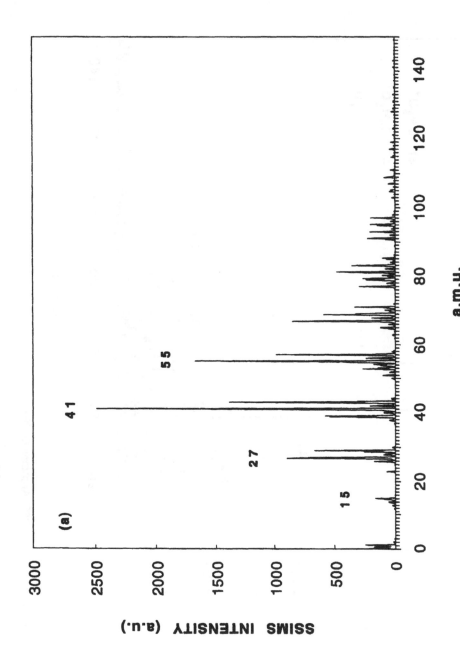

Figure 5a. Positive SSIMS spectra of $C_{36}H_{74}$: untreated.

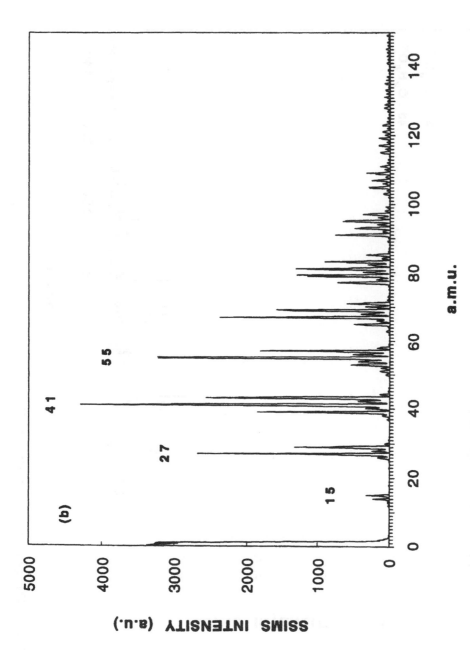

Figure 5b. Positive SSIMS spectra of $C_{36}H_{74}$: O_2 plasma for 10 s.

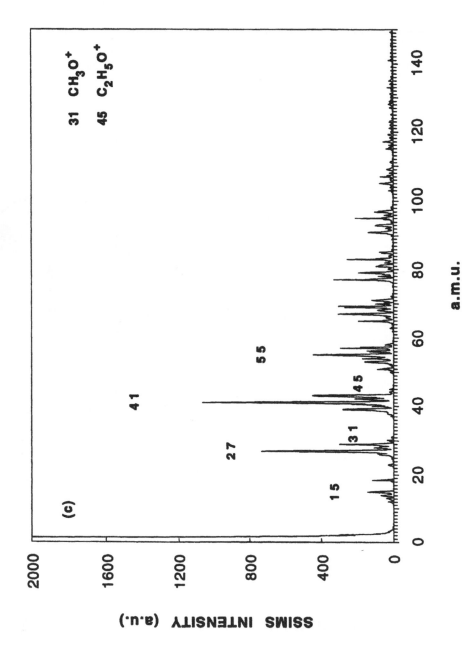

Figure 5c. Positive SSIMS spectra of $C_{36}H_{74}$: Ar plasma for 20 s.

Table 5.

SIMS results obtained on $C_{36}H_{74}$ and HDPE treated in argon (20 s) and oxygen (10 s) plasmas (errors: 20%)

	$C_{36}H_{74}$			HDPE		
	Untreated	Ar plasma treated	O₂ plasma treated	Untreated	Ar plasma treated	O₂ plasma treated
Σ arom.	5.4	5.9	7	4.6	7.3	7.3
Σ C8/Σ C2	23	15	28	21	37	30
Σ > 100	5.8	7.9	8.5	5.5	13	14
Σ fct	0.3	0.6	0.4	1.2	1.4	1.4

The data presented in Table 5 were calculated from the normalized histograms of SSIMS spectra (see comments at the end of the Experimental section). The data are presented as a percentage of the overall secondary-ion intensity. Σ arom. is the sum of normalized masses 76, 77, 91, 105, 115, 117, 128, 141, 165, and 178 (aromatic structures). An increase in the intensity of aromatic structures is representative of the unsaturation [11, 13]. Σ C8 is the sum of normalized masses 103–113, and Σ C2 is the sum of normalized masses 24–29. An increasing Σ C8/Σ C2 ratio may be representative of branching and crosslinking at the surface of polyolefins [13, 14]. Although van Ooij and Brinkhuis [14] insist that an increase in the intensity of masses higher than 100 is due to crosslinking, we must emphasize that the oxidized fragments also contribute to Σ > 100. In any event, all these parameters must be treated very carefully. Briggs [13] showed that this type of calculation is really dependent on the instrument and the experimental conditions used. Σ fct is the sum of normalized masses 31, 45, 47, 59, 61, 73, 75, 87, 89, and 101 (peaks associated with $^+C_nOH_{2n+1}$, $^+C_nO_2H_{2n-1}$, and $^+C_nO_2H_{2n+1}$ ions).

The effect of the plasma treatment is shown by a parallel increase in all of these values, indicating that double-bond formation, branching, and crosslinking seem to occur in oxygen as well as in argon plasma treatment and on $C_{36}H_{74}$ as well as on HDPE. Nevertheless, some differences can be pointed out.

The changes in Σ arom. and Σ C8/Σ C2 are very close for $C_{36}H_{74}$ and HDPE after oxygen plasma treatment but larger for HDPE than for $C_{36}H_{74}$ after argon plasma treatment. The value of Σ C8/Σ C2 is surprisingly low on $C_{36}H_{74}$ treated in argon. Branching and crosslinking are caused by the recombination of radicals. In an argon plasma, the radicals are not consumed by the reactions with molecular oxygen and, therefore, the formation of branches should be favoured, as is observed for HDPE. In $C_{36}H_{74}$, this phenomenon may be hindered because the mobility of the crystalline chains is less than in the amorphous regions of HDPE.

The increase of Σ > 100 appears to be closely related to the nature of the sample, whatever the plasma atmosphere: it is higher for HDPE than it is for hexatriacontane. The molecular weight distributions are drastically different, and after plasma-induced chain scissions, the molecular weight distributions of fragments will be different too. In the case of HDPE, the heavier oxidized fragments will contribute more to Σ > 100 than the lighter oxidized fragments formed hexatriacontane. Moreover, heavier fragments will diffuse more slowly towards the bulk than the lighter ones.

This effect may explain the absence of oxygen on the $C_{36}H_{74}$ surface as observed by ISS.

The increase of the values of Σ fct, which is representative of oxygen incorporation, is quite close for both treatments. An important point should be mentioned: the peaks used in Σ fct correspond without ambiguity to oxygenated (for example, 31 amu is CH_2OH^+ or CH_3O^+) but peaks such as 43, which can be an oxygenated fragment ($CH_3C=O^+$) or a hydrocarbon one ($CH_3CH_2CH_2^+$), are excluded. This data treatment can explain the apparently low functionalization detected by SSIMS.

The combined results of the contact angle measurements, ISS, SSIMS, and XPS show that (1) the oxygen incorporation decreases in the order HDPE treated in an oxygen plasma $\geqslant$ HDPE treated in an argon plasma $\geqslant$ $C_{36}H_{74}$ treated in an argon plasma $>$ $C_{36}H_{74}$ treated in an oxygen plasma; (2) the level of oxygen incorporation in HDPE does not depend on the nature of the gas, while for $C_{36}H_{74}$ the higher degradation in an O_2 plasma corresponds to a smaller concentration of incorporated oxygen; and (3) in $C_{36}H_{74}$, the diffusion of lower-molecular-weight oxygenated fragments towards the bulk seems to induce the absence of oxygen in the uppermost layer.

Because of traces of O_2 in the argon plasma, the incorporation of oxygen cannot be clearly attributed to a post-discharge reaction with the atmosphere after an argon treatment. Working with isotope labelling is a way to answer more precisely this question and to go beyond the limited mass resolution of the quadrupole spectrometers by shifting the peaks corresponding to the oxidized fragments to $M+2$ or $M+4$ amu, depending on whether they contain one or two isotopically labelled oxygen plasma.

3.3. Comparison between plasma treatments in $^{18}O_2$ and $^{16}O_2$

Hexatriacontane, HDPE, and polystyrene (PS, chosen because its high reactivity in an oxygen plasma is well known) samples were treated in an $^{18}O_2$ plasma under the same operating conditions as those used for the $^{16}O_2$ plasma (see Section 3.2). Moreover, the $^{18}O_2$ plasma-treated surfaces were deactivated in an $^{16}O_2$ atmosphere. Consequently, the functionalization resulting from reactions during the plasma treatment will be ^{18}O-labelled and those resulting from 'post-plasma' reactions will be ^{16}O-labelled (contact with deactivating gas or air). The contribution of $^{16}O_2$ traces in the reactor ($<5\%$) to the functionalization during the plasma treatment should not be higher than its relative concentration in air because its reactivity is equivalent to that of $^{18}O_2$. Therefore, the contribution due to $^{16}O_2$ present in the plasma should not exceed 5%.

Table 6.
XPS results obtained for $^{18}O_2$ and $^{16}O_2$ plasma treatments on $C_{36}H_{74}$, HDPE, and PS

Samples	Atomic per cent oxygen		
	Untreated	$^{16}O_2$ plasma treated	$^{18}O_2$ plasma treated
PS	1.0	33	35
HDPE	0.4	18	17
$C_{36}H_{74}$	0.5	13	11

First, XPS analyses were performed in order to check the reproducibility of the functionalization obtained for the same working conditions of the plasma reactor. The results obtained for the $^{18}O_2$ and $^{16}O_2$ plasma treatments are similar (Table 6). The oxygen atomic concentration incorporated on the surfaces is quite similar regardless of which oxygen was used, but depends markedly on the reactivity of the considered samples (in terms of degradation and functionalization), in agreement with the literature [28, 29]:

$$polystyrene > polyethylene > C_{36}H_{74}.$$

The ISS spectra presented in Figs 6, 7, and 8 show that the oxygen incorporation in the upper layer is higher in PS than in HDPE and cannot, as discussed in Section 3.2, be detected on $C_{36}H_{74}$. The overall level of oxidation achieved does not differ between the two plasmas and is the result of both ^{18}O and ^{16}O oxidation (see Table 7). This fact has already been underlined by Occhiello *et al.* [9]. As the $^{18}O_2$ plasma also contains traces of $^{16}O_2$, it is difficult to assert that the ^{16}O incorporation takes place only after plasma treatment has been completed. The $^{18}O/^{16}O$ ratio calculated from ISS spectra acquired with higher ion doses (6×10^{14} ions $^3He^+/cm^2$; working with such ionic doses yields results for sampling depths comparable to those probed by SSIMS) shows that the contributions of ^{16}O are equal to 37% and 57% for PS and HDPE, respectively. These values are higher than that expected from the $^{16}O_2$ traces in the reactor (5%). As no sign of preferential reaction with $^{16}O_2$ or $^{18}O_2$ was observed, it appears that even in an oxygen plasma treatment some oxygen is incorporated after the treatment.

The oxidation $C_{36}H_{74}$ and HDPE, as detected in the uppermost layer, is very low compared with PS. These results are consistent with a surface rearrengement favoured in the following order according to the degradation rates:

$$PS < HDPE < C_{36}H_{74}.$$

Identification of the peaks observed in the SSIMS spectra of the $^{18}O_2$-plasma-treated samples is comparable to that described above, except that new peaks are present or the intensity of the characteristic peaks is increased. Figures 9, 10, and 11 display the 28–52 amu range of the positive spectra of $^{18}O_2$- and $^{16}O_2$-plasma-treated PS, $C_{36}H_{74}$, and HDPE. After $^{16}O_2$ plasma treatment, $^+C_nOH_{2n+1}$ ions appear

Table 7.
ISS results obtained on $^{18}O_2$- and $^{16}O_2$-plasma-treated PS, HDPE, and $C_{36}H_{74}$. The O/C ratios were obtained for the ionic dose 1.5×10^{14} ions $^3He^+/cm^2$, except for * which was 6×10^{14} ions/cm^2

Samples	Corrected O/C atomic ratio				
	Untreated	$^{16}O_2$ plasma	$^{18}O_2$ plasma deactivated in $^{16}O_2$		
	$^{16}O/C$	$^{16}O/C$	$^{18}O/C$	$^{16}O/C$	$^{18}O/^{16}O$*
PS	0.04 (± 0.02)	2.06 (± 0.16)	1.26 (± 0.21)	0.69 (± 0.21)	1.7
HDPE	—	0.15 (± 0.03)	—	0.12 (± 0.02)	0.77
$C_{36}H_{74}$	—	—	—	—	—

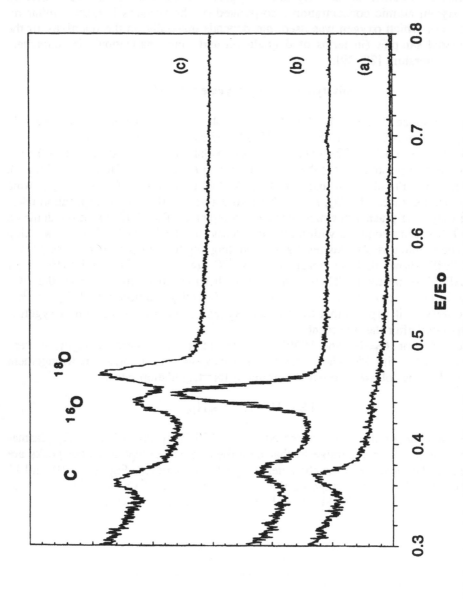

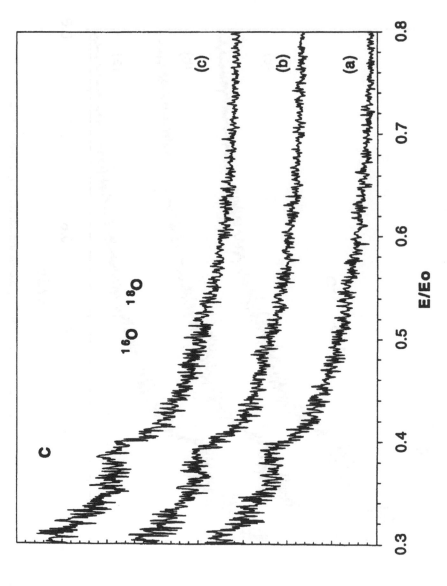

Figure 7. ISS spectra of C$_{36}$H$_{74}$ (ion dose: 1.5×10^{14} ions/cm^2). (a) Untreated; (b) ^{16}O$_2$ plasma treated; (c) ^{18}O$_2$ plasma treated.

86   F. Clouet et al.

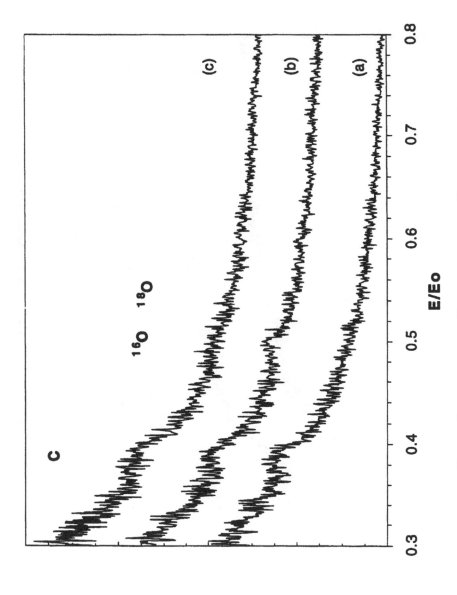

Figure 8. ISS spectra of HDPE (ion dose: 1.5×10^{14} ions/cm^2). (a) Untreated; (b) $^{16}O_2$ plasma treated; (c) $^{18}O_2$ plasma treated.

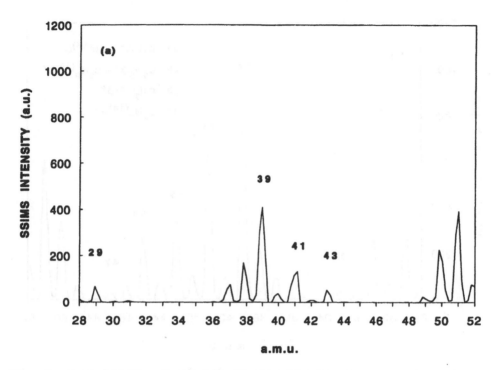

Figure 9a. Positive SSIMS spectra (range 28–52 amu) of PS: untreated.

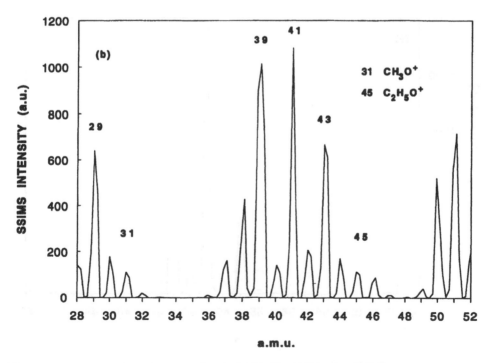

Figure 9b. Positive SSIMS spectra (range 28–52 amu) of PS: ^{16}O$_2$ plasma treated.

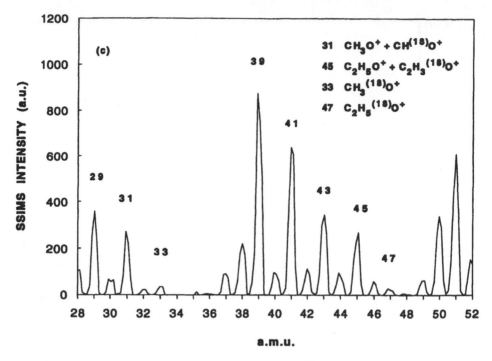

Figure 9c. Positive SSIMS spectra (range 28–52 amu) of PS: $^{18}O_2$ plasma treated.

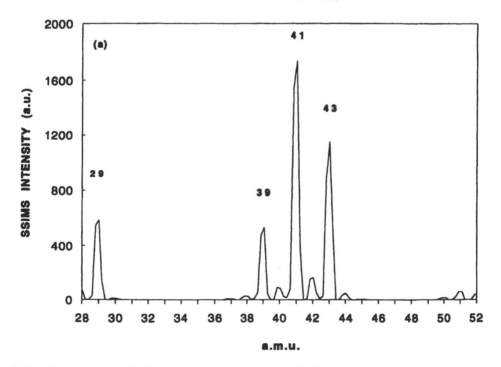

Figure 10a. Positive SSIMS spectra (range 28–52 amu) of $C_{36}H_{74}$: untreated.

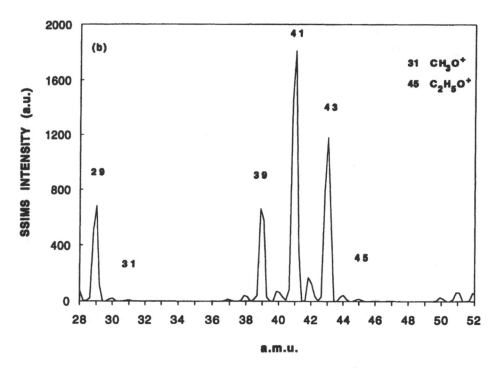

Figure 10b. Positive SSIMS spectra (range 28–52 amu) of $C_{36}H_{74}$: $^{16}O_2$ plasma treated.

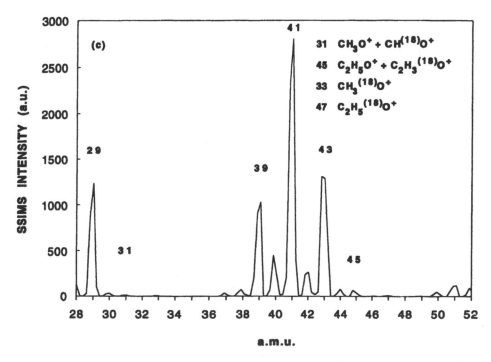

Figure 10c. Positive SSIMS spectra (range 28–52 amu) of $C_{36}H_{74}$: $^{18}O_2$ plasma treated.

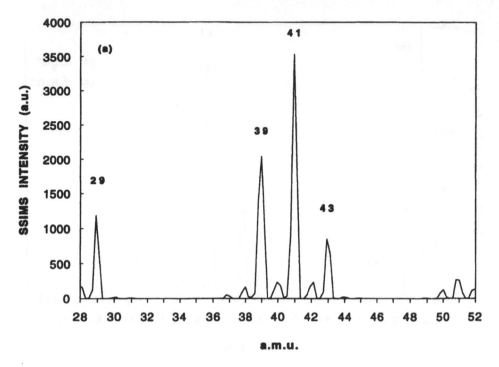

Figure 11a. Positive SSIMS spectra (range 28–52 amu) of HDPE: untreated.

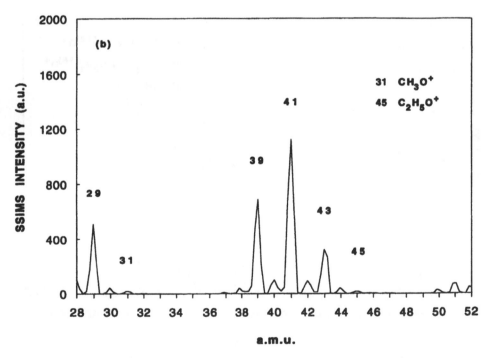

Figure 11b. Positive SSIMS spectra (range 28–52 amu) of HDPE: $^{16}O_2$ plasma treated.

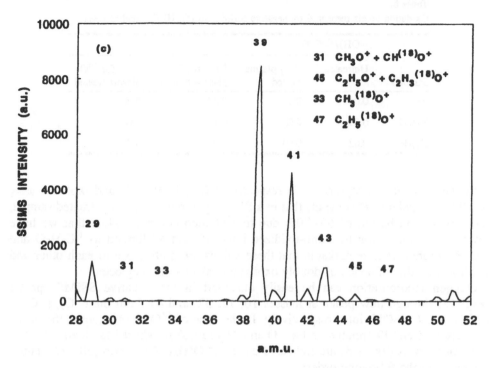

Figure 11c. Positive SSIMS spectra (range 28–52 amu) of HDPE: $^{18}O_2$ plasma treated.

(at 31 and 45 masses). After $^{18}O_2$ plasma treatment, these ions still increase ($^+C_n{}^{16}OH_{2n+1}$ ions and $^+C_{n-1}{}^{18}OH_{2(n-1)-1}$ ions) but new peaks (33 and 47) appear and are essentially identified with $^+C_n{}^{18}OH_{2n+1}$ ions. This evolution is observed clearly for PS but more weakly for HDPE and $C_{36}H_{74}$. For PS, we tried to use these shifts in a more quantitative way. Indeed, it was possible to calculate that part of the positive SSIMS spectra of $^{18}O_2$-plasma-treated films due to ^{16}O incorporation during, or after, plasma treatment. The hypothesis was to use the positive SSIMS spectra of $^{16}O_2$-plasma-treated films as a reference for the ^{16}O incorporation and thus use an equation such as:

$$M_x^{O18} = M_x^{O16} r + [M_{x-2}^{O16} - M_{x-2}^{NT}](1 - r).$$

M_x^T is the normalized intensity of an oxygenated fragment for mass x in the T treatment (NT for untreated, O16 for the $^{16}O_2$ plasma treatment, and O18 for the $^{18}O_2$ plasma treatment). The intensity of any one purely oxygenated fragment at x amu in an $^{18}O_2$-treated sample spectrum is identified as the sum of two contributions: an $^{16}O_2$-labelled ion contribution (r) and an $^{18}O_2$-labelled ion one ($1 - r$). The first contribution can be calculated from the corresponding $^{16}O_2$-treated-sample spectrum. The second one can also be calculated from the same spectrum because it must be equal to the shift of the oxygenated contribution from the peak at $x - 2$ amu. The contribution from the same peak in the untreated-sample spectrum must be subtracted to take into account the fact that the hydrocarbon part of the peak must not be shifted.

Table 8.
Oxidation levels measured by negative SSIMS on PS, HDPE, and $C_{36}H_{74}$

	(−OH)/(−CH)			
Samples	Untreated	$^{16}O_2$ plasma treated	$^{18}O_2$ for $^{18}O_2$ plasma treated	$^{16}O_2$ for $^{18}O_2$ plasma treated
PS	0.04	0.6	0.28	0.26
HDPE	0.03	0.23	0.1	0.08
$C_{36}H_{74}$	0.02	0.09	0.07	0.08

In the case of polystyrene, r is respectively 0.6, 0.59, 0.7, and 0.65 for $x = $ 31, 33, 45, and 47. This means that in a SIMS spectrum of an $^{18}O_2$-treated sample, there is a contribution of 60–70% due to ^{16}O incorporation. Note that we have limited our calculation to mono-oxidized ions (45 and 47 limited to $^+C_2H_5O$ and $^+C_2H_7O$ ions). It is remarkable that these values are quite close to each other and confirm that the influence of doubly oxidized peaks can be neglected.

Oxygen incorporation can be easily observed in the negative SSIMS spectra (Fig. 12) through new peaks at masses 16, 17, 32, and 41 from the (^-O), (^-OH), ($^-O_2$), and (^-C_2OH) ions, respectively. In the case of ^{18}O incorporation, the peaks at masses 16 and 17 (relative to the ^-O and ^-OH ions) are shifted to 18 and 19. The oxidation levels (Table 8) are indicated by the (^-OH)/(^-CH) ratio [30]. Oxidation increases in the following order:

$$\text{polystyrene} > \text{HDPE} > \text{hexatriacontane } (C_{36}H_{74}).$$

The ^{18}O contents are very similar to the values calculated from positive SSIMS and ISS spectra acquired with 6×10^{14} ions/cm^2. The relative concentrations of ^{18}O and ^{16}O are equivalent in the depth probed by SSIMS. The absence of ^{18}O for HDPE and $C_{36}H_{74}$, revealed by the ISS spectra acquired with low ion doses, seems to be limited to the uppermost surface.

In brief, Table 9 summarizes the total oxidation level revealed for PS, PE, and $C_{36}H_{74}$ by the different techniques characterized by various surface sensitivities.

Table 9.
Comparison of oxygen incorporation levels measured by corrected O/C ratios from ISS (ion dose: 1.5×10^{14} ions/cm^2), corrected O_{1s}/C_{1s} ratios from XPS, and (−OH)/(−CH) ratios from SSIMS on PS, PE, and $C_{36}H_{74}$

	Oxygen incorporation					
	$^{16}O_2$ plasma treated			$^{18}O_2$ plasma treated		
	PS	PE	$C_{36}H_{74}$	PS	PE	$C_{36}H_{74}$
ISS (O/C)	2.06	0.15	0	1.96	0.12	0
SIMS (−OH/−CH)	0.6	0.23	0.09	0.54	0.18	0.15
XPS (O/C)	0.33	0.18	0.14	0.35	0.18	0.11

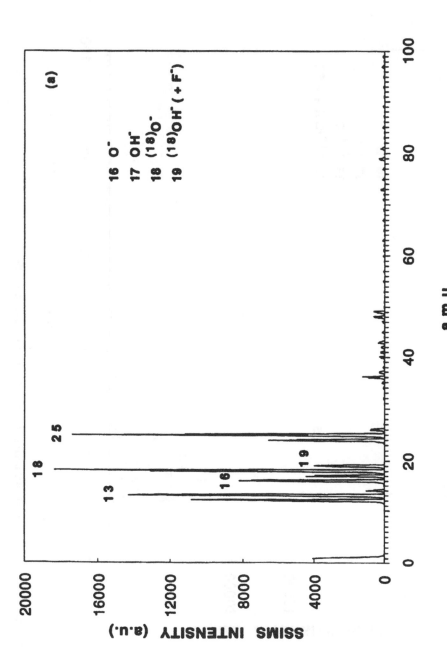

Figure 12a. Negative SSIMS spectra (range 0–100 amu) of PS treated in an ^{18}O$_2$ plasma for 10 s.

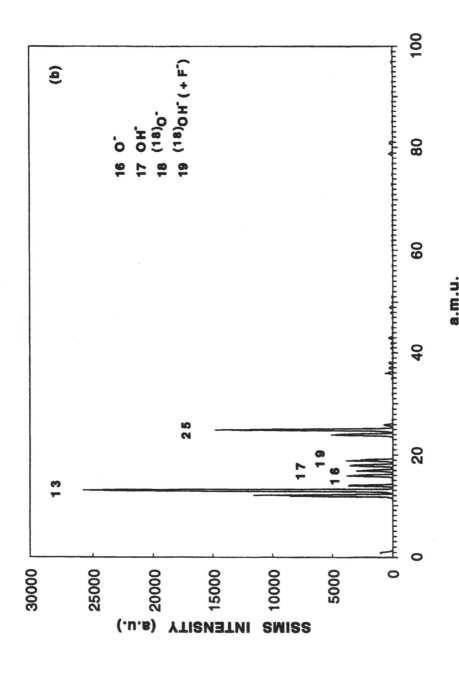

Figure 12b. Negative SSIMS spectra (range 0–100 amu) of PE treated in an $^{18}O_2$ plasma for 10 s.

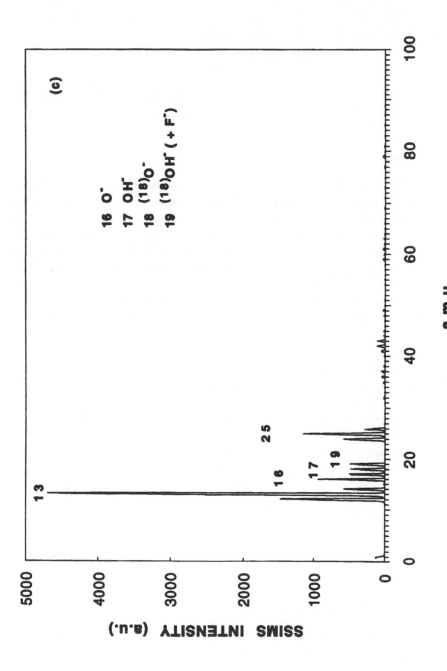

Figure 12c. Negative SSIMS spectra (range 0–100 amu) of C$_{36}$H$_{74}$ treated in an ^{18}O$_2$ plasma for 10 s.

Note that the O/C values obtained from the ISS results are not really quantitative. The influence of elastic and inelastic shadowing cannot be easily quantified and leads to an overestimation of the O/C ratios. The isotope labelling with ^{18}O shows the good reproducibility of the plasma treatments. The oxygen incorporation level is in the opposite order of the degradation rates, that is:

$$PS > HDPE > C_{36}H_{74}.$$

The contribution of ^{16}O in the oxygen level detected is higher than expected. This effect indicates that some oxygen is incorporated after the treatment.

4. CONCLUSIONS

This study by ISS, SSIMS, XPS, and contact angle measurements of the functionalization induced by the oxygen plasma and argon plasma treatments of the surface of hexatriacontane, HDPE, and PS shows the following:

(1) In the case of the model molecule hexatriacontane, the migration of the low-molecular-weight oxygenated fragments towards the bulk is favoured by the small size of fragments. The low oxygen level in $C_{36}H_{74}$ may also be due to the expulsion of the volatile oxidized species. In the case of polymer samples, oxygen is always observed but the lack of really quantitative measurements does not allow us to highlight differences with respect to the various sampling depths probed.

(2) In a plasma, the functionalization is in the opposite order of the degradation rate. But treated HDPE has the same oxygen level in both argon and oxygen plasmas, even though the degradation rate is six times higher in the oxygen plasma than it is in the argon plasma. Therefore, the degradation rates and the reactions leading to the oxygen incorporation should be taken into account when comparing the behaviour of different materials in different plasmas.

(3) The $^{18}O_2$ plasma treatment shows that the global oxygen concentration is independent of the isotope effect. It appears that a higher concentration of $^{16}O_2$ than expected is detected on the samples treated in $^{18}O_2$. This result means that even in an oxygen plasma treatment, some of the surface oxygen is introduced after the treatment has been completed.

These results emphasize the efficiency and the complementarity of the techniques used here in the characterization of polymer surfaces treated by cold plasmas.

REFERENCES

1. F. Clouet and M. K. Shi, J. Appl. Polym. Sci. 46, 1955–1966, 2063–2074 (1992).
2. M. Anand, R. E. Cohen and R. F. Baddour, Polymer 22, 361–371 (1981).
3. S. Nowak, H. P. Haerri, L. Scalpbach and J. Vogt, Surface Interface Anal. 16, 418–423 (1990).
4. J.-L. Dewez, A. Doren, Y.-J. Schneider, R. Legras and P. G. Rouxhet, Surface Interface Anal. 17, 499–502 (1991).
5. J.-L. Dewez, E. Humbeek, E. Everaet, A. Doren and P. G. Rouxhet, in: Polymer–Solid Interfaces, J. J. Pireaux, P. Bertrand and J. L. Bredas (Eds), pp. 463–474. IOP Publishing Ltd., Bristol (1992).
6. H. Yasuda, E. J. Charlson, E. M. Charlson, T. Yasuda, M. Miyama and T. Okuno, Langmuir 7, 2394–2400 (1991).

7. R. Foerch and D. Johnson, *Surface Interface Anal.* **17**, 847–854 (1991).
8. E. Occhiello, F. Garbassi and M. Morra, *Surface Sci.* **211–212**, 218–226 (1989).
9. E. Occhiello, M. Morra, F. Garbassi, D. Johnson and P. Humphrey, *Appl. Surface Sci.* **47**, 235–242 (1991).
10. J. Lub, F. C. B. M. van Vroonhoven, E. Bruninx and A. Benninghoven, *Polymer* **30**, 40–44 (1989).
11. Y. De Puydt, D. Leonard and P. Bertrand, in: *Metallized Plastics 3: Fundamental and Applied Aspects*, K. L. Mittal (Ed.), pp. 225–241. Plenum Press, New York (1992).
12. W. J. van Ooij and R. S. Michael, in: *Metallization of Polymers*, E. Sacher, J. J. Pireaux and S. P. Kowalczyk (Eds), pp. 60–87. Am. Chem. Soc. Symposium Series No. 440 (1990).
13. D. Briggs, *Surface Interface Anal.* **15**, 734–745 (1990).
14. W. J. van Ooij and R. H. G. Brinkhuis, *Surface Interface Anal.* **11**, 430–440 (1988).
15. J. A. Gardella, Jr and D. M. Hercules, *Anal. Chem.* **53**, 1879–1884 (1981).
16. Y. De Puydt, D. Leonard, P. Bertrand, Y. Novis, M. Chtaib and P. Lutgen, *Vacuum* **42**, 811–817 (1991).
17. P. Bertrand and Y. De Puydt, *Nucl. Instrum. Meth., Phys. Res. B* (in press).
18. P. Marie and Y. Rouault, *Macromolecules* (submitted).
19. Y. Rotenberg, L. Borukva and A. W. Neumann, *J. Colloid Interface Sci.* **93**, 169–183 (1983).
20. P. Cheng, D. Li, L. Borukva, Y. Rotenberg and A. W. Neumann, *Colloids Surfaces* **43**, 151–167 (1990).
21. S. Wu, *J. Polym. Sci.* **C34**, 19–30 (1971).
22. M. K. Shi, Y. Holl, Y. Guilbert and F. Clouet, *Makromol. Chem., Rapid Commun.* **12**, 277–283 (1991).
23. D. T. Clark and A. Dilks, *J. Polym. Sci., Polym. Chem. Ed.* **17**, 957–976 (1989).
24. M. Suzuki, A. Kishida, H. Iwata and Y. Ikada, *Macromolecules* **19**, 1804–1808 (1986).
25. F. Rouzhebi, B. Catoire, M. Goldmann and J. Amouroux, *Rev. Int. Hautes Temp. Refract. Fr.* **23**, 211–220 (1986).
26. H. Niehus and R. Spitzl, *Surface Interface Anal.* **17**, 287–307 (1991).
27. D. Briggs, A. Brown and J. C. Vickerman, in: *Handbook of Static Secondary Ion Mass Spectrometry (SIMS)*, pp. 86–87. John Wiley, Chichester (1989).
28. D. T. Clark and R. Wilson, *J. Polym. Sci., Polym. Chem. Ed.* **21**, 837–853 (1983).
29. S. J. Moss, A. M. Jolly and B. J. Tighe, *Plasma Chem. Plasma Process.* **6**, 401–416 (1986).
30. A. Chilkoti, B. Ratner and D. Briggs, *Anal. Chem.* **63**, 1612–1620 (1991).

Plasma Surface Modification of Polymers, pp. 99–111
M. Strobel, C. Lyons and K. L. Mittal (Eds)
© VSP 1994

Plasma surface modification of polyethylene: short-term vs. long-term plasma treatment

R. FOERCH,* G. KILL and M. J. WALZAK†

Surface Science Western, Western Science Centre, The University of Western Ontario, London, Ontario, Canada N6A 5B7

Revised version received 27 April 1993

Abstract—A remote plasma reactor, with air as the plasma gas, has been used for in-line surface modification of linear low-density polyethylene tape (LLDPE) passing 10 cm below the main plasma zone. Line speeds of up to 0.70 m/s were tested, allowing the study of 0.014 s exposure times to the plasma. Oxygen to carbon (O/C) ratios averaging 0.11 were observed on a reproducible basis. The reactor was also used for static plasma treatment under similar experimental conditions. This allowed a comparative study of short-term (milliseconds) vs. long-term (several seconds) plasma treatment. High-resolution X-ray photoelectron spectroscopy (XPS) analysis of the treated polymer surface suggested the formation of hydroxyl (C—OH), carbonyl (C=O) and carboxyl (O—C=O) groups, even after short plasma treatment. The intensities of these components were seen to increase in approximately equal quantities with increasing O/C ratio. Water washing of polyethylene surfaces with high O/C ratios showed a loss of oxygen, apparent as a decrease in O—C=O groups in the C 1s spectra. A smaller loss in oxygen was observed when washing samples that had been plasma-treated for milliseconds. A surface ageing study revealed that polyethylene surfaces that had been plasma-treated for short time periods showed only a negligible loss of oxygen on prolonged exposure to air. Surfaces treated for longer time periods showed a loss of up to 50% of the total oxygen on the surface within a few days of treatment. Static secondary ion mass spectrometry has provided some supporting evidence for surface damage of the treated films.

Keywords: Plasma; polyethylene; surface modification; ageing; XPS.

1. INTRODUCTION

The main gas-phase techniques presently in use for the modification of polymer surfaces are processes involving excited-state chemistry such as corona discharge [1–5] and glow-discharge treatments [6–10]. Glow-discharge plasma treatment has been under intense study and some commercial processes have been introduced, most of these relying on batch rather than in-line treatment. Batch processing generally involves plasma exposure of at least several seconds to alter the surface properties [11]. Since treatment times of this length have been believed necessary, in-line plasma treatment of polymers is still fairly uncommon as an industrial process within the polymer industries. Thus, very little has been reported in the literature on the effect of short-term plasma treatment on polymer surfaces, making this study timely.

It is well known that most practical polymer surfaces are covered by a thin

---

*Present address: Institut für Microtechnik Mainz, Physical Technology Division, Carl Zeisstr. 18-20, P.O. Box 2440, 6500 Mainz 42, Germany.

†To whom correspondence should be addressed.

surface layer of hydrocarbon contamination, antioxidants, and slip agents. When such a polymer surface is plasma-treated, three main processes occur:

(1) *surface cleaning:* usually an oxidative degradation of the contaminating hydrocarbon species (dust, fingerprints, process residues, etc.);
(2) *surface modification:* insertion of oxygen at active sites on the polymer surface to form $C-O$, $C=O$, $O-C=O$, ethers, and epoxy and ester groups [2, 3, 12, 13]. This is usually accompanied by surface cross-linking;
(3) *surface degradation:* excessive chain scission leading to a layer of short-chain oxidized material on the surface. This layer has previously been termed LMWOM (low-molecular-weight oxidized materials) [4]. With longer treatment, or stronger process conditions, ablation of the surface may occur.

Thus, on a molecular level, the surface experiences a complex set of reactions during plasma treatment leading to a mixture of low and high molecular weight material on the surface. After the plasma has been turned off, the surface will tend to stabilize itself to reduce its surface energy. It is believed that this occurs mainly by macromolecular motion and a mixing of the shorter-chained material with the polymer bulk [14, 15]. This can also lead to the reorientation of functional groups into the polymer bulk. Some reports have also suggested migration of additives from the bulk to the surface [16, 17], thus further affecting the oxygen to carbon (O/C) values observed on XPS analysis of the surface.

Over the past few years, researchers have been very active in determining the effect of the short-chained material formed during surface modification to explain industrial problems such as adhesion failure, surface ageing, and the loss of surface properties [4, 5, 11, 13, 16, 18]. This has predominantly concentrated on corona-discharge-treated surfaces. However, glow-discharge plasma treatment is believed to cause similar effects. Gerenser *et al.* [3] established, through a series of surface washing experiments on polyethylene treated by corona discharge, that water washing of the plasma-treated surface always gave an O/C ratio of 0.10–0.12 regardless of the initial O/C value. They attributed this to the loss of water-soluble, short-chained species on the surface. Strobel *et al.* [4] showed that after dissolution of the LMWOM formed on a corona-discharge-treated polypropylene surface, no further LMWOM was formed with subsequent ageing in air. Also, the extent of formation of LMWOM was reported to depend on the type of polymer and process conditions. Occhiello *et al.* [19] reported the effect of the plasma modification of polypropylene (PP) on the strength of a modified PP/epoxy joint. Two interfaces were identified: one at the modified PP/epoxy border, and the other at the modified PP/bulk PP border. Adhesion failure was shown to occur predominantly at the modified PP/bulk PP interface. Brewis [20], Brewis and Briggs [21], and Strobel *et al.* [4], however, point out that the presence of LMWOM may actually assist adhesion properties in cases where a mechanism exists for incorporation of the LMWOM into the adherate. The identification of the short-chained material, or LMWOM, is unfortunately very difficult, because the layer is very thin and contains the same elements and functional groups as the modified polymer bulk. Thus, analysis using standard spectroscopic techniques has not been very successful.

Static secondary ion mass spectroscopy (SSIMS) is rapidly developing into a major surface analytical tool for the analysis of polymeric substances [22–24].

The SSIMS experiment typically analyses a depth of 1–2 nm and can thus be regarded as a complementary technique to XPS, which probes typical depths of 3–10 nm. Because of the shallow depth analyzed, the SSIMS experiment can potentially reveal additional information specific to LMWOM species.

Recently, two independent SSIMS studies [16, 25] of polymers treated in a corona and glow-discharge plasmas have reported the presence of high-intensity mass fragments corresponding to aromatic structures. It was suggested that the aromatic fragments were associated with the surface LMWOM. Further spectroscopic evidence of plasma-induced surface damage was observed in a detailed study of the valence band spectra [26] of polymers plasma-treated to very high levels of oxygen and nitrogen; these studies suggested the presence of branching by the formation of a peak typical for dangling $-CH_3$ groups.

Earlier studies have concentrated on static plasma treatment, during which the polymer surface is exposed to the plasma for several seconds. In most cases, it is difficult to control process conditions at exposure times less than 1 s and data have suggested [26] that the O/C ratio is close to the maximum even after 1 s of plasma treatment. Thus, there is a need to study the effect of shorter treatment times to investigate the early stages of the plasma/polymer surface reaction. In order to do this, a system has been built [27] which allows the in-line treatment of polymer tapes at various speeds, thus enabling process times to be reduced to milliseconds. In the present study, surfaces of LLDPE were exposed to short-term ($t < 1$ s) and long-term ($t > 1$ s) remote air plasmas and investigated using XPS and SSIMS analyses, in combination with the more traditional surface-washing experiments. Some comparisons are made with the previous long-term plasma treatment [16] ($t > 10$ s) to investigate surface modification and surface degradation occurring at the various exposure times.

2. EXPERIMENTAL

Linear low-density polyethylene (LLDPE) (51 μm gauge, 6 mm width Sclair film) was supplied by DuPont Canada, Kingston, Ontario. Because this is a commercial film, there are low concentrations of antioxidants and stabilizers present. The XPS spectra of the control sample show either no oxygen concentration or oxygen concentrations of less than 1%, which can arise from minor oxidation of the surface.

The plasma system employed uses a 120 W, 2.45 GHz microwave generator which applies microwaves to gases passing through a 10 mm diameter quartz plasma-drift tube. The microwave energy is applied to a volume of approximately 10 cm^3 positioned approximately 10 cm above the sample. This distance is not fixed and can be varied depending on the extent of modification required. The gas used was Liquid Carbonic extra-dry air. Typical pressures near the sample surface during the static experiments were 1.3 Pa with no gas flow and 270 Pa when the gas flow was 1000 sccm [9, 10]. There is some heating of the sample on prolonged exposure. Typical film temperatures reached after 3 s of plasma treatment were 60–70°C. For the in-line treatment of tapes, the reactor was modified by the addition of two sets (inlet and outlet) of differentially pumped chambers separated by a series of Teflon slits [27]. No contamination was transferred from the apparatus to the moving tape surface, as evidenced by a lack of change in the

XPS spectra of film samples run through the system in the absence of plasma. The outer chambers were differentially pumped by a high-speed single-stage mechanical pump, and a two-stage mechanical pump was used to further reduce the pressure in the plasma reaction chamber. The tape speeds available ranged from 0.18 to 0.70 m/s. The pressures during the in-line experiments were 300 Pa with no gas flow and 600 Pa with 1000 sccm of air. For in-line treatment, approximate exposure times (t) were calculated assuming that the plasma stream covered an area of 1 cm^2 on the sample surface. Thus, a line speed of 0.70 m/s represents an exposure time of approximately 0.014 ± 0.005 s; 0.40 and 0.18 m/s represent exposure times of 0.025 ± 0.005 and 0.056 ± 0.005 s, respectively. The errors in the exposure times arise predominantly from variations in the web speed.

Surface-washing experiments were carried out on the samples prepared in-line by cutting three pieces from a 15 cm length of film and submerging them in deionized water for 5 min. The washed samples were dried in a vacuum desiccator for 24 h. The treated, unwashed samples were checked by XPS immediately and 24 h later (at the same time as the washed samples were analyzed) to ensure that ageing did not play a major role in the washing experiment. The zero-day analyses were done between 30 min and 4 h after reaction. XPS studies did not show any ageing effects during the 24 h. Samples in the ageing studies were stored at room temperature (22°C) in small containers with the treated surface untouched. Samples for XPS analysis were cut from the centre of the sample to accommodate minor variations in the oxygen uptake over the width of the film.

X-ray photoelectron spectroscopy was performed using an SSL SSX-100 X-ray photoelectron spectrometer, which use monochromatized Al K_a X-rays to analyze the sample. The binding energy reference for the C 1s of the C—C and C—H peak was set at 285.0 eV. For the elemental analysis a spot size of 1000 or 600 μm and a pass energy of 150 eV were utilized. For the high-resolution spectra, the X-ray spot size and pass energy were reduced to 150 μm and 50 eV, respectively. During the analysis the electron take-off angle (ETOA) was 37°. Charging was controlled using the flood gun/screen technique [28]. Angle-resolved XPS analysis was performed on a Scienta ESCA 300 X-ray photo-electron spectrometer, which combines a high-power rotating anode X-ray source with a focusing quartz crystal monochromator, high transmission/imaging electron optics, and a multichannel detector. This instrument has been described in detail elsewhere [26, 29]. During the analysis, the ETOA was varied from 20° to 70°. At each angle, the C 1s, O 1s, and valence band spectra were recorded.

Static SIMS analysis was performed after plasma treatment on the same samples as above. The SSIMS spectra were run on a VG ESCALAB MkI. The ion beam source was an AG61 providing a 4 keV beam of Xe$^+$. The ion current was approximately 0.4 nA, corresponding to an ion beam density of ~ 1 nA/cm^2. Spectra were obtained in less than 30 min after first exposing the sample to the Xe$^+$ beam and therefore were carried out within the static SIMS limit. Charge compensation was achieved by an LEG31 electron flood gun. Mass analysis was done using a VG MM12-1200 quadrupole mass spectrometer equipped with an energy pre-filter. Both positive and negative ion spectra were collected for all samples for the mass range of $m/z = 0$–200.

3. RESULTS AND DISCUSSION

3.1. XPS elemental and core level analysis

The uptake of oxygen by a polymer surface is dependent on a number of factors such as the pressure, the microwave power, the gas flow rate, and the position of the sample with respect to the plasma [9, 10, 27]. O/C ratios were chosen to indicate the level of oxidation or the extent of modification of the surface. The effect of treatment time on the uptake of oxygen on LLDPE is shown in Table 1. During the dynamic plasma treatment, the O/C ratio did not change between the fastest and the medium speeds (O/C = 0.11) but did increase slightly for the slowest speed (O/C = 0.12). The shortest treatment time possible with a static treatment was 1 ± 0.5 s, which typically gave an O/C ratio of 0.16 ± 0.03. Air plasma treatment usually caused uptakes of small amounts of nitrogen (< 2%), which are not recorded in the table.

Table 1.
XPS data for a series of LLDPE samples treated under various experimental condition showing increasing O/C ratio

Experimental conditions[a]	Treatment time (s)	Power (W)	O/C
Dynamic	0.014	30	0.11 ± 0.02
Dynamic	0.025	30	0.11 ± 0.01
Dynamic	0.056	30	0.12 ± 0.01
Static	2	30	0.17 ± 0.02
Static	20	30	0.24 ± 0.01
Static	20	40	0.25 ± 0.03
Static	30	30	0.26 ± 0.03

[a]All experiments were performed with the sample 10 cm from the plasma at an air flow rate of 1000 sccm.

The high-resolution C $1s$ spectra for this series of experiments (Fig. 1) indicated the presence of C—O—C (or C—OH), C=O and O—C=O groups on all surfaces. These species become apparent in the C $1s$ spectrum at typical binding energy shifts from the main hydrocarbon peak of 1.5 ± 0.1, 3.0 ± 0.2, and 4.4 ± 0.3 eV, respectively [9, 10]. Even with the short exposure times encountered during dynamic plasma treatment, the C $1s$ spectrum showed the presence of highly functionalized groups. The relative intensities of the high-binding-energy components (more highly oxidized carbon) were seen to increase with increasing O/C ratio. This suggests that there is a trend for oxygen to add preferentially to specific carbon sites until no further C—O bonds can be formed. Further oxygen then adds to that carbon which has already reacted, eventually leading to the formation of the O—C=O species commonly observed in the C $1s$ spectra of oxygen- or air-plasma-treated LLDPE samples.

Valence band spectra were recorded for samples with increasing exposure times to an air plasma. The valence band spectra obtained here showed no appreciable differences with increasing O/C ratio. No evidence for surface branching was apparent at an O/C of 0.20. This may suggest that surface branching can be observed by XPS valence band analysis only when the surface is more highly functionalized than in this work.

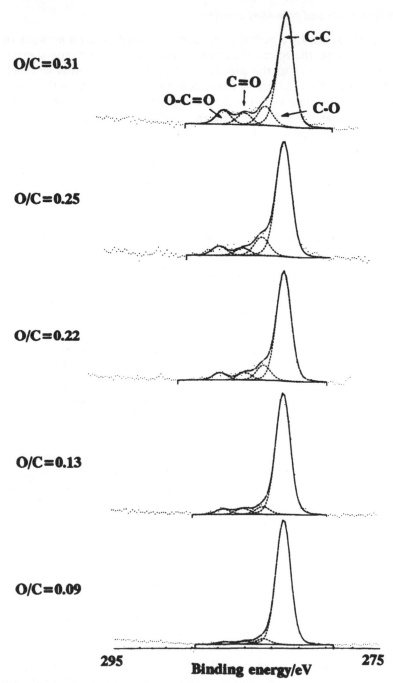

Figure 1. C 1*s* spectra of air-plasma-treated LLDPE as a function of increasing O/C ratio.

3.2. Surface ageing

The surface ageing phenomenon has been studied by a number of groups [4–14] and it has generally been found that glow-discharge-modified polymer surfaces lose a significant portion of the newly acquired functional groups within a few days after treatment. This is generally accompanied by a loss of surface adhesion properties. LLDPE specimens exposed to static plasma conditions for 3 s had an O/C ratio of 0.20 (see Table 2). Such specimens were found to lose ~ 40% of this oxygen when left in air for a period of 45 days. Previous work on LLDPE [10, 26] has shown that after static plasma treatment (O/C = 0.25) the surface loses ~ 30% of the initial oxygen on the surface within 5 days. These static plasma experiments were performed as part of a separate set of experiments and so are not done under exactly the same conditions as the dynamic plasma treatments. From our work with the ageing of these samples and those treated in a similar manner in an oxygen plasma, we are confident that the behaviors are comparable. Similar ageing studies were performed on LLDPE exposed to a high-speed dynamic plasma treatment resulting in O/C ratios of 0.11. Little or no ageing was observed for these samples.

Table 2.
Ageing data for dynamic air-plasma treatments (t = 0.014 s) and static air-plasma treatment (t = 3 s) of LLDPE over a period of 45 days.

Exposure to air after treatment (days)	O/C, dynamic	O/C, static[a]
0	0.11 ± 0.02	0.20 ± 0.03
1	0.09 ± 0.03	—
2	0.09 ± 0.03	—
6	0.09 ± 0.01	—
8	—	0.16 ± 0.03
15	0.10 ± 0.01†	0.11 ± 0.03[b]
20	0.10 ± 0.02	—
28	—	0.15 ± 0.03
45	—	0.12 ± 0.032

[a]Conditions for static treatment—2000 sccm, 22 cm, 60 W.
[b]Angle-resolved XPS measurements of these samples show no apparent change in the O/C ratio as a function of the sampling depth.

This behavior could be rationalized if it is assumed that the ageing phenomenon is mostly a measure of the degree of chain motion occurring as the surface attempts to stabilize after plasma treatment. Thus, the greater the increase in O/C ratio with treatment, the greater the tendency for chain mobility. At low O/C ratios, chain mobility may be reduced because (a) less LMWOM is present on the surface and (b) the change in surface energy may be less, thus reducing the migration of the low molecular weight species. Thus, sample surfaces exposed to shorter treatment times, typical of the dynamic plasma treatment, appear to be less affected by ageing than those surfaces originally containing more oxygen.

3.3. Surface washing

Table 3 shows the XPS O/C ratio observed for a series of washing experiments on samples treated at various times of plasma exposure at constant gas flow (1000 sccm), power (30 W), distance from the plasma (10 cm), and pressure (220 Pa). The values quoted in Table 3 are averages of nine independent experiments. LLDPE specimens treated in the dynamic plasma all showed a loss of ~ 25% of the surface oxygen upon washing. Changes in the line speed did not affect the losses on washing. Increasing the plasma exposure time to 2 s (O/C = 0.17) showed an increase in the quantity of water-soluble material, which represented 40% of the initially observed surface oxygen. With longer exposure times, the average decrease in the O/C atomic ratio upon washing was 45–52%.

Table 3.
XPS analysis of a series of LLDPE samples treated in an air plasma—to achieve increasing O/C values—showing the effect of surface washing in deionized water on the surface oxygen concentration

Exposure time (s)	Film line speed (m/s)	O/C ratio (unwashed)	O/C ratio (washed)	% O removed
Control	—	0	0.03 ± 0.01	—
0.014	0.70	0.11 ± 0.02	0.08 ± 0.02	27
0.023	0.40	0.11 ± 0.01	0.08 ± 0.03	27
0.056	0.18	0.12 ± 0.01	0.09 ± 0.02	25
2	—	0.17 ± 0.02	0.10 ± 0.01	41
10	—	0.27 ± 0.05	0.13 ± 0.01	52
20	—	0.24 ± 0.01	0.13 ± 0.03	46
30	—	0.26 ± 0.03	0.14 ± 0.01	46
> 60	—	0.27 ± 0.01	0.15 ± 0.02	44

The O/C ratio for unwashed surfaces was noted to vary considerably at longer exposure times. Washing the surfaces, and thus removing the LMWOM, results in lower and more consistent O/C values than before washing. It is possible that the inconsistency in the O/C ratios before washing indicates uneven rates of oxygen reaction and/or different reaction products. The relative rates of modification and degradation on the LLDPE surface appear extremely sensitive to local plasma conditions. During much shorter plasma exposure times ($t = 0.014$–0.056 s), surface degradation appears to be less significant. Oxygen incorporated into the polyethylene surface (O/C = 0.11) appears to be very stable; some of the oxygen can be removed by immediate water washing (25%) but there is no significant loss on prolonged exposure to air. If samples treated for milliseconds are stored for 14 days before washing, there is no loss of oxygen. This indicates that the small amounts of LMWOM formed can be stabilized on the surface with time. For plasma exposure times greater than 1 s, this behavior is seen to change and surface degradation (or the formation of LMWOM) seems to dominate.

A change in the C 1s peak profiles was observed for all polyethylene surfaces studied. Washing of the surface (Fig. 2) shows a decrease of the higher-binding-energy peak components, representing acidic or ester groups, especially at the longer treatment times. Thus, with increasing treatment time there is a trend towards more highly oxidized carbon species and a greater loss of oxygen on washing.

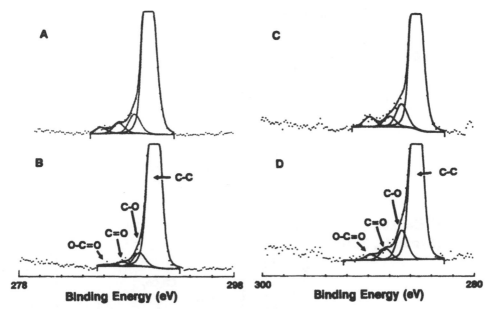

Figure 2. Effect of washing on the C $1s$ spectra of LLDPE after short ($t = 0.014$ s) or long ($t = 20$ s) air-plasma treatment. (A) and (B) represent short treatment times, (A) unwashed and (B) washed; (C) and (D) represent long treatment times, (C) unwashed and (D) washed.

3.4. Static SIMS analysis

Static SIMS (SSIMS) was performed 2 weeks after plasma treatment on the same samples as above. Negative-ion SSIMS analysis showed several differences on the surfaces of untreated and plasma-treated materials (Fig. 3). Untreated LLDPE showed evidence of low levels of surface oxidation by peaks at $m/z = 16^-$ (O^-), 17^- (OH^-), 45^- ($C_2H_5O^-$) and 54^- ($CH_3CO_2^-$). After dynamic plasma treatment, peaks at $m/z = 16^-$, 17^-, and 54^- increased in relative intensity, while that at $m/z = 45^-$ decreased. Upon increasing the power from 30 to 50 W during dynamic treatment, the peaks due to the hydrocarbon components at $m/z = 13$ (CH^-) and 25 (C_2H^-) were seen to increase in relative intensity. These peaks were noted to increase even further for samples exposed to a static plasma for 6 s (30 W). The relative intensity increase in the C_1 and C_2 hydrocarbon components is believed to represent greater chain scission and degradation of the surface with longer (or higher power) plasma treatment.

The positive SSIMS spectrum of untreated LLDPE showed evidence of surface oxidation by low itensity peaks appearing at $m/z = 31^+$, 45^+, 59^+, etc. corresponding to $CH_3(CH_2)_nO^+$, $n = 0, 1, 2, \ldots$ (see Fig. 4). Peaks at $m/z = 77^+$, 91^+, 105^+, 115^+, and 128^+ were observed in the spectra and represent aromatic fragments from the surface. High levels of aromatic fragments in SSIMS spectra have previously been observed [16] for various other polymers at high O/C and N/C ratios after remote oxygen- or nitrogen-plasma treatment.

It is believed that the formation of LMWOM occurs during the early stages of plasma treatment and is therefore present on most plasma-treated polymer surfaces discussed in the literature. The low molecular weight species are formed during scission of the polymer chains and are known to affect the surface

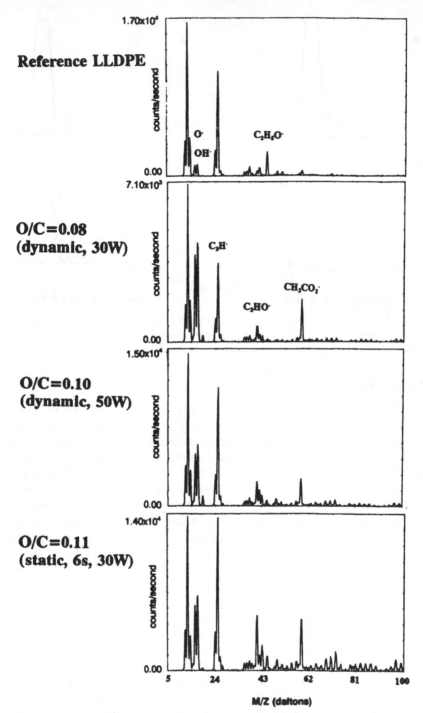

Figure 3. Negative SSIMS spectra of LLDPE before and after air-plasma treatment as a function of increasing O/C ratio. The treatment time for the dynamic sample was 0.014 s; the static system treatment time was 6 s.

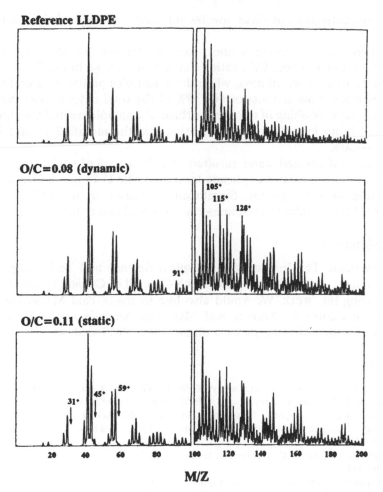

Figure 4. Positive SSIMS spectra of LLDPE before and after air-plasma treatment as a function of increasing O/C ratio. The treatment time for the dynamic system was 0.014 s; the static system treatment time was 6 s.

properties. This work has suggested that very short exposure times to a plasma will effectively modify the polymer to produce a surface that is stable with respect to ageing; longer exposure times to create high O/C ratios will result in increased ageing and high oxygen loss on exposure to water.

4. CONCLUSIONS

XPS and SSIMS analyses of the surface of polyethylene treated in an air plasma for various time durations has shown the following:

(1) An XPS O/C ratio equal to 0.11 can be reached after an exposure time of only 0.014 s. This uptake of oxygen does not increase proportionally with the exposure time.
(2) XPS C 1s spectra indicate the formation of C—O—C (or C—OH), C=O and O—C=O groups on the surface even after treatment times of only 0.014 s.

The concentrations of these species increase equally with increasing O/C ratio.

(3) Based on surface ageing studies, samples treated for short ($t = 0.014$ s) treatment times (lower O/C ratio) showed no change in the O/C ratio with exposure to air over 20 days, while those samples plasma-treated for longer time periods showed a loss of 40–50% of the total oxygen from the surface over 45 days. Washing of the aged surface of a sample treated for short times resulted in no loss of oxygen, indicating that the modified surface is stable after 14 days.

(4) Washing in deionized water resulted in a 25% loss of surface oxygen even with the short ($t = 0.014$ s) plasma treatment; the C 1s spectra showed a loss of acidic or ester groups. Significant decreases in the O/C ratio were observed for samples washed immediately after longer plasma treatments.

Acknowledgements

We wish to thank The Institute for Chemical Science and Technology and the Natural Sciences and Engineering Research Council of Canada for their financial support during this work. We would also like to thank Paul Marsh and Martin Hearn of the Centre for Surface and Materials Analysis at the University of Manchester for their kind assistance with the static SIMS analysis.

REFERENCES

1. D. Briggs, D. G. Rance, C. R. Kendall and A. R. Blythe, *Polymer* 21, 895–905 (1980).
2. D. Briggs, C. R. Kendall, A. R. Blythe and A. B. Wootton, *Polymer* 24, 47–51 (1983).
3. L. J. Gerenser, J. F. Elman, M. G. Mason and J. M. Pochan, *Polymer* 26, 1162–1166 (1985).
4. M. Strobel, C. Dunatov, J. M. Strobel, C. S. Lyons, S. J. Perron and M. C. Morgan, *J. Adhesion Sci. Technol.* 3 321–335 (1989).
5. J. M. Strobel, M. Strobel, C. S. Lyons, C. Dunatov and S. J. Perron, *J. Adhesion Sci. Technol.* 5, 119–130 (1991).
6. L. J. Gerenser, *J. Adhesion Sci. Technol.* 1, 303–318 (1987).
7. L. J. Gerenser, *J. Vac. Sci. Technol. A* 6, 2897–2903 (1988).
8. M. Morra, E. Occhiello and F. Garbassi, *J. Colloid Interface Sci.* 132, 504–508 (1989).
9. R. Foerch, J. Izawa, N. S. McIntyre and D. H. Hunter, *J. Appl. Polym. Sci.: Appl. Polym. Symp.* 46, 415–437 (1990).
10. R. Foerch and D. H. Hunter, *J. Polym. Sci.: Polym. Chem. Ed.* 30, 279–296 (1992).
11. E. M. Liston, *J. Adhesion* 30, 199–218 (1989).
12. D. Briggs and C. R. Kendall, *Int. J. Adhesion Adhesives* 2, 11–17 (1982).
13. J. M. Pochan, L. J. Gerenser and J. F. Elman, *Polymer* 27, 1058–1062 (1986).
14. F. Garbassi, M. Morra, E. Occhiello, L. Barino and R. Scordamaglia, *Surface Interface Anal.* 14, 585–589 (1989).
15. J. D. Andrade (Ed.), *Polymer Surface Dynamics*. Plenum Press, New York (1988).
16. R. Foerch and D. Johnson, *Surface Interface Anal.* 17, 847–854 (1991).
17. J. Adelsky, *TAPPI* 72, 181 (1989).
18. D. Briggs, in: *Surface Analysis and Pretreatment of Plastics and Metals*, D. M. Brewis (Ed.), Chap. 9, pp. 199–226. Macmillan, New York (1982).
19. E. Occhiello, M. Morra, G. Morini and F. Garbassi, *Mater. Res. Soc. Symp. Proc.* 153, 199–204 (1989).
20. D. M. Brewis, *Prog. Rubber Plastic Tech.* 1, 1–21 (1985).
21. D. M. Brewis and D. Briggs, *Polymer* 22, 7–16 (1981).
22. D. Briggs, *Surface Interface Anal.* 15, 734–738 (1990).
23. D. Briggs and A. B. Wootton, *Surface Interface Anal.* 4, 109–115 (1982).
24. D. Briggs, *Surface Interface Anal.* 9, 391–404 (1986).

25. V. Andre, Y. DePuydt, F. Arefi, J. Amouroux, P. Berrand and J. F. Silvain, in: *ACS Symposium Series 440*, E. Sacher, J. J. Pireaux and S. Kowalczyk (Eds). 423–432. American Chemical Society, Washington, DC (1990).
26. R. Foerch, G. Beamson and D. Briggs, *Surface Interface Anal.* **17**, 842–846 (1991).
27. R. Foerch, N. S. McIntyre and D. H. Hunter, *Kunststoffe/German Plastics* **81**, 47–48 (1991).
28. C. Bryson, *Surface Sci.* **189/190**, 50–58 (1987).
29. G. Beamson, D. Briggs, S. F. Davies, I. W. Fletcher, D. T. Clark, J. Howard, U. Gelius, B. Wannberg and P. Barker, *Surface Interface Anal.* **15**, 541–549 (1990).

Plasma Surface Modification of Polymers, pp. 113–121
M. Strobel, C. Lyons and K. L. Mittal (Eds)
© VSP 1994

Plasma oxidation of polystyrene vs. polyethylene

R. K. WELLS,[1] J. P. S. BADYAL[1]*, I. W. DRUMMOND,[2] K. S. ROBINSON[2] and F. J. STREET[2]

[1] *Department of Chemistry, Science Laboratories, Durham University, Durham DH1 3LE, U.K.*
[2] *Kratos Analytical, Barton Dock Road, Urmston, Manchester M31 2LD, U.K.*

Revised version received 11 June 1993

Abstract—Polyethylene and polystyrene film surfaces have been plasma-oxidized and subsequently characterized by X-ray core level and valence band spectroscopies. The extent of polyethylene surface oxidation was found to be dependent on the power of the oxygen glow discharge employed and the length of time that the treated sample was left exposed to air prior to analysis. In marked contrast to these observations, plasma-oxidized polystyrene surfaces were much less dependent on the oxygen glow discharge power and were also found to retain their oxygenated character over much longer periods of ageing. These differences in oxidative behaviour are explained in terms of the molecular structures of the respective polymers.

Keywords: Plasma oxidation; ageing; polyethylene; polystyrene; polypropylene; XPS.

1. INTRODUCTION

Surface modification of a polymer by a glow discharge is a useful means for increasing the surface energy of polymers and improving their adhesion properties [1, 2]. Plasma oxidation comprises degradation of the substrate and reaction with the ions, atoms, ozone, and metastables of atomic and molecular oxygen as well as electrons and a broad electromagnetic spectrum. Investigation by surface-sensitive techniques such as X-ray photoelectron spectroscopy (XPS) [3] and secondary ion mass spectrometry (SIMS) [4, 5] have revealed that the extent and stability of such treatments are critically dependent on the glow discharge characteristics (e.g. reactor configuration, electromagnetic excitation frequency, substrate location, etc.). However, a very poor understanding still remains concerning exactly what occurs during the exposure of a polymer substrate to an oxygen glow discharge, and how the surface subsequently behaves during ageing [6]. In this paper, we compared the relative reactivities of polyethylene and polystyrene towards an oxygen plasma under identical experimental conditions. The surface chemistry of interest was monitored by X-ray core level and valence band spectroscopies; these techniques are widely recognized as being complementary tools for the surface analysis of polymers [7]. The former is routinely used for determining the presence of heteroatoms within a polymeric structure [8]. However, this technique is incapable of identifying the different types of hydrocarbon segments associated with a specific polymeric backbone. In fact, the XPS valence band region provides a much more informative picture [7]. Typically, the valence electronic levels can serve as a unique fingerprint for a specific polymeric structure.

---

*To whom correspondence should be addressed.

2. EXPERIMENTAL

Additive-free low-density polyethylene (ICI, 2 cm^2 × 0.1 mm, M_w = 250 000) and polystyrene (ICI, 2 cm^2 × 0.1 mm, M_w = 100 000) pieces were cleaned in an ultrasonic bath with isopropyl alcohol and subsequently dried in air. Research-grade-quality oxygen (BOC) was used without any further purification.

A 13.56 MHz RF generator was inductively coupled to a cylindrical glass reactor (4.5 cm diameter, 490 cm^3 volume) via an externally wound copper coil; this arrangement was used for the plasma treatments [9]. A strip of polymer was positioned in the centre of the coils (i.e. within the glow region). The reaction vessel was initially evacuated by a two-stage rotary pump to a base pressure of less than 1×10^{-2} Torr. Then oxygen was passed through the chamber at a pressure of 0.2 Torr and a constant flow rate of 2.0 cm^3 min^{-1} for 10 min, followed by plasma modification for 5 min (longer exposures resulted in no further changes to the polymer surface as determined by XPS). The substrate remained at ambient temperature during the glow discharge treatment.

Two different electron spectrometers were used for surface analysis of the treated films. Low-resolution core level X-ray photoelectron spectra were acquired on a Kratos ES300 surface analysis instrument, which collected electrons in the fixed retarding ratio (FRR) analyser mode. Unmonochromatized Mg K_α radiation (1253.6 eV) was used as the excitation source. These measurements were taken with an electron take-off angle of 30° from the surface normal. Data accumulation and component peak analysis were performed on an IBM PC. All binding energies are referenced to the hydrocarbon component ($-\underline{C}_xH_y-$) at 285.0 eV [10], and the instrumentally determined sensitivity factors are such that for unit stoichiometry the C(1s):O(1s) intensity ratio is ~ 0.55.

High-resolution core level and valence region spectra were taken on a Kratos AXIS HS instrument. This instrument was equipped with a monochromatic Al K_α (1486.6 eV) X-ray source and a novel magnetic immersion lens system that yields high resolution and high sensitivity. Uniform charge neutralization with little energy shift was effected with a low-energy electron source.

3. RESULTS

3.1. Untreated polyethylene and polystyrene

Only one C(1s) XPS peak is seen for untreated polyethylene at 285.0 eV, which can be associated with the $-\underline{C}H_2-$ linkages present in the polymer [11]. In the valence band region, overlap of the C(2s) orbitals along the hydrocarbon backbone yields bonding (~ 19 eV) and anti-bonding (~ 13 eV) molecular orbitals [12] (Fig. 1a). The C(2p) and H(1s) orbitals involved in the C—H bond contribute towards the weak, broad structure seen below ~ 10 eV. The relative intensities are dependent on the ionization cross-sections of the various molecular orbitals.

In addition to a hydrocarbon component at 285.0 eV, the XPS spectrum of clean polystyrene displays a distinctive satellite structure at ~ 291.6 eV, which arises from low-energy $\pi \rightarrow \pi^*$ shake-up transitions accompanying core level ionization [3] (Fig. 2a). The alkyl backbone in polystyrene makes only a minor contribution to the XPS valence band region because the density of states for these levels spans a rather broad structure, and the number of such levels is much

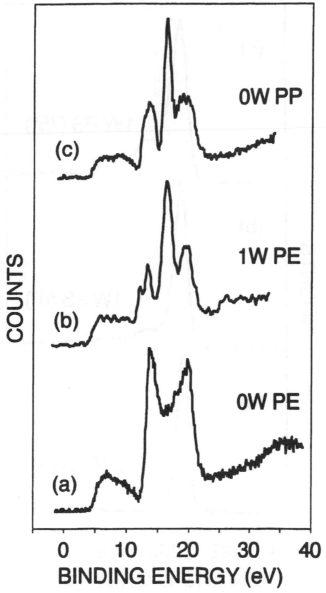

Figure 1. XPS valence band spectra of (a) clean polyethylene, (b) 1 W plasma-oxidized/aged polyethylene (10^5 min), and (c) clean polypropylene. The surface normal was tilted 30° away from the axis of the spectrometer.

smaller in comparison with those associated with the phenyl groups. As a result, the sharp features of the benzene spectrum predominate [12, 13] (Fig. 3a).

3.2. *The extent of surface oxidation as a function of the glow discharge power*

Detailed chemical information about the modified polymer surfaces was obtained by fitting the C(1s) XPS spectra to a range of carbon functionalities: carbon adjacent to a carboxylate group ($\underline{C}-CO_2 \sim 285.7$ eV), carbon singly bonded to

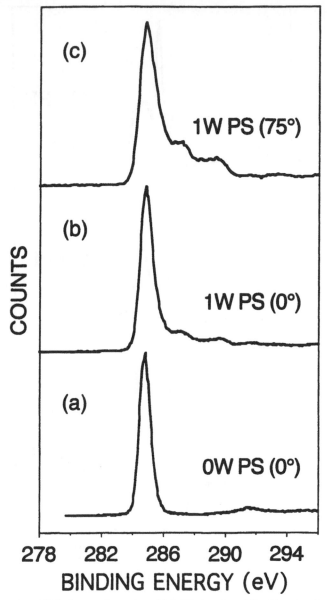

Figure 2. C($1s$) region of (a) clean polystyrene, (b) 1 W plasma-oxidized/aged polystyrene (10^5 min; the surface normal was aligned with the axis of the spectrometer), and (c) 1 W plasma-oxidized polystyrene/aged (10^5 min; the surface normal was tilted 75° away from the axis of the spectrometer).

one oxygen atom ($\underline{C}$—O ~ 286.6 eV), carbon singly bonded to two oxygen atoms or carbon doubly bonded to one oxygen atom (O—$\underline{C}$—O or $\underline{C}$=O ~ 287.9 eV), carboxylate groups (O—$\underline{C}$=O ~ 289.0 eV), and carbonate atoms (O—$\underline{C}$O—O ~ 290.4 eV) [14]. The C($1s$) envelope for plasma-oxidized polystyrene is shown in Fig. 2b. Loss of aromaticity in the polystyrene film could be monitored by the decrease in the intensity of the $\pi \rightarrow \pi^*$ shake-up satellite.

The O($1s$)/C($1s$) ratios of these samples were determined immediately after

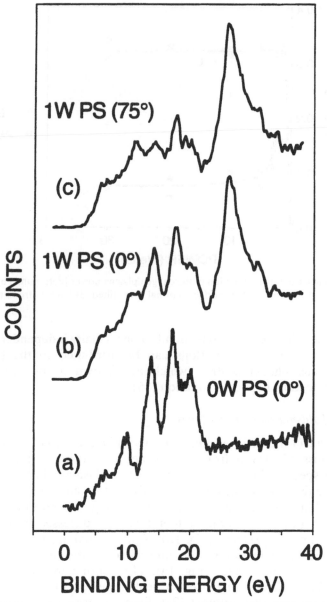

Figure 3. XPS valence band spectra of (a) clean polystyrene, (b) 1 W plasma-oxidized/aged polystyrene (10^5 min; the surface normal was aligned with the axis of the spectrometer), and (c) 1 W plasma-oxidized/aged polystyrene (10^5 min; the surface normal was tilted 75° away from the axis of the spectrometer).

plasma oxidation as a function of the glow discharge power (Fig. 4). A significant quantity of oxidized functionalities was measured at the polyethylene surface following exposure to an oxygen plasma. The O/C ratio passes through a maximum with increasing wattage; presumably this trend reflects a greater amount of sputtering occurring at higher plasma powers.

Polystyrene undergoes a relatively higher degree of oxidation in comparison

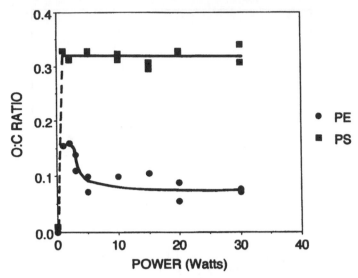

Figure 4. O/C ratio ($\pm$ 0.01) as a function of the oxygen plasma power [determined by measuring the O($1s$) and C($1s$) peak areas]. The surface normal was tilted 30° away from the axis of the spectrometer.

with polyethylene. The O/C ratio is found to be virtually independent of the glow discharge power. These observations can be attributed to the fact that this polymer contains phenyl centres, which are known to be reactive, e.g. as chromophores leading to photo-oxidation [15].

3.3. Ageing of plasma-oxidized surfaces

The O($1s$) and C($1s$) core level spectra of plasma-oxidized polyethylene indicate a significant loss of oxygen from the surface on storing in air (Fig. 5). XPS valence band spectra (Fig. 1b and 3b) of aged plasma-oxidized polymers are shown because the changes with respect to the untreated surfaces are so marked and clear-cut. The variation in the XPS valence region with time during ageing appears to be highly complex and is therefore not addressed in this paper. The limiting valence band spectrum (Fig. 1b) of plasma-oxidized polyethylene (aged over 10^5 min) is markedly different in appearance to the characteristic spectrum of a clean polyethylene surface (Fig. 1a). The major new feature in the C($2s$) vicinity overlaps exactly with the extra peak reported for polypropylene. The relative intensities of the outer C($2s$) features are also consistent with a polypropylene-type structure (i.e. the attachment of a methyl side-chain to the main polyethylene backbone introduces an extra feature in the middle of the C($2s$)–C($2s$) region at ~ 16 eV [12], Fig. 1c). A slight splitting is evident in the polyethylene band at ~ 13 eV; this could be a manifestation of head-to-head and head-to-tail linkages in the polypropylene structure [12], or a mixture of iso- and syndiotactic polypropylene [16]. O($2s$) features are known to appear at ~ 25–27 eV, whereas O($2p$) electrons associated with oxygen lone-pairs in π and non-bonding molecular orbitals, O($2p$)–H($1s$), and O($2p$)–C($2p$) bonds, all occur below 10 eV [12]. A small amount of O($2s$) is still evident for the plasma-oxidized/aged polyethylene surface.

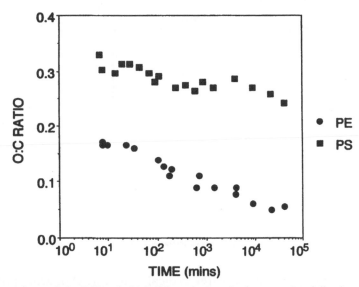

Figure 5. Dependence of the O/C ratio (± 0.01) on the length of storage time following a 1 W plasma oxidation treatment [determined by measuring the O(1s) and C(1s) peak areas]. The surface normal was tilted 30° away from the axis of the spectrometer.

The C(1s) and O(1s) core-level XPS spectra show that the 1 W-treated polystyrene film ages at a much slower rate than the corresponding 1 W plasma-oxidized polyethylene surface. Angle resolved core-level and valence band measurements divulge that there is a greater degree of oxidation towards the surface (Fig. 2c); in fact, very little polystyrene valence band structure is detectable at highly grazing electron-emission angles (Fig. 3c).

4. DISCUSSION

In this study, the extent of polyethylene surface modification appears to result in a more well-defined set of resultant molecular orbitals (i.e. selectivity) compared with other valence-band investigations [17, 18]. This discrepancy is probably due to different experimental parameters and reactor configurations (e.g. remote oxygen plasma treatment [17, 19]). Also, polystyrene is found to undergo a greater degree of oxidation than polyethylene [14, 20].

The major advantage of studying XPS rather than UPS valence band spectra is that the photo-ionization cross-section for molecular orbitals with major 2s character is greater than that for molecular orbitals with dominant 2p character [21]; this is of significant benefit, because the 2s bands are easier to interpret than their 2p counterparts [18, 22]. The C(2s)–C(2s) band can serve as a fingerprint for the electronic structure of a polymeric system. The very strong valence band spectral features observed for plasma-oxidized/aged polyethylene can be attributed to either extensive crosslinking or to rearrangement taking place at the polymer surface to form a polypropylene-type structure as depicted overleaf:

The plasma would be expected to extensively rupture the polyethylene chains to form primary hydrocarbon carbocations and free radicals, which are energetically unstable with respect to secondary isomers [23]. Therefore these species will undergo rearrangement, and subsequent recombination may occur

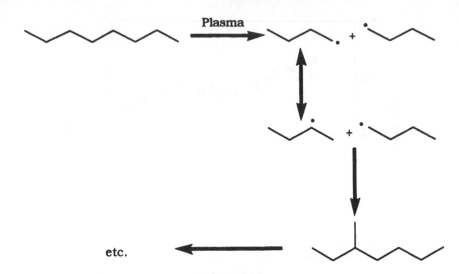

etc.

between a secondary carbocation/free radical with a neighbouring polyethylene chain which has just undergone fragmentation. This sequence of steps results in the generation of a polyethylene backbone to which methyl side-groups are attached at random. To a first approximation, the valence band spectrum of a polymer based on a polyethylene backbone to which a short chain is fixed can be considered to be a simple summation of the individual constituents, i.e. polyethylene + attached group [24]. Therefore, the XPS valence band spectrum of a polyethylene chain with randomly attached methyl side-groups would be expected to bear a strong resemblance to the valence region spectrum reported for polypropylene. The loss of oxygen from the surface with ageing may be a direct consequence of the creation of this new extended structure. Ageing may occur via migration of the oxygenated moieties towards the subsurface, conglomeration into very localized pockets at the surface, or simply desorption into the gas phase. Examination of the valence band spectra of a whole range of other hydrocarbon polymers supports this assignment [25] (the closest is found to be a polyethylene backbone with two methyl side-groups on each carbon centre, i.e. polyisobutylene; however, the polyisobutylene binding energies differ significantly, but their minor presence could explain the origins of the slight splitting seen in the ~ 13 eV band).

In the case of polystyrene, activation of the phenyl centres by the oxygen plasma offers numerous reaction pathways which can lead to a considerable breakdown of the parent polymer and a greater degree of oxidation [26]. Electromagnetic radiation from the glow discharge can penetrate below the surface to cause photo-excitation of the phenyl groups, and extensive crosslinking reactions may subsequently occur within the subsurface region. This may be the reason why plasma-oxidized polystyrene is much more resistant to ageing than its polyethylene counterpart.

5. CONCLUSIONS

These systematic studies have shown that the plasma oxidation of polyethylene followed by ageing either results in a polypropylene-type surface structure, or else implies that some kind of crosslinking has taken place. It should be noted that these results are the first example of such a highly selective reaction occurring at a polymer surface under the influence of a glow discharge. Furthermore, the plasma modification of polyethylene surfaces is found to be much more critically dependent on the process parameters employed than is the case for polystyrene. Polystyrene experiences a greater degree of oxidation and is much more resistant to subsequent loss of surface oxygen. This can be attributed to the unsaturated/chromophoric phenyl groups in polystyrene being highly vulnerable towards plasma activation.

Acknowledgements

R. K. W. thanks the SERC and BP International Ltd. for a CASE Studentship.

REFERENCES

1. J. M. Burkstrand, *J. Vac. Sci. Technol.* **15**, 223–226 (1978).
2. W. L. Wade, R. J. Mammone and M. Binder, *J. Appl. Polym. Sci.* **43**, 1589–1591 (1991).
3. D. T. Clark and A. Dilks, *J. Polym. Sci. Polym. Chem. Ed.* **15**, 15–30 (1977).
4. J. Lub, F. C. M. van Vroonhoven, D. van Leyen and A. Benninghoven, *Polymer* **29**, 998–1003 (1988).
5. D. Briggs, *Br. Polym. J.* **21**, 3–15 (1989).
6. M. Morra, E. Occhiello, F. Garbassi, D. Johnson and P. Humphrey, *Appl. Surface Sci.* **47**, 235–242 (1991).
7. J. J. Pireaux, J. Riga, P. Boulanger, P. Snauwaert, Y. Novis, M. Chtaib, C. Gregoire, F. Fally, E. Beelen, R. Caudano and J. J. Verbist, *J. Electron Spectrosc. Relat. Phenom.* **52**, 423–445 (1990).
8. D. T. Clark and D. Shuttleworth, *J. Polym. Sci. Polym. Chem. Ed.* **18**, 27–46 (1980).
9. A. G. Shard, H. S. Munro and J. P. S. Badyal, *Polym. Commun.* **32**, 152–154 (1991).
10. G. Johansson, J. Hedman, A. Berndtsson, M. Klasson and R. Nilsson, *J. Electron. Spectrosc. Relat. Phenom.* **2**, 295–317 (1973).
11. M. H. Wood, M. Barber, I. H. Hillier and J. M. Thomas, *J. Chem. Phys.* **56**, 1788–1789 (1972).
12. J. J. Pireaux, J. Riga and J. J. Verbist, in: *Photon, Electron and Ion Probes for Polymer Structure and Properties*, D. W. Dwight, T. J. Fabish and H. R. Thomas (Eds), ACS Symposium Series No. 162, pp. 169–201. Am. Chem. Soc., Washington, DC (1981).
13. A. Chilkoti, D. Castner and B. D. Ratner, *Appl. Spectrosc.* **45**, 209–217 (1991).
14. D. T. Clark and A. Dilks, *J. Polym. Sci. Polym. Chem. Ed.* **17**, 957–976 (1979).
15. A. G. Shard and J. P. S. Badyal, *J. Phys. Chem.* **95**, 9436–9438 (1991)
16. J. Delhalle, R. Montigny, C. Demanet and J. M. Andre, *Theor. Chim. Acta.* **50**, 343–349 (1979).
17. R. Foerch, G. Beamson and D. Briggs, *Surface Interface Anal.* **17**, 842–846 (1991).
18. L. J. Gerenser, *J. Adhesion Sci. Technol.* **1**, 303–318 (1987).
19. R. Foerch, N. S. McIntyre and D. H. Hunter, *J. Polym. Sci. Polym. Chem. Ed.* **28**, 193–204 (1990).
20. P. M. Triolo and J. D. Andrade, *J. Biomed. Mater. Res.* **17**, 129–147 (1983).
21. U. Gelius in: *Electron Spectroscopy*, D. A. Shirley (Ed.), p. 311. North-Holland, Amsterdam (1972).
22. W. R. Salaneck, *CRC Crit. Rev. Solid State Mater. Sci.* **12**, 267–296 (1985).
23. J. McMurray, *Organic Chemistry*, 2nd edn, pp. 180–207. Brooks/Cole, Belmont, USA (1988).
24. J. J. Pireaux, J. Riga, R. Caudano, J. J. Verbist, J. Delhalle, J. M. Andre and Y. Gobillon, *Phys. Scr.* **16**, 329–338 (1977).
25. G. Beamson and D. Briggs, *High Resolution XPS of Organic Polymers: The Scienta ESCA300 Database*. John Wiley, Chichester (1992).
26. A. G. Shard and J. P. S. Badyal, *Macromolecules* **25**, 2053–2054 (1992).

implies that some kind of cross-linking has taken place. It should be noted that these results are the first example of such a highly selective reaction occurring at a polymer surface under the influence of a glow discharge. Furthermore, the plasma modification of polyethylene surfaces is found to be much more difficult than, say, the process normally employed when in the case of polypropylene polystyrene, example, a given dose of radiation and a much more restricted population of available surface oxygen. This can be attributed to the unsaturated nature of the polymer groups in polypropylene, being highly vulnerable to such glass transition.

Plasma Surface Modification of Polymers, pp. 123–146
M. Strobel, C. Lyons and K. L. Mittal (Eds)
© VSP 1994

Evolution of the surface composition and topography of perfluorinated polymers following ammonia-plasma treatment

THOMAS R. GENGENBACH, XIMING XIE, RONALD C. CHATELIER
and HANS J. GRIESSER*

Division of Chemicals and Polymers, CSIRO, Private Bag 10, Rosebank MDC, Clayton 3169, Australia

Revised version received 16 August 1993

Abstract—Treatment of fluorinated ethylene propylene (FEP) and polytetrafluoroethylene (PTFE) in ammonia plasmas produced surfaces with very high wettability by water, but on storage in air at ambient temperature, the air/water contact angles increased markedly. The evolution of the surface composition and topography was studied by angle-dependent X-ray photoelectron spectroscopy (XPS), derivatization of amine groups with fluorescein isothiocyanate, scanning tunelling microscopy (STM), and atomic force microscopy (AFM). XPS demonstrated a continuous increase in the oxygen content over periods of weeks; this was assigned to oxidation of trapped radicals and subsequent secondary reactions. In addition, the fluorine content also changed markedly on storage; the XPS fluorine signal suggested that there was a substantial amount of fluoride in the freshly treated surfaces, and this component disappeared rapidly on storage. STM and AFM showed no changes in topography with aging but suggested surface hardening on plasma treatment. The events following treatment of FEP and PTFE in ammonia plasmas are not adequately described by a model involving plasma-induced, instantaneous chemical modification followed by surface restructuring; the surface and sub-surface compositions evolve over a period of several weeks due to the occurrence of oxidative reactions, and these chemical changes interact with the physical process of surface restructuring.

Keywords: Plasma surface treatment; XPS; contact angles; derivatization; fluoropolymers; ammonia plasma; oxidation; surface rearrangement.

1. INTRODUCTION

Plasma surface treatments are potentially very attractive for the covalent incorporation into polymer surfaces of extraneous reactive groups suitable for participation in further, conventional chemical reactions at the interface. The attachment of amine groups to commodity polymer surfaces is particularly attractive because amines are relatively reactive nucleophiles useful either for the covalent binding of larger chemical compounds that can be used to control interfacial interactions, or for the provision of reactive sites for interfacial adhesive bonding. A convenient and rapid means for the surface amination of polymers is treatment with a plasma of ammonia. A number of reports have described the modification of hydrocarbon polymers with ammonia plasmas with the intention of placing amine groups on the

---

*To whom correspondence should be addressed.

surface [1–11]. Plasma surface treatments, however, typically produce multifunctional surfaces. On ammonia-plasma-treated polystyrene, for instance, amine groups were not the only species attached, as shown by the detection of oxygen by X-ray photoelectron spectroscopy (XPS) [7, 9–11].

Of particular interest to us are plasma-treated perfluorinated polymers, which have been studied to a lesser extent than plasma-treated hydrocarbon polymers. In particular, the chemical consequences of the ammonia plasma treatment of fluorocarbon polymers have, to our knowledge, not been described. The chemical results of ammonia plasma treatments of fluoropolymers may differ considerably from those of hydrocarbon polymers. The electron-withdrawing effect of adjacent fluorines should make the amine groups attached to a fluorocarbon backbone considerably more acidic than the amine groups on a hydrocarbon polymer segment.

The capacity for surface-rearrangement motions in modified polymers also has to be considered. Polymer surfaces generally are highly mobile [12, 13], and this also applies to the commonly available perfluorinated polymers polytetrafluoroethylene (PTFE) and fluorinated ethylene propylene (FEP) copolymer. The mobility of polymer chain segments allows surface restructuring to occur, and this typically leads to a considerable decrease over time of the effects conferred on the surface by non-depositing plasma treatments [14–19]. Surface restructuring results from an energetically unfavourable situation that arises when polar groups are located at the interface with the non-polar air environment; chain segments carrying polar groups migrate into the polymer, unless prevented by adjacent crosslinks, and hydrophobic polymer chains re-emerge. The partial or complete disappearance of the intended reactive groups is a concern in the context of using modified surfaces for subsequent chemical reactions, but characterization of the aging effects can assist in avoiding adverse consequences, for instance, by specifying a maximum delay between plasma treatment and chemical reaction.

Such movement of some of the polar groups away from the interface occurs also in the surface layers of FEP and PTFE treated in ammonia plasmas when stored in air: the wettability of the modified surfaces decreased substantially over the course of a few days after treatment [20]. Detailed further analysis has shown, however, that surface restructuring does not fully describe the changes that occur on storage of ammonia-plasma-treated fluoropolymers; the composition of the modified surface layers also undergoes substantial post-treatment changes. Here we report data on these compositional changes on storage, by surface analysis of ammonia-plasma-treated FEP and PTFE by XPS and derivatization. In addition, we have utilized scanning tunnelling microscopy (STM) and atomic force microscopy (AFM) to probe the surface topography of the modified polymers. Samples were assessed over extended periods of time in order to gain an improved understanding of the processes which lead to long-term changes in the surface wettability and composition. The evolution of the modified surfaces is interpreted in terms of several processes which occur on a similar time scale on storage in air after ammonia plasma treatment.

2. EXPERIMENTAL

2.1. Plasma surface treatments

Ammonia plasma treatments were performed on the substrate materials FEP and PTFE, which were obtained in the form of tapes of 12.7 mm width. The former was obtained commercially (Du Pont FEP 100 Type A), while the PTFE tape was a gift from C. S. Lyons, 3M Corporate Research Labs, St. Paul, MN, USA. Ammonia was obtained from Matheson.

The general procedure that we use for plasma surface treatments has been reported earlier [20], and the custom-built plasma apparatus has also been described previously [21]. Briefly, the reactor chamber (volume $= 7600$ cm^3) contained two vertically oriented copper electrodes of dimensions 90×18 mm placed in a vertical glass cylinder. The electrodes were spaced apart by 16 mm. Plastic and glass fittings defined a controlled path for the flow of incoming process gas, extending from the gas inlet to the electrodes. Ammonia gas was supplied from a cylinder via a stainless steel line and a mass flow controller (MKS). The pressure in the reactor was adjusted via the throttle valve at the reactor outlet. The reactor also incorporated a tape transport system for semi-continuous treatment of 12.7 mm wide tape moving through the plasma at a controlled speed. Experiments were performed either with short (75 mm) sections of fluorocarbon tape attached by thin double-sided adhesive tape to the face of the electrodes, or with extended lengths of tape moving at a constant speed through the plasma zone. The treatment of extended lengths of tape moving at a constant speed enabled the fabrication of a large number of nominally identical specimens which, after plasma treatment, were stored in tissue-culture polystyrene (TCP) dishes at ambient temperature (22 ± 1°C) and assessed periodically. A fresh piece of tape was used for each contact angle measurement and each XPS analysis. The storage of samples in TCP dishes, which are explicitly designed to avoid contamination by polymer additives, evidently avoids problems with adventitious contamination; we have previously discussed our evidence, obtained on a range of samples of different surface energies, against the significant build-up of hydrocarbon and other contamination on storage [20].

The plasma discharge was powered by a commercial generator (ENI ACG-3) operating at 13.56 MHz and equipped with a matching network. The plasma treatment conditions were as follows: pressure, 0.2 to 0.5 Torr; power, 20 to 120 W; treatment time, 10 to 120 s; flow rate, 1 to 6 sccm. Altogether, over 20 combinations of parameters were used; however, since all samples showed a very similar behaviour in the evolution of the contact angles on storage [20], only a limited number of samples were assessed in detail by XPS and derivatization. The initial and final values of the contact angles differed somewhat, although not greatly, with the plasma conditions. These quantitative differences are of no significance for the present purposes because, qualitatively, the underlying processes governing the evolution of the treated surfaces appeared to be the same. For consistency, we will report XPS data recorded on one representative series of identical specimens taken from a length of FEP tape treated while moving through the plasma. These particular experimental conditions were pressure, 0.28 Torr; plasma power, 30 W; flow rate, 2 sccm; exposure to the plasma for 63 s while the tape was moving through the glow zone at a tape transport speed of 8.6 cm/min.

2.2. XPS analysis

The unit used for most of the work was a VG Escalab V spectrometer using non-monochromatic Al K$_\alpha$ radiation at a power of 200 W. The pressure in the analysis chamber typically was 2×10^{-9} mbar. The binding energy scale was calibrated using sputter-cleaned foils of nickel, silver, gold, and copper. The C $1s$ signal for CF_2 groups, with a binding energy of 292.2 eV, was used as an internal standard for the correction of the charging of the samples. Elements present were identified by survey spectra. High-resolution spectra were recorded from individual peaks at a 30 eV pass energy in the fixed analyser transmission mode. The elemental composition of the surface was determined based on a first principles approach [22]; atomic ratios were calculated from integral peak intensities using a non-linear Shirley-type background and published values for photoionization cross sections [23]. The inelastic mean free path of the photoelectrons was assumed to be proportional to $E^{0.5}$, where E is the kinetic energy [24]. The transmission function of the analyser was determined to be proportional to $E^{-0.5}$. The random error in the quantitative analysis of elemental compositions is 5–10% (usually 7–8%) on this XPS unit. A value of 1.94 was obtained for the F/C ratio of cleaned, untreated FEP.

Additional data were obtained on a Surface Science Instruments (SSI) X-Probe 100 spectrometer with monochromatic Al K_α radiation in the NESAC/BIO Laboratory (Professor B. Ratner, University of Washington, Seattle, WA, USA), to check the data obtained on the Escalab and to improve spectral resolution. Using the same cleaning procedure, an FEP specimen was measured on the SSI instrument, which uses calibrated sensitivity factors (calibration with PTFE tape) to quantify data. An F/C ratio of 1.99 was measured. The F/C ratios obtained on the VG Escalab unit, given below, were not corrected for the evident slight underestimation of the F/C ratio.

A new specimen was used for each XPS analysis in order to avoid the accumulation of effects arising from the inevitable, slow degradation of the sample under the non-monochromatic X-ray irradiation in the Escalab spectrometer. The effects of sample decomposition during analysis were minimized by exposing specimens to radiation for less than 1 min before the start of data acquisition and limiting the total analysis time to less than 30 min. Samples were assessed after various periods of storage, in parallel with known control samples. The first analysis was performed within 20–30 min after plasma treatment and the venting of the reactor.

Angle-dependent XPS (ADXPS) analysis was performed on treated FEP samples following assessment by STM of the surface topography of the FEP and the PTFE. STM showed the latter to have a topography with relatively steep slopes, whereas the former was sufficiently smooth to allow the use of glancing photoelectron emission. The XPS signal at low emission angles is a superposition of electrons escaping at a range of angles relative to the local, microscopic surface plane. In view of the limited signal-to-noise ratio at low emission angles in our instrument, the finite collection angle, and the problems associated with a mathematical description of the undulating surface topography, construction of depth profiles was not attempted and only qualitative ADXPS was performed. An upper limit of 75° was chosen for the emission angle (θ, measured from the surface normal) and data were collected at the two emission angles of 0° (normal) and 75°, which correspond to an analysis depth of approximately 10 nm and 2–3 nm, respectively, assuming an attenuation length of the photoelectrons of 3 nm.

2.3. Derivatization of amine groups with fluorescein isothiocyanate

Primary and secondary surface amine groups on ammonia-plasma-treated FEP were reacted with fluorescein isothiocyanate (FITC, Isomer I, Molecular Probes, Junction

City, OR, USA) using a variation of a published method [25]. To couple FITC onto amine groups, sample strips (12.7 × 25 mm) were immersed in 5 ml of a freshly prepared solution of 100 μg/ml of FITC in 50 mM sodium phosphate buffer (pH 8.0) for 2 h at room temperature in the dark. The strips were removed, washed thoroughly with distilled water, and immersed in the buffer for another 2 h at room temperature in the dark. The strips were then immersed in 2.5 ml of 0.1 M NaOH for 30 min; in this alkaline solution, the thiourea linkage is hydrolysed to detach FITC from the surface. The solution absorption of FITC was determined using a Hewlett-Packard UV/Vis absorption spectrometer and the amount of surface-attached FITC was calculated using the equation

$$\sigma = \frac{\text{Abs} V N_A}{\varepsilon l A}, \tag{1}$$

where σ is the surface density of FITC dye in molecules/nm^2, Abs is the optical density of the solution at 492 nm, V is the volume in litres, N_A is Avogadro's number, ε is the extinction coefficient (7.5×10^4 M^{-1}cm^{-1}), l is the path length in cm, and A is the area of the sample in nm^2. In each assay, a piece of unmodified FEP tape was treated in the same way to enable determination of the amount of FITC that is non-specifically bound to the surface, that is, not by covalent attachment, but adsorbed strongly enough to resist removal by washing. Assuming that the non-specific adsorption component was identical on both modified and unmodified FEP, the density of reactive amine groups was calculated from the difference between the amount of FITC assayed on the ammonia-plasma-treated surface and that determined on the untreated FEP strip.

2.4. Contact angle measurements

The apparatus and methods used for the determination of the advancing and receding air/water contact angles (CAs) were the same as those detailed in our previous report [20]. Measurements were performed in quadruplicate; standard deviations typically were 1–2°.

2.5. Surface topography

Scanning tunnelling microscopy (STM) was performed using a custom-built air STM unit incorporating an inverted piezo-tube scanner design. The electronics and data acquisition consist of a PI controller and a computer-driven XY raster system. The sensitivity of the STM head was calibrated to within ±10% using highly oriented pyrolytic graphite (x,y-sensitivity) and a nickel compact-disc stamper with a pit height of 140 nm, equivalent to 1/4 of the wavelength of the pick-up laser (z-sensitivity). The insulating specimens were mounted on stubs using conducting cement, with some of the cement lapped around the specimen edge. Subsequently they were coated with a RF plasma-sputtered layer of platinum 3–5 nm thick. One specimen of untreated FEP and one of ammonia-treated FEP were studied by atomic force microscopy (AFM) using a Nanoscope II unit (Digital Instruments, Santa Barbara, CA, USA). The specimens used for AFM were not overcoated, and were mounted on standard sample stubs. In the case of the ammonia-treated FEP specimen, the attractive force to the SiN tip was 13 nN, and some sticking was observed as the tip was moved away from the surface. The untreated specimen was studied with a force of 4 nN. Any further decrease in the force led to detachment in areas where the sample sloped away from the tip.

3. RESULTS AND DISCUSSION

The cleanliness of the FEP and PTFE materials before treatment was assessed by XPS as described previously [20]. Following correction for the satellite contribution arising from the non-monochromatic source [18], the XPS C 1s region for both FEP and PTFE showed a signal characteristic of fluorocarbons only, with no residual contribution from hydrocarbon at 285 eV. The materials thus appeared clean as supplied; ultrasonic washing in ethanol produced no detectable changes in the contact angles or the XPS F/C elemental ratio. For experiments with moving, extended lengths of tape, washing was impractical, and XPS was used to ascertain the cleanliness of samples cut from tape adjacent to those lengths used for the moving-tape treatments.

Over a range of conditions, the ammonia plasma treatments produced highly modified surfaces with air/water contact angles which were unusually low compared with those obtained with treatments in oxygen, argon [18], or water vapour [20] plasmas. However, the CAs of ammonia-plasma-treated fluorocarbon surfaces increased substantially in the course of a week of storage at ambient conditions. This behaviour, reported in our previous communication [20], has been ascribed to surface reorientation: unfavourable interfacial energetics associated with the presence of polar groups at the interface with air provides a driving force that displaces modified polymer chains from the surface into deeper regions of the polymer, with concomitant emergence of hydrophobic material. This surface restructuring does not proceed to completion; some of the attached polar groups are prevented from migrating away from the interface, presumably by adjacent crosslinks. Polar groups are, of course, also not in an energetically favourable situation when inside a fluoropolymer, but according to XPS they exist in the sub-surface region; we have speculated previously that the polar groups can, in principle, lower their energy by associating with each other [20], forming, for instance, hydrogen-bonded dimers or micro-micellar clusters. XPS is incapable of testing this hypothesis. An alternative driving force for surface reorientation may be provided by the entropy increase which would result from randomization of the surface groups.

The surface restructuring of ammonia-plasma-treated fluorocarbon surfaces has been studied further, as reported in the following, using a range of techniques. XPS analysis has indicated that the post-treatment evolution of plasma-treated fluoropolymer surfaces is not so simple as to be fully described by the surface-restructuring model; the physical process of chain segmental reorientation is accompanied by changes in the chemical composition, and both have an influence on the evolution of the CAs. In this paper we present data on the long-term changes in the chemistry of the surface and sub-surface layers of ammonia-plasma-treated fluorocarbon polymers and discuss the effects of these changes on the CAs. The evolution of the CAs of ammonia-plasma-treated FEP and PTFE surfaces has been reported in our earlier communication [20], but one representative set of data is reproduced in Fig. 1 for ease of comparison with the following surface analytical data obtained after various periods of storage in air.

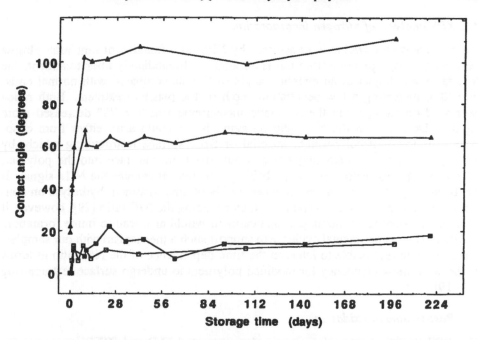

Figure 1. Advancing (triangles) and receding (squares) air/water contact angles of FEP (filled symbols) and PTFE (empty symbols) treated in an ammonia plasma using conditions as given in the text. The values for the untreated materials are advancing contact angle and receding contact angle of 117° and 98°, respectively, for FEP, and 162° and 98° of PTFE. Standard deviations of quadruplicate measurements are in the range 1–2°.

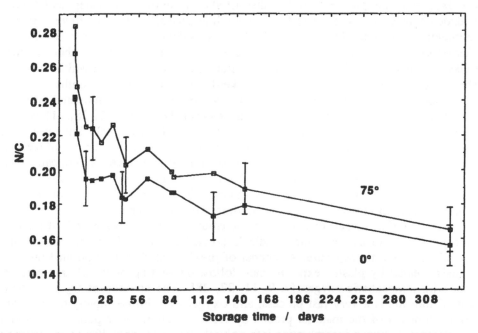

Figure 2. The XPS N/C ratio at emission angles of 75° and 0° of ammonia-plasma-treated FEP, as a function of the storage time in air. The error bars represent one standard deviation.

3.1. XPS analysis of nitrogen incorporation

Figure 2 shows the N/C ratio as assessed by XPS, as a function of time after plasma treatment, at two photoelectron emission angles. Immediately after treatment, the N/C ratio was higher at an emission angle of 75° as compared with normal emission (0°), indicating a low penetration depth of the plasma treatment. Both ratios decreased on storage, but the N/C ratio measurement at $\theta = 75°$ decreased more rapidly. Decreases in the N/C ratio on aging in air could arise either from evaporation of low-molecular-weight material or from surface restructuring which, by moving the nitrogen-containing groups away from the interface into the polymer, reduces the photoelectron escape probability and thus attenuates the N $1s$ signal. It has been suggested that imines produced by N_2 plasma treatment hydrolyse on contact with atmospheric water vapour, thereby reducing the N/C ratio [29]; however, it is not clear whether ammonia-plasma treatment would also lead to imine formation, and there is no experimental evidence to support such a mechanism with our samples. It seems more reasonable to interpret the time dependence of the N/C ratio in terms of the well-known tendency for modified polymers to undergo surface restructuring [12–19].

3.2. Post-treatment oxidation

The XPS survey spectra of ammonia-plasma-treated FEP and PTFE specimens invariably contained a signal assignable to oxygen, even after careful leak testing and thorough evacuation of the reactor. Furthermore, the oxygen content of the samples increased with time as the samples were stored in air after treatment. Figure 3 documents, for NH_3-treated FEP, the XPS O/C ratio as a function of the storage time. The oxygen uptake proceeded in two distinct stages: an initial, fast step, and a more extended process that continued for weeks while gradually slowing down. For experimental reasons, the first point in all of our XPS data was obtained $\approx$ 20–30 min after exposure to air. At that time, the O/C ratio measured at normal photoelectron emission was $\approx$ 0.05. The O/C ratio then increased about four-fold within 10 days. The oxygen uptake then slowed down markedly but continued for months.

A number of previous studies have detected oxygen on plasma-treated surfaces even when the plasma gas did not contain oxygen [7, 9–11, 27, 28]. However, the existence of two distinct steps, and the slow, continuing uptake have not been discussed before. The existence of an extended oxygen uptake and the rate of oxygen incorporation are of fundamental interest as well as of relevance for applications of plasma-treated surfaces. Such long-term changes may substantially affect the surface properties of materials on storage, in particular the nature and density of reactive groups capable of covalent interfacial bonding with an adhesive.

The fast, initial oxygen uptake can be attributed to the addition of in-diffusing, atmospheric O_2 to carbon-centred radicals created in the surface and sub-surface layers by the plasma exposure. A number of previous studies have pointed out that radicals created by plasma exposure can, following venting of the plasma reactor, react with atmospheric oxygen [7, 9–11, 27, 28]. The addition of O_2 to carbon-centred radicals is a very fast process, and oxygen can diffuse rapidly through the few nanometres of the material probed by XPS: Gerenser has reported substantial post-treatment oxygen incorporation into polyethylene after only 30 s of exposure to the atmosphere [29].

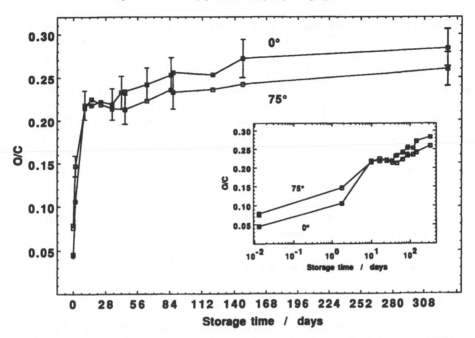

Figure 3. The XPS O/C ratio at emission angles of 75° and 0° of ammonia-plasma-treated FEP, as a function of the storage in air. The error bars represent one standard deviation. The inset shows the same data on a logarithmic time scale.

The slow, extended oxygen uptake is likely to involve secondary reactions analogous to those involved in the oxidative degradation of polyolefins, which proceeds by complex sets of radical reactions [31–34]. The gradual decay of metastable peroxides can maintain oxidative reaction cycles over extended periods of time, thus producing a continuing, long-term oxygen uptake. A corresponding study of the long-term oxidative reactions in a hydrocarbon plasma polymer after deposition showed that the reaction pathways and cycles involved are complex [35]. A similar degree of complexity can be expected in the case of plasma-treated fluorocarbon polymers. With the small amounts of material involved and the extreme demands on sensitivity, detailed analysis of the oxidative reactions occurring in the surface and sub-surface layers of plasma-treated polymers is challenging. It is, however, not our intention at present to elucidate the reaction steps of this post-treatment oxidation but merely to point out that a scheme based on metastable intermediates can account for the continuing, slow oxygen uptake of ammonia-plasma-treated fluorocarbon polymers.

Comparison of the XPS O/C ratios obtained at 0° and 75° emission angles (Fig. 3) presents a further point of interest: while for a few days after the plasma treatment the oxygen content is higher at the surface as compared with the sub-surface region, the situation is reversed later on. After 3–4 weeks, XPS consistently detected more oxygen at normal emission than at 75°. This is emphasized in Fig. 3 (inset), where the O/C ratio is displayed on a logarithmic time scale. These results indicate that two different processes occur simultaneously: the continuous incorporation of carbon-oxygen functionalities, and rearrangement motions that transport some of these polar groups away from the surface, thus shifting the peak of the oxygen distribution below the surface (but still within the analysis depth of XPS).

Contact angle data show major changes within the first 10 days after sample fabrication (an enlarged display of the initial period has been given in our earlier communication [20]), and very little subsequent change (Fig. 1). The major increases in the contact angles and the rapid initial oxygen uptake thus occur within the same period of time. Evidently, surface restructuring and oxidation take place simultaneously and, in counteracting ways, define the evolution of the surface composition within the first 10 days. Additional polar groups are continually incorporated into the surface layers, but the contact angles continue to increase, indicating that surface restructuring competes quite successfully with oxidation and rapidly removes polar groups from the immediate surface to the sub-surface region where XPS, but not contact angles, can still report the presence of the oxygen-containing groups (air/water contact angles are believed to probe a depth of ~ 0.5 nm [30], whereas the probe depth of XPS is several nanometres). However, not all of the polar groups are removed from the surface; the contact angles do not revert to values characteristic of pure fluorocarbon surfaces. This system thus represents an excellent opportunity for studying the interplay between surface restructuring and post-treatment oxidation.

3.3. Fluorine abstraction

So far, data are consistent with a model involving the transport of part of the surface population of polar groups away from the polymer surface into the 'bulk' (deeper than ~ 0.5 nm) with the simultaneous production of additional oxygen-containing groups both at the surface and in the bulk. Additional complexity is, however, indicated in the XPS C 1s and F 1s spectral regions, and by the F/C ratios, which

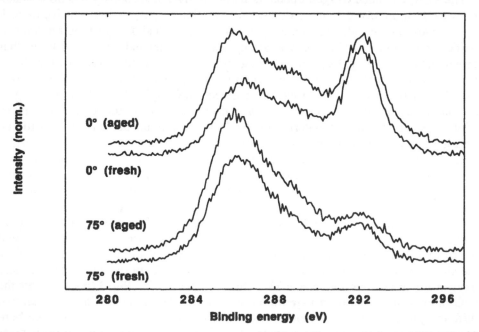

Figure 4. XPS C 1s spectra of ammonia-plasma-treated FEP, recorded immediately after treatment and after 4 months of storage at two different emission angles (75° and 0°). The intensity of the spectra at one particular emission angle was normalized with respect to the CF$_2$ component.

showed marked compositional variations with the emission angle and time. An example of a C $1s$ spectrum is shown in Fig. 4. At $\theta = 75°$, the intensity in the perfluorocarbon component at 292.2 eV was remarkably small, indicating a highly carbonized surface. The spectra and F/C ratios of different samples varied somewhat with the treatment conditions but were always in agreement with substantial fluorine abstraction.

Particularly at the 75° take-off angle, however, the C $1s$ spectra contained a small fluorocarbon component, yet the F/C ratio (Fig. 5) exceeded 0.5. Analysis of the F $1s$ region showed that the peak consists of *two* components (Fig. 6). The main peak was clearly from perfluorocarbon fluorine as evidenced by its binding energy, which was the same as that of the F $1s$ peak of untreated FEP. The second, smaller component appeared at a binding energy of ≈ 686 eV, i.e. slightly more than 3 eV below the position of the main peak. This component was assigned to fluoride ion (F^-), possibly adjacent to an $R\text{-}NH_3^+$. Its contribution was substantial only in spectra recorded on freshly treated specimens and it vanished on storage (Fig. 6). Hence, it appeared that a considerable fraction of the fluorine atoms cleaved from C–F bonds were still present in the surface and sub-surface layers, but had been converted to fluoride ion by the plasma exposure by mechanisms which are not understood at present. The ratio of fluoride-F to perfluorocarbon-F was determined by NLLS curve-fitting to be approximatelly 1/2 at the 75° take-off angle and 1/3 at 0°. In Fig. 5, both components of the F $1s$ signal, and the total F/C ratio, are plotted as a function of the storage time. The fluoride component at the 75° take-off angle is not shown for clarity; it likewise decays rapidly with time. The mechanism behind the disappearance of the fluoride ions from the plasma-treated surface layers is not clear; one possibility is the loss of HF, with the proton coming from attached amine groups or water vapour. Fluoride ions might also participate in secondary reactions.

The F/C ratio was much smaller close to the surface (75° take-off angle) than in deeper regions (normal emission); this is consistent with the higher nitrogen incorporation close to the surface (Fig. 2). However, the amount of fluorine abstracted was considerably larger than the amount of nitrogen incorporated. It appears reasonable to assume that the first step in the ammonia-plasma modification is the homolytic cleavage of C–F bonds. Hence, the difference between fluorine abstraction and nitrogen incorporation suggests that a substantial fraction of the carbon-centred radicals created by the plasma exposure did not react with an ammonia molecule, or a nitrogen-containing fragment thereof, arriving from the plasma phase. Some of the remaining radicals reacted with atmospheric oxygen, as attested to by the oxygen uptake. Even the sum of the N and the O uptake, however, is clearly not sufficient to account for all of the loss of F, indicating that the majority of the radicals presumed to have been created by C–F bond scission during the plasma treatment underwent reactions other than addition of an extraneous species. The fate of these radicals, and the reactions involved, is not clear; possible reactions are the combination of radicals to form crosslinks, and the disproportionation of radicals, which forms double bonds [17]. The relative efficiency of such reactions cannot be predicted, nor are techniques available at present to determine the extent of crosslink and double-bond formation. The suggested inefficient quenching of carbon-centred radicals by oxygen may seem somewhat surprising given the relatively high oxygen solubility and permeability of perfluorinated polymers, but it has been pointed out that the reaction

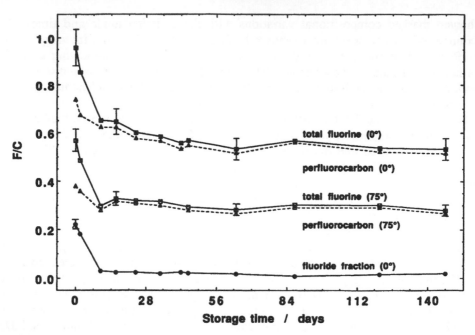

Figure 5. The XPS F/C ratio of ammonia-plasma-treated FEP, as a function of the storage time in air. Plotted are the fluoride component at 0° emission, the perfluorocarbon component, and the total fluorine, the latter two at 75° and 0°.

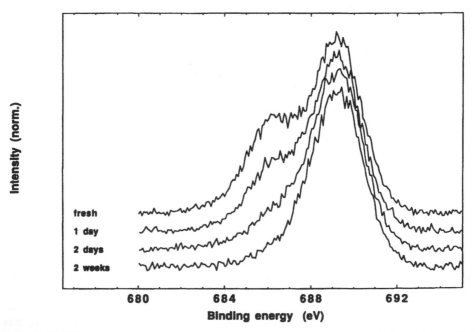

Figure 6. XPS F 1s spectrum of ammonia-plasma-treated FEP, freshly treated, and after various periods of storage. All spectra were recorded at an emission angle of 0°.

of oxygen with perfluoroalkyl radicals is decelerated by several orders of magnitude relative to alkyl radicals [31]. Secondly, the plasma-treated tape is under vacuum for several minutes following treatment, and hence not immediately exposed to oxygen. It is possible that a substantial fraction of the C• radicals can undergo other reactions before oxygen addition occurs, and the formation of additional crosslinks by the combination of neighbouring radicals would seem a probable major reaction that could occur relatively rapidly given the high mobility of perfluorocarbon polymer segments.

The cleavage of C–F bonds was not confined to the initial plasma exposure: in addition to the more pronounced loss of fluoride-F, some fluorine is also lost from the perfluorocarbon component (Fig. 5). One possibility is the hydrolysis of acid fluorides, which would also be consistent with the observed increase in the O/C ratio, but spectral overlaps in both the C $1s$ and the F $1s$ peaks prevent direct study of the importance of this reaction. The relative intensity of the CF_2 component of the F $1s$ peak decreased at a rate slower than that of the increase in the oxygen content. This suggests that the post-treatment cleavage of C–F bonds occurs as a result of secondary reactions in the oxidation pathway; the hydrolysis of acid fluorides is one of a number of possible reactions. Another reaction sequence that may proceed on such a time scale as observed here to reduce the F/C ratio is the decomposition of metastable peroxides into alkoxy radicals, with the subsequent liberation of small, volatile fragments rich in fluorine, for example by a radical-shift reaction such as:

$$\overset{\overset{\displaystyle \dot{O}}{\displaystyle \|}}{\text{\textasciitilde\textasciitilde\textasciitilde CF}_2\text{-C-CF}_2\text{-}\dot{\text{C}}\text{F}_2} \quad \longrightarrow \quad \text{\textasciitilde\textasciitilde\textasciitilde CF}_2\text{-C} \overset{\displaystyle \nearrow O}{\underset{\displaystyle \text{CF}_3}{\diagdown}} \quad + \quad \text{CF}_2\text{=CF}_2$$

XPS analysis thus shows that the chemical effects of the ammonia-plasma treatment of these perfluorocarbon polymers are much more complex than one might surmise. The ammonia plasma does not simply achieve a replacement of some F atoms by NH_2 groups; the term surface amination is not an accurate description. In addition, the surface composition continues to evolve over a considerable period of time, with changes to the content of both oxygen and fluorine, and by complex reactions which are presently not understood.

The possible role of diffusion in the aging process must be considered. The out-diffusion of dispersed additives can be a problem in the analysis of modified polymer surfaces; in the case of FEP, however, we found no evidence for the presence of additives. It is also possible that unmodified fluorocarbon chains of relatively low molecular weight could out-diffuse to the surface, driven by interfacial energetics, and lead to the observed increases in the CAs. The reduction in the F/C ratio on aging is, however, not in accordance with such a scenario. The in-diffusion of modified low-molecular-weight material (produced by C–C bond scission during the plasma exposure and then driven away from the surface by unfavourable interfacial energetics arising from the presence of polar groups) would likewise have to be accompanied by an increase in the F/C ratio.

The above data and discussion have focused on the XPS analysis of ammonia-plasma-treated FEP. The surface topography of PTFE is too rough to be suitable for

ADXPS work [20]; however, analogous data (not shown), recorded at an emission angle of 0° only and thus lacking the depth information, showed a qualitatively identical behaviour of ammonia-plasma-treated PTFE to that of treated FEP, and it appears reasonable to assume that qualitatively similar phenomena proceed on aging.

3.4. Surface derivatization

The density of surface amine groups can be determined by the covalent reaction between amines and isothiocyanates. FITC is a suitable probe molecule as it enables quantitation both by absorption spectroscopy and by fluorescence [25, 26]. FITC does not react significantly with weaker nucleophiles such as hydroxyl groups within the reaction time used [25]. Some adsorption, however, usually also occurs onto surfaces, such as cleaned FEP, which definitely have no functional group capable of reacting with FITC. Accordingly, the amount of FITC attached to ammonia-plasma-treated surfaces must be corrected by subtracting the amount of FITC non-specifically adsorbed to untreated surfaces. This procedure assumes that the covalent binding of FITC onto reactive amines does not influence the adsorption of FITC molecules onto untreated surface patches in-between the amine groups, and that the presence of other, unreactive, polar surface groups on the patches in-between the amines does not affect the adsorption affinity of FITC. The applicability of these assumptions is not established and the magnitude of systematic errors is unknown at present, but we accept the limited accuracy of quantitative determinations and apply the method for *comparing* the ability for covalent binding of FITC of fresh and aged samples.

Solution derivatization reactions can be subject to considerable uncertainties arising from surface restructuring caused by the solvent; dissolution of surface material; and migration of the probe into the polymer, perhaps assisted by swelling. When samples of ammonia-plasma-treated FEP aged in air were immersed in water, no change in the contact angles was observed [20], suggesting that surface restructuring is irreversible and unimportant in the course of FITC derivatization using water as a solvent. Washing experiments performed in conjuction with contact angle and XPS analyses showed that part of the freshly modified surface structures could be removed, probably by dissolution of low-molecular-weight polymer fragments, but the effect was minor, and the washed surfaces then showed marked increases, almost to the same extent as unwashed samples, in the contact angles on aging [20]. On aged samples, ultrasonic washing in ethanol produced no increase in the CAs, indicating the absence of soluble low-molecular-weight fragments [20]. Thus, we consider that the dissolution of low-molecular-weight fragments from freshly treated surfaces may be a minor effect, perhaps affecting the accuracy of the FITC derivatization of amine groups, but it appears not to be the limiting factor at present given the larger uncertainties arising from non-specific adsorption. Finally, angle-resolved XPS analysis of an FITC-derivatized surface (to be reported elsewhere) has shown a very steep depth dependence of signals assigned to the FITC, indicating that significant migration into the polymer does not occur, as expected with such a large probe molecule and particularly when the polymer surface becomes partly crosslinked during the plasma modification. Hence, only amine groups at the very surface are believed to react with FITC within the reaction time used.

The amounts of FITC attached to and redissolved from the surface of ammonia-plasma-treated FEP samples, freshly treated and after storage, are listed in Table 1.

Table 1.

Advancing and receding air/water contact angles, and the density of attached FITC dye on ammonia-plasma-modified FEP samples treated for various plasma exposure times, for freshly treated and aged samples

Treatment time (s)	Storage time (days)	Contact angle (degrees)[a]		Density of FITC (molecules/nm^2)
		ACA	RCA	
10	0	55	5	0.4
	140	84	20	0
20	0	45	6	0.6
	139	75	8	0.1
30	0	37	6	0.5
	138	78	14	0.1
60	0	12	0	0.5
	136	71	20	0.1
120	0	12	0	0.4
	134	71	22	0

[a] Typical standard deviations of quadruplicate measurements are 1–2°.

The values determined here are considerably below the maximum packing density of FITC, which is ≈ 4 molecules/nm^2 [25], and thus should not be affected by possible inefficient derivatization arising from steric crowding. Although the values have a relatively large experimental uncertainty of ±0.1 molecules/nm^2 (and possibly systematic errors as well, as discussed above), the comparison of fresh and aged samples suggests substantial losses of reactive surface-amine groups.

An unresolved question is whether the amines disappear mainly by surface restructuring or by oxidation to amides (following which some of the amides would have to move away from the interface in order to allow for the observed increase in the CAs). Amide groups do not react with FITC, which is a probe for nucleophiles. We cannot separate the amine and amide contributions to the XPS N 1s peak due to the very small difference in binding energy between the two groups.

3.5. Modelling of contact angles

The fact that aging does not cause the contact angles to reach the values of the untreated materials (Fig. 1) indicates that there are *two* populations of attached polar groups: one population is immobile, and the other is mobile. These terms are used in a functional sense to describe the ability of chain segments to move the attached polar groups away from the interface to a depth exceeding the probe depth of contact angle measurements. Immobile groups are likely to be adjacent to interchain crosslinks. Thus, the surface is pictured in terms of microscopic domains of varying crosslink densities; some domains are immobile, some are mobile. As oxygen-containing groups are incorporated by post-treatment oxidation, they likewise will be subjected to the same mobility considerations. We will use an interpretation of the time dependence of contact angles to estimate the importance of immobilized patches.

The information content and problems in contact angle measurements have been the subjects of thorough discussion by Morra *et al.* [36]. Meaningful for the present purposes are the advancing (ACA) and the receding (RCA) contact angles, in that they probe for hydrophobic and hydrophilic components of the surface, respectively. The distribution and precise nature of the polar and dispersive surface sites are not known, but it would appear reasonable, as a first approximation, to make use of both the ACA and the RCA as they both would be affected by a time-dependent redistribution of the relative amounts of these surface components. In the absence of a more detailed understanding of the contact angle analysis of heterogeneous, mobile surfaces, we will attempt an approach to the study of surface restructuring based on a mean contact angle, θ_M, defined by

$$\cos(\theta_M) = \frac{\cos(RCA) + \cos(ACA)}{2}. \tag{2}$$

We will use a description of a heterogeneous surface based on the Cassie equation [37]:

$$\cos \theta_M = f_p \cos \theta_M^p + (1 - f_p) \cos \theta_M^{np}, \tag{3}$$

where f_p is the fraction of surface area covered by polar segments, θ_M^p is the air/water contact angle of a surface consisting only of polar (non-dispersive) segments, and θ_M^{np} analogously applies to a pure non-polar (dispersive) surface. We depict the plasma-treated FEP surface as comprising segments with dispersive character and consisting of CF_2 and CF_3 structural elements, and polar segments consisting of amine, amide, and various oxygen-containing groups. The plasma-treated surfaces are not likely to possess any significant degree of order, and thus, a random distribution of such groups and structural elements is likely to prevail. The individual segments of polar and dispersive characters are therefore likely to be of random size and very small on the average, with some comprising only one chemical group. An advancing or receding water drop will experience an interfacial force that is the sum of the forces emanating from a large number of such individual surface patches. The Cassie equation was established for surfaces of heterogeneous structures on a larger size scale, and it is not known whether one can use this approach to describe surfaces which are heterogeneous on the size scale of individual chemical groups. It appears reasonable to us that if the Cassie approach is useful for heterogeneous surfaces consisting of larger patches, then the analogous summation over a larger number of smaller patches should also be applicable.

At zero storage time, we define the polar segments in terms of a fraction covered by immobile polar groups, f_{im}, and a fraction covered by mobile polar groups, f_m. We then assume that the mobile groups are lost in a single exponential process with a time constant τ, and that, as they are lost, their place on the surface is taken by non-polar species that migrate up from sub-surface layers. Hence, we extend the Cassie equation to describe θ_M as a function of the storage time, t, as follows:

$$\cos \theta_M(t) = (f_{im} + f_m e^{-t/\tau}) \cos \theta_M^p + (1 - f_{im} - f_m e^{-t/\tau}) \cos \theta_M^{np}. \tag{4}$$

As an approximation, we assume that the non-polar species have the same values of ACA and RCA as FEP, 117° and 98°, respectively, yielding a $\cos \theta_M^{np}$ value of -0.2966. The treated surface is much poorer in fluorine than FEP, so that θ_M^{np} might be somewhat smaller, but this is a minor effect. For the hydrophilic domains, we assume that all of the polar groups are fully wettable, so that the ACA and the RCA are both $0°$, yielding a $\cos \theta_M^p$ value of unity. A value of θ_M^p of $0°$ appears reasonable given the very low CAs measured immediately after treatment (Fig. 1). With the amine groups protonated, ACA and RCA values of $0°$ would apply [38]. On storage, however, oxygen-containing groups increasingly contribute to the surface polarity (Fig. 3), and θ_M^p becomes a sum of various contributions which cannot be calculated in the absence of data on the nature and density of all of the groups present on the surfaces. We ignore this potential time dependence and use $\theta_M^p = 0°$ throughout.

An example of the application of equation (4) to fit the experimental data is shown in Fig. 7. The value of f_m was set equal to $(1 - f_{im})$, so that there were only two fit parameters. After performing a weighted non-linear least-squares fit, the values obtained were $f_{im} = 0.78 \pm 0.02$ and $\tau = 2.1 \pm 0.8$ days. The contact angles of various samples differed somewhat with the plasma treatment conditions, and hence the value for f_{im} also varied between samples, but the general considerations applied equally to all samples studied. The above figures do not imply that 78% of the surface which existed immediately after plasma treatment was immobile; f_{im} is determined by the CAs at long times and thus contains contributions from polar groups attached both by the plasma and by post-treatment oxidation. The information in f_{im} is that, after long storage, the density of polar surface groups is 78% of that immediately after treatment, and that these remaining polar groups, irrespective of whether they originated from plasma exposure or post-treatment oxidation, are immune to surface energetics attempting to drive them away from the interface.

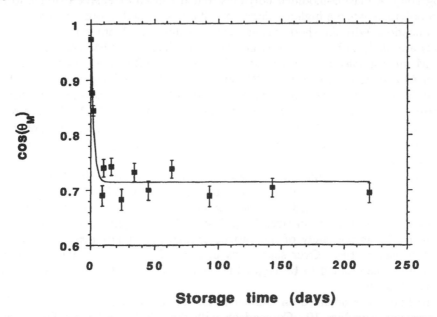

Figure 7. Experimental contact angle data (■) and the best fit using equation (4) (see text). The error bars represent one standard deviation.

The FITC derivatization results report a much larger decrease in the density of reactive amine groups with time (although there are considerable uncertainties in both methods). Immediately after plasma treatment, amines are likely to be the major polar species on the surface, but with time, oxygen-containing groups increasingly contribute to the wettability, and finally seem responsible for more than half of the surface polarity, as suggested by the fact that the O/C ratio exceeds the N/C ratio at long times. There is also likely to be considerable oxidative conversion of amines to amides by oxidation at the carbon atom adjacent to an amine group. Thus, on both counts one expects the wettability and the amine surface density not to correlate, and the density of amines at the surface to reduce markedly with time.

In the present case, surface restructuring is not very efficient at removing polar groups; a considerable surface density remains after extended periods of time. Such inefficiency suggests substantially reduced mobility of the surface layers, due most likely to a considerable extent of surface crosslinking, an interpretation that is consistent with the large decrease in the F/C ratio. Conversely, the limited reduction in surface polarity also attests to the efficiency of oxidative processes in inserting polar groups; the oxidative processes considerably oppose the effects of surface-rearrangement motions.

3.6. Surface topography by STM and AFM

Air/water contact angles can be affected by surface topography [36]. It is possible that the plasma treatment could alter the surface topography, and that such topographical changes would affect the CAs. In addition, a modified surface topography could alter or relax on storage, thus producing time-dependent topographical effects on the CAs. Scanning electron microscopy can be used for the study of the surface topography of plasma-modified polymers, but it caused extensive damage to these fluoropolymers when a high accelerating voltage was used in order to access higher magnification. With the short plasma treatment times used, any topography/etching effects are likely to be on a nanoscale. We have thus used AFM and STM.

AFM micrographs are reproduced in Fig. 8. The difference between the treated (panel A) and the untreated (panel B) samples is striking and suggests etching by the ammonia-plasma treatment to produce fine, worm-like ridges and additional structure superimposed on the gentle undulations native to untreated FEP. Line scans were taken at several random locations across the areas; Fig. 9 displays two representative examples. These line scans clearly show more detailed topography of the ammonia-treated sample.

For STM analysis, samples were coated with a thin layer of Pt (thickness $\approx$ 3 nm). An RF-sputtered Pt coating deposited onto cleaved mica under the same conditions had a structure characterized by a mean grain diameter of $\approx$ 4 nm and a relatively narrow distribution of grain sizes [39]. The surface roughness contribution by the Pt coating limits the study of the surface topography of polymers to a resolution of a few nanometres. Over scan areas of the order of 1 μm^2, the Pt coating is sufficiently fine-grained to be negligible, and STM and AFM information can be compared directly.

STM line traces obtained on untreated FEP and on ammonia-plasma-treated FEP are reproduced in Fig. 10. Comparison with the AFM line traces (Fig. 9) shows a striking difference for the traces from the *untreated* FEP sample, whereas STM

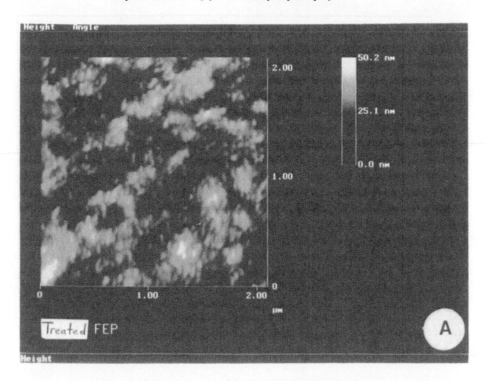

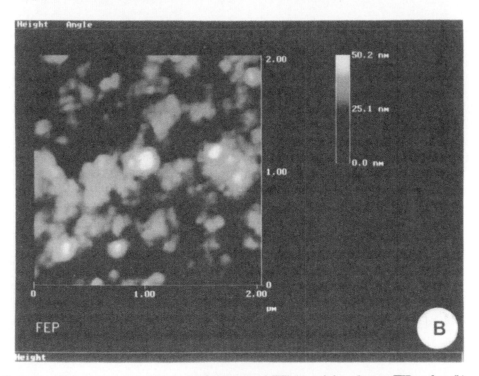

Figure 8. AFM micrographs of ammonia-plasma-treated FEP (a) and the reference FEP surface (b).

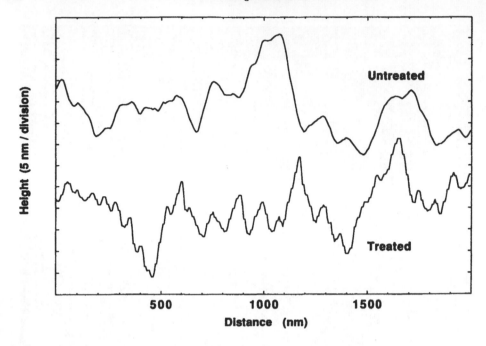

Figure 9. AFM line traces of ammonia-plasma-treated FEP and reference FEP surface.

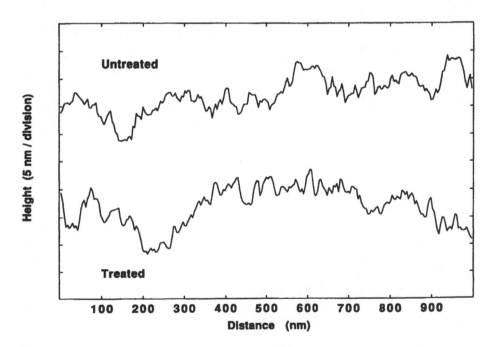

Figure 10. STM line traces of ammonia-plasma-treated FEP and the reference FEP surface.

and AFM gave the same fine structure in the case of the ammonia-plasma-treated FEP sample. Also, the STM line traces were qualitatively identical for the two surfaces. We therefore attribute the AFM results obtained on the untreated FEP surface to an experimental artefact: if the surface were sufficiently soft, then the AFM tip might deform the surface and thus smooth out structural detail. In STM, on the other hand, such deformation forces do not apply under good scanning conditions. In addition, the metallic overlayer can be expected to stabilize the soft polymer. We believe that the reason why only the reference surface, but not the ammonia-plasma-treated FEP surface, was affected by the presumed smoothing is related to the observed fluorine abstraction. With crosslinks produced by the plasma treatment, the treated FEP surface should have a higher resistance to deformation by the AFM tip. The AFM observations thus agree with the concept of surface crosslinking suggested by both the reduction in the F/C ratio and the contact angle data, which show that a considerable fraction of the polar surface groups is prevented from participating in surface restructuring.

3.7. Surface restructuring models and long-term chemical changes

Modified polymer surfaces generally are unstable; surface restructuring typically leads to a decrease over time of the treatment effects. The extent of disappearance of the treatment effects on storage is variable [14–20, 40–42], and the reasons for such differences in behaviour are not understood. Morra *et al.* found that the XPC O/C ratio correlated inversely with the extent of hydrophobic recovery in three plasma-treated hydrocarbon polymers, and postulated that a high extent of oxygen incorporation into the surface was necessary to achieve a marked reduction in the hydrophobic recovery [40]. However, this may not apply to all classes of surface-modified polymers; water-plasma-treated FEP and PTFE surfaces had a much lower oxygen content [20] than the ammonia-plasma-treated surfaces of the present study; yet, the latter reoriented while the former did not show any changes in the contact angles on storage [20]. Alternative mechanisms for surface stabilization must be considered.

The extent of surface restructuring might be related to the degree of crosslinking conferred by the plasma treatment. However, water-plasma-treated FEP and PTFE surfaces did not undergo surface restructuring at all, although they showed a much smaller decrease in the F/C ratio on treatment [20] and thus presumably are much less crosslinked than the present ammonia-plasma-treated materials. Hence, it is doubtful whether surface crosslinking is the dominant factor affecting the stability of a modified surface stored in air.

Alternatively, the stability of a surface could be related to the penetration depth of the treatment or, in other terms, the steepness of the compositional gradient; a shallow compositional gradient would cause only a relatively small driving force for reorientation. This interpretation is supported by the observation of a considerable depth of penetration of water-plasma treatments of FEP and PTFE [20], whereas most other plasma treatments appear to penetrate to a much smaller depth and thereby cause steeper compositional gradients. The present ammonia-plasma-treated FEP samples also are modified over only a few nanometres, and this is consistent with instability according to the above interpretation. We are currently investigating the relationship between the stability and the penetration depth of a wider range of plasma modifications.

The above models implicitly describe the surface restructuring of *compositionally invariant* surfaces; only the depth distribution of (a constant total amount of) polar groups is considered to vary with time. Our data, however, now demonstrate that there may occur an evolution of the chemical composition of the surface and near-surface layers on a time scale comparable to that of surface restructuring, which in turn should be affected by the changing composition. It is thus clear that current models for the surface restructuring of modified polymers are in need of refinement. We cannot at present offer a well-supported model, and it is not clear at this stage whether it is indeed warranted to envisage a unified model for the surface evolution on aging of the various combinations of polymers and plasma-treatment gases studied by us and other workers. The effects of plasma exposure and post-treatment oxidative reactions on segmental mobility probably vary considerably with reactivities determined by the original chemical composition of the polymer, its inherent segmental mobility, the nature and density of the groups and radicals introduced by the plasma treatment, and the treatment and aging conditions.

A key issue in the discussion of polymer surface mobility is the density of crosslinks Plasma treatments generally are thought to form crosslinks; in the present case, considerable abstraction of fluorine occurs, and crosslinks are likely to be major products. The density of crosslinks may, however, also evolve on aging. Crosslinks are known to form in the course of the conventional oxidation of polyolefins; on the other hand, chain-scission processes also occur [31–34]. The oxidative pathways operating in plasma-treated polymers might also affect the density of crosslinks, but the net balance between crosslinking and scission reactions cannot be predicted and may vary considerably with the polymer, the treatment gas, and the aging conditions. Therefore, the process of oxidation not only increases the density of polar surface groups that the surface restructuring process is trying to reduce, but also likely affects the rate of reorientation by altering the crosslink density with time. Moreover, substantial changes to the mobility of the surface and near-surface layers could lead to changes, in the course of aging, of not only the rate of ongoing oxidation reactions, but even the relative importance of various competing reactions, thus altering the product distribution with time. Interpretation of the interplay between oxidation and surface restructuring requires further detailed analytical data. A major obstacle is the absence of a method for the determination of crosslink densities in treated polymer surface layers. Indirect evidence for a substantial density of crosslinks in ammonia-plasma-treated FEP and PTFE is provided by the lack of mobility of a considerable fraction of the polar surface groups, the reduction in the XPS F/C ratio, and the apparent surface hardening observed by comparison of AFM and STM results, but these observations do not give quantitative information, nor do they distinguish crosslinks formed during the plasma treatment from those formed on aging by oxidative reactions.

Modelling of the superimposed effects of surface restructuring and oxidation is thus not simple. The evolution of modified polymer surfaces by the interplay of chemical processes (reactions continuing for extended periods of time) and physical processes (polymer segmental mobility) is complex and dependent on a number of factors. Since changes in surface composition with time are difficult to predict *a priori*, we recommend that the aging of plasma-treated polymers be monitored in future studies, and that compositional data be accompanied by information on storage times and conditions.

4. CONCLUSIONS

The aging ammonia-plasma-treated FEP and PTFE surfaces does not only involve hydrophobic recovery by a physical process based on segmental mobility. Contact angle data are deceptively simple in suggesting an asymptotic surface-restructuring process whereby polar groups are moved away from the interface; however, considerable compositional changes that occur after the ammonia-plasma treatment are evident in XPS.

The chemical effects of ammonia-plasma treatment of fluoropolymers are complex: the surface layers undergo considerable defluorination in the treatment, accompanied by the incorporation of nitrogen. XPS results indicate that some of the remaining fluorine is in the form of fluoride ion. After the plasma treatment, the surface composition continues to evolve on aging: polar carbon-oxygen functional groups are formed as a result of post-deposition oxidation, and an additional decrease in the fluorine content is observed. The fluoride ion component is reduced drastically within short times after treatment. The aged surfaces contain more oxygen than nitrogen. Thus, chemical changes are superimposed on surface-rearrangement motions. There is, on the other hand, no evidence for changes in the surface *topography* on aging.

The term 'surface amination' is therefore not an accurate description of the results of the ammonia-plasma treatments and does not describe the time-dependent character of the resultant surfaces. Attached amine groups disappear on storage, probably both by surface restructuring and by oxidation to amides, in spite of a considerable degree of immobilization of the surface.

Acknowledgements

We are indebted to Professor B. Ratner, NESAC/BIO and the Division of Research Resources (N. I. H. Grant RRO1296) for support and access to the SSI X-Probe spectrometer, and K. Kjoller of Digital Instruments for performing the AFM measurements.

REFERENCES

1. J. R. Hollahan, B. B. Stafford, R. D. Falb and S. T. Payne, *J. Appl. Polym. Sci.* 13, 807–816 (1969).
2. M. R. Wertheimer and H. P. Schreiber, *J. Appl. Polym. Sci.* 26, 2087–2096 (1981).
3. A. S. Chawla and R. Sipehia, *J. Biomed. Mater. Res.* 18, 537–545 (1984).
4. D. L. Cho and H. Yasuda, *J. Vac. Sci. Technol.* A4, 2307–2316 (1986).
5. B. Jansen, H. Steinhauser and W. Prohaska, in: *Biological and Biomechanical Performance of Biomaterials*, P. Christel, A. Meunier and A. J. C. Lee (Eds), p. 207. Elsevier, Amsterdam (1986).
6. Y. Iriyama and H. Yasuda, *J. Appl. Polym. Sci.: Appl. Polym. Symp.* 42, 97–124 (1988).
7. Y. Nakayama, T. Takahagi, F. Soeda, K. Hatada, S. Nagaoka, J. Suzuki and A. Ishitani, *J. Polym. Sci., Part A: Polym. Chem.* 26, 559–572 (1988).
8. C. Munkholm, D. R. Walt, F. P. Milanovich and S. M. Klainer, *Anal. Chem.* 58, 1427–1430 (1986).
9. J. Lub, F. C. B. M. van Vroonhoven, E. Bruninx and A. Benninghoven, *Polymer* 30, 40–44 (1989).
10. N. Inagaki, S. Tasaka and H. Kawai, *J. Adhesion Sci. Technol.* 3, 637–649 (1989).
11. P. J. C. Chappell, J. R. Brown, G. A. George and H. A. Willis, *Surface Interface Anal.* 17, 143–150 (1991).

12. F. J. Holly and M. F. Refojo, *J. Biomed. Mater. Res.* **9**, 315–326 (1975).
13. J. D. Andrade (Ed.), *Polymer Surface Dynamics.* Plenum Press, New York (1988).
14. H. Yasuda, A. K. Sharma and T. Yasuda, *J. Polym. Sci., Polym. Phys. Ed.* **19**, 1285–1291 (1981).
15. F. Garbassi, M. Morra, E. Occhiello, L. Barino and R. Scordamaglia, *Surface Interface Anal.* **14**, 585–589 (1989).
16. M. Morra, E. Occhiello and F. Garbassi, *J. Colloid Interface Sci.* **132**, 504–508 (1989).
17. D. Youxian, H. J. Griesser, A. W. H. Mau, R. Schmidt and J. Liesegang, *Polymer* **32**, 1127–1130 (1991).
18. H. J. Griesser, D. Youxian, A. E. Hughes, T. R. Gengenbach and A. W. H. Mau, *Langmuir* **7**, 2484–2491 (1991).
19. E. Occhiello, M. Morra, P. Cinquina and F. Garbassi, *Polymer* **33**, 3007–3015 (1992).
20. X. Ximing, T. R. Gengenbach and H. J. Griesser, *J. Adhesion Sci. Technol.* **6**, 1411–1431 (1992).
21. H. J. Griesser, *Vacuum* **39**, 485–488 (1989).
22. J. T. Grant, *Surface Interface Anal.* **14**, 271–283 (1989).
23. J. H. Scofield, *J. Electron Spectrosc. Relat. Phenom.* **8**, 129–137 (1976).
24. M. P. Seah and W. A. Dench, *Surface Interface Anal.* **1**, 2–11 (1979).
25. H. J. Griesser and R. C. Chatelier, *J. Appl. Polym. Sci., Appl. Polym. Symp.* **46**, 361–384 (1990).
26. T. G. Vargo, D. J. Hook, K. S. Litwiler, F. V. Bright and J. A. Gardella, Jr, *Polym. Mater. Sci. Eng.* **62**, 259–263 (1990).
27. H. Yasuda, H. C. Marsh, S. Brandt and C. N. Reilley, *J. Polym. Sci., Polym. Chem. Ed.* **15**, 991–1019 (1977).
28. D. S. Everhart and C. N. Reilley, *Anal. Chem.* **53**, 665–676 (1981).
29. L. J. Gerenser, *J. Adhesion Sci. Technol.* **1**, 303–318 (1987).
30. C. D. Bain and G. M. Whitesides, *J. Am. Chem. Soc.* **110**, 5897–5898 (1988).
31. M. Lazar and J. Rychly, *Adv. Polym. Sci.* **102**, 189–221 (1992).
32. R. Arnaud, L. J. Lemaire and A. Jevanoff, *Polym. Degrad. Stabil.* **15**, 205–218 (1986).
33. F. Gugumus, *Polym. Degrad. Stabil.* **27**, 19–34 (1990).
34. J. Lacoste, D. J. Carlsson, S. Falicki and D. M. Wiles, *Polym. Degrad. Stabil.* **34**, 309–323 (1991).
35. T. R. Gengenbach, Z. R. Vasic, R. C. Chatelier and H. J. Griesser, *J. Polym. Sci., Part A: Polym. Chem.* (in press).
36. M. Morra, E. Occhiello and F. Garbassi, *Adv. Colloid Interface Sci.* **32**, 79–116 (1990).
37. A. B. D. Cassie, *Discuss. Faraday Soc.* **3**, 11–16 (1948).
38. R. J. Ruch and L. S. Bartell, *J. Phys. Chem.* **64**, 513–519 (1960).
39. T. R. Gengenbach, Z. R. Vasic and H. J. Griesser, Unpublished results.
40. M. Morra, E. Occhiello and F. Garbassi, in: *Polymer-Solid Interfaces*, J. J. Pireaux, P. Bertrand and J. L. Bredas (Eds), pp. 407–428. IOP Publishing Ltd., Bristol, UK (1992).
41. E. M. Cross and T. J. McCarthy, *Macromolecules* **23**, 3916–3922 (1990).
42. H. C. van der Mei, I. Stokroos, J. M. Schakenraad and H. J. Busscher, *J. Adhesion Sci. Technol.* **5**, 757–769 (1991).

Plasma Surface Modification of Polymers, pp. 147–165
M. Strobel, C. Lyons and K. L. Mittal (Eds)
© VSP 1994

Surface fluorination of polyethylene films by different glow discharges. Effects of frequency and electrode configuration

Y. KHAIRALLAH,[1,*] F. AREFI,[1] J. AMOUROUX,[1] D. LEONARD[2] and P. BERTRAND[2]

[1]*Laboratoire des Réacteurs Chimiques en Phase Plasma, Université Paris VI, ENSCP, 11 Rue P. et M. Curie, 75231 Paris Cedex 05, France*
[2]*UCL-PCPM-Bâtiment Boltzmann, 1 Place Croix du Sud, B 1348 Louvain-la-Neuve, Belgium*

Revised version received 11 November 1993

Abstract—Polyethylene has been treated in a CF_4 plasma using two configurations for the electrodes: parallel-plate electrodes at 13.56 MHz, and a nonsymmetrical configuration of electrodes at 70 kHz. The chemical species involved in the plasmas and the subsequent surface modifications were characterized. Comparison of the emissions in the range 200–450 nm was made with optical emission spectroscopy. The excited radicals' CF_2 bands and the CF^+ broad continuum were detected and a comparison of their relative intensities indicated a more energetic aspect to the low-frequency discharge. This was confirmed by comparing the ratio of N_2^+/N_2 in the two cases. Analysis of the surface properties by means of X-ray photoelectron spectroscopy (XPS), static secondary ion mass spectrometry (SSIMS), and contact angle measurements showed rapid fluorination of the surface ($t = 0.1$ s) for the nonsymmetrical electrodes. Longer treatment times (up to 15 s) led to more fluorinated surfaces, thereby decreasing the wettability of the treated substrates; fluorine incorporation was, however, more significant at equivalent treatment times in the diode reactor. SSIMS analysis indicated the formation of more complex fluorine compounds with an increase in the treatment time with no notable modification other than replacement of H by F. Further increases in the treatment time for low-frequency treated samples in the nonsymmetrical configuration of electrodes caused an increase in the surface roughness, as observed by SEM (scanning electron microscopy) analysis. RBS (Rutherford backscattering spectroscopy) measurements showed that beyond 0.1 s, which corresponds to the time required for rapid fluorination, fluorine diffuses through the polymeric matrix and its concentration is enhanced with increasing treatment time.

Keywords: Surface fluorination; electrode configuration; frequency; plasma diagnostics; polyethylene; XPS; SSIMS.

1. INTRODUCTION

Electrical discharges in gases such as carbon tetrafluoride (CF_4) are commonly used to etch a variety of materials during the fabrication of microelectronic circuits [1]. They are also used as a potential method for modifying polymer surface characteristics such as wettability and adhesion [2, 3]. Parallel-plate radio-frequency (RF) reactors operating at 13.56 MHz are generally used for this purpose. Optical emission spectroscopy [4, 5] and actinometry [6, 7] are widely used to monitor the excited

---

*To whom correspondence should be addressed.

species and the energetic aspects of the discharge. The effect of frequency has been discussed in the case of $H_2 + CH_4$ discharges [8]. The electron temperature at low frequencies ($T_e = 16\,000$ K at 1 kHz) was greater than that in high-frequency plasmas ($T_e = 8200$ K at 13.56 MHz). An increase in ion production with decreasing frequency [9] in a chlorine RF plasma showed that the high-energy 'tail' of electrons is smaller at 13.56 MHz than at lower frequencies for the case of a parallel-plate reactor.

The interaction of plasmas generated by fluorinated monomers such as CF_4, C_2F_4, or SF_6 with polymers gives rise to the fluorination of the treated surfaces. This can be obtained by two principal pathways: plasma polymerization and plasma grafting [10]. The predominant mechanism depends on the plasma operating conditions and the composition of the feed gas (CF_x/F ratio). In fluorinated plasmas, CF_x radicals are considered to be the building blocks for polymerization [11, 12]. A low CF_x/F ratio, such as in CF_4, produces fluorination through a direct grafting of F atoms to the polyolefin by the replacement of H atoms by F atoms. Simultaneously, F atoms, which are known to be etching agents, can give rise to the ablation of the fluorinated layer. In such discharges, the suppression of polymerization reactions is due to the predominance of the ablation phenomenon occurring under ion bombardment.

Different techniques such as static secondary ion mass spectrometry (SSIMS) [13–17], X-ray photoelectron spectroscopy (XPS) [18], and contact angle measurements [19] have been used as complementary analytical tools for the study of the physicochemical changes induced by the surface treatment of polyolefin samples. The molecular specificity of SSIMS and its high surface sensitivity (about 1 nm against 5 nm for XPS) have already been proven in the study of polymer surfaces [13, 14]. Contact angle measurements allow one to follow the variation of the surface energy as a function of the working parameters.

This paper reports on the characteristics of plasmas at two different frequencies: 70 kHz and 13.56 MHz, and two different configurations of electrodes: a nonsymmetrical configuration of electrodes (hollow-blade type cylinder) and parallel-plate electrodes. Optical emission spectroscopy was used to study the energetic aspects of the discharge. The modifications of the polymer surface properties were also studied by different analytical techniques: XPS, SSIMS, contact angle measurements, scanning electron microscopy (SEM), and Rutherford backscattering spectroscopy (RBS).

2. EXPERIMENTAL

Schematic diagrams of the two plasma used in this study are shown in Figs 1 and 2.

The first reactor was a low-pressure plasma reactor with a nonsymmetrical configuration of electrodes (Fig. 1) consisting of a hollow electrode-grounded cylinder (diameter = 0.07 m) and using an excitation frequency of 70 kHz. The hollow electrode was made of stainless steel and the polymer (22×22 cm^2) to be treated was rolled onto the grounded cylinder. The electrical characteristics of the discharge were measured with a Lecroy 9400 digital oscilloscope with a sampling frequency of 100 MHz. A typical power of 150 W and a total flow rate of 100 sccm were maintained during the experiments.

The second reactor was a diode reactor with an excitation frequency of 13.56 MHz (Fig. 2). The plasma discharge chamber was made of stainless steel and pumped

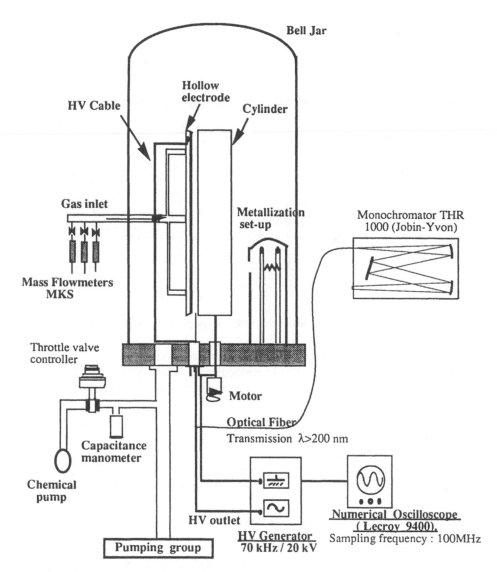

Figure 1. Schematic diagram of the bell jar reactor with a nonsymmetrical configuration of electrodes used for the surface treatment of polyethylene films by a CF_4 low-pressure plasma.

to a background pressure of 10^{-3} Pa. The stainless steel electrodes were 60 mm in diameter and the gap was maintained at 1 cm. The polymer film (diameter = 60 mm) was placed on the grounded electrode for treatment. The total flow rate was maintained at 30 sccm and the power at 80 W.

CF_4 provided by Setic Labo (France) at a purity of 99.7% was used without further purification. The gas flow was metered by MKS flow-meters/controllers which accurately measured and controlled the flow rate of the gas which was introduced through the powered electrode, allowing its direct excitation in the inter-electrode gap. The pressure was measured with MKS 127 A capacitance manometers and

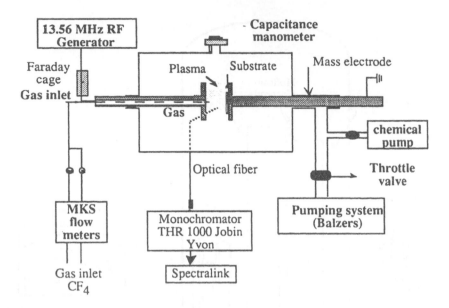

Figure 2. Schematic diagram of the diode reactor with internal parallel-plate electrodes.

controlled by MKS 252 exhaust-throttle-valve controllers. The main chamber was evacuated by a TPH 170 (Balzers) turbomolecular pumping system and a base pressure of 10^{-3} Pa was obtained. The operating pressure was maintained at 100 Pa by a 2012 AC chemical pump in both reactors.

Plasma emission was collected by an optical fiber inserted into the plasma. The emission signal was selected by a Jobin-Yvon spectrometer having a 1 m focal length and a grating of 3600 grooves/mm.

Low-density polyethylene (LDPE) film (50 μm thick) supplied by Sodap company was used. It contained no additives except 1500 ppm of an SiO_2 anti-blocking agent.

2.1. Surface analysis: analytical techniques

2.1.1. XPS. XPS analyses were performed a few weeks after the treatment at Laboratoire ITODYS (University of Paris VII). XPS spectra were recorded using a VG Scientific ESCALAB MK I system interfaced to a Cybernetix Data Acquisition System based on a personal computer. The spectrometer was operated in the constant analyzer energy (CAE) mode at a pass energy of 20 eV to insure good spectral resolution and a high signal-to-noise ratio. The electron take-off angle was 90° with respect to the surface. A non-monochromatic MgK_α source, operating at a power of 200 W, was used. Peak fitting of the high-resolution XPS spectra was done using software written at the ITODYS laboratory. Quantification of the XPS data was achieved using peak areas and experimental sensitivity factors. The full width at half-maximum (FWHM) was constrained so that it was the same for all singlet peaks within a spectrum. A linear background was assumed and the peak shape was 100% Gaussian. The spectra were calibrated with respect to the C 1s (C—C, C—H) components at 285 eV.

2.1.2 SSIMS. SSIMS analyses were performed at Louvain-la-Neuve on treated samples a few weeks after they were prepared. Details of the geometry of the experimental system have been given elsewhere [15]. In this case, it was necessary to use an electron flood gun (EG2 from VSW Scientific Instruments Ltd.) for charge compensation during the SSIMS analysis of the PE samples. In the case of hydrocarbon polymers, static conditions, i.e. conditions under which degradation of the samples during the analysis is avoided, correspond to ion doses definitely lower than 10^{13} ions/cm^2 [20, 21]. In the positive mode, typical values for the filament current and the electron kinetic energy are 2 A and 200 eV, respectively. In the negative mode, it was helpful, for charge compensation, to flood the surfaces using low-energy electrons (10 eV) emitted from a heated tungsten filament (2.5 A).

2.1.3. RBS. RBS analyses were run at Louvain-la-Neuve on the treated samples a few weeks after the treatment. Details of the RBS equipment and the simulation of the experimental spectra by using the RUMP routine of the GENTLOT program (Computer Graphic Service, USA) are given elsewhere [15]. Note that in this case the ion current is 15 nA (measured in the Faraday cup) and the total ion dose is 3×10^{15} ions/cm^2.

2.1.4. SEM. Microscopic characterization of the treated samples was carried out by SEM. The samples were coated with a 5 nm thick gold layer deposited by sputtering, and examined under low acceleration voltage conditions (5 kV).

2.1.5. Contact angle measurements. Contact angle measurements were performed in our laboratory with the help of an image processing system [22], using two liquids: water and formamide. The reported values correspond to the average of five measurements of the advancing angle, performed on different parts of the sample. Typical shifts of $\pm 3°$ with respect to this average value were observed. The polar and dispersive components of the surface energy were calculated using the Kaelble method [23] and are to be considered only as relative values.

3. RESULTS AND DISCUSSION

3.1. Characterization of the discharge by optical emission spectroscopy

The emission spectra resulting from the interaction of a CF$_4$ plasma, obtained by different glow discharges, with a polyethylene (PE) substrate were investigated. The emission spectrum in the range 200–450 nm is reported in Fig. 3 for the nonsymmetrical configuration of electrodes at an excitation frequency of 70 kHz. The broad continuum centered at 290 nm was assigned to CF$_2^+$ [7].

It can be observed that the emission intensities of the CF$_2$ lines due to the vibronic transition (020–030) of the $(\tilde{A}^1B_1 - X^1A_1)$ system are much smaller than the intensity of the continuum assigned to excited CF$_2^+$. This suggests that in the 70 kHZ discharge, the population of electrons in the low-energy range (4.6 eV) responsible for the emission of CF$_2$ lines is not significant compared with that of

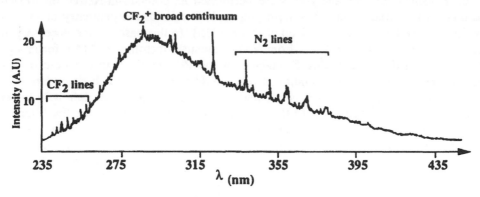

Figure 3. Emission spectrum in the range of 235–450 nm for a pure CF_4 discharge. $P = 100$ Pa; flow rate, $Q = 100$ sccm; $P_w = 150$ W; $f = 70$ kHz; nonsymmetrical configuration of electrodes.

Table 1.
Values of the N_2^+/N_2 ratio as a function of the injected power

W/FM (J/g) $\times 10^{-4}$	2.1	3.1	4.1	5.2
N_2^+/N_2 ratio	3.1	3.4	4.2	4.5

$P = 100$ Pa; $Q = 100$ sccm; $f = 70$ kHz; non-symmetrical configuration of electrodes.

the energetic electrons (16.4 eV) responsible for the ionization of CF_2 by direct electronic impact from the ground state [24].

The ratio of the first negative system of N_2^+ line at 394.1 nm and that of the second positive system of N_2 line at 394.3 nm was used to follow the energetic aspect of the discharge [5]. These lines result from the excitation of the residual nitrogen always present as a trace contaminant in the reactor. The ionization of N_2 (threshold energy $E_{th} \approx 15$ eV) is assumed to occur mainly by direct electronic impact of the ground state. The results are reported in Table 1 for different values of the ratio W/FM (where W is the discharge power and FM is the mass flow of the monomer), considered by Yasuda *et al.* as a very efficient parameter for controlling plasma processes in polymerization reactions [25] as well as in non-depositing plasmas [12].

The emission spectra in the case of a 13.56 MHz RF parallel-plate reactor were recorded for different powers (Fig. 4). The intensities of the CF_2 lines as compared with that of the excited CF_2^+ continuum are significantly greater than in the case of the nonsymmetrical configuration of electrodes (Fig. 3). The N_2^+/N_2 ratio varied between 0.9 and 1.2 with increasing power (W/FM $= 0.9 \times 10^4$ and 5.0×10^4 J/g, respectively). The lower values of this ratio, as compared with those listed in Table 1, confirm our previous assumption that the discharge is less energetic in this case than in the case of the nonsymmetrical electrode geometry with an excitation frequency of 70 kHz.

In low-frequency excitation systems (20–400 kHz), the transition time of the ions in the sheath in less than the half period of the applied field. Therefore the ions have a tendency to disappear during every period on one electrode or the other [26]. This

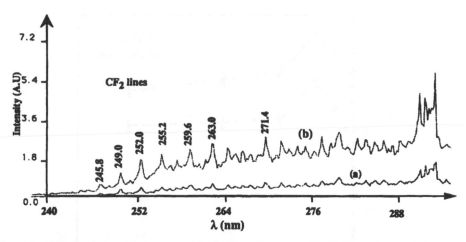

Figure 4. Emission spectra in the range of 240–300 nm for a pure CF_4 discharge in a parallel-plate configuration of electrodes. $P = 100$ Pa; $Q = 30$ sccm; $f = 13.56$ MHz; (a) $P_w = 18$ W; (b) $P_w = 50$ W.

ionic impact brings about the emission of secondary electrons from the electrodes, which is considered to be the predominant mechanism for sustaining the discharge. The nonsymmetrical geometry of the electrodes could also contribute to increase the energetic aspect of this discharge [26].

3.2. Surface characterization of CF_4-plasma-treated PE

Different complementary analytical techniques were used to characterize the treated surfaces and understand the physicochemical modifications brought about by the CF_4 plasma treatment. The main parameter investigated was the treatment time in both reactors. In the nonsymmetrical configuration of electrodes, the real treatment time (i.e. the period during which the polymer is exposed to the plasma) is calculated by multiplying the total rotation time by the ratio of the plasma width (0.5 cm) on the polymeric surface to the total perimeter of the cylinder (22 cm).

3.2.1. Contact angle measurements. Measurements of the contact angles using two liquids, water and formamide, showed, in the case of the nonsymmetrical configuration of electrodes, that the surface wettability increases (θ decreases) for treatment times less than 0.2 s and for those exceeding 15 s (see Table 2), whereas it decreases between the two values (Fig. 5).

This feature could be explained by the fact that very short treatment times ($t < 0.2$ s) lead to the incorporation of polar groups on the treated PE surface. This was accompanied by an increase in the polar component of the surface energy (see Table 5 below). The decrease of the surface wettability (increase in θ) for longer treatment times ($0.2 < t < 15$ s) indicates the substantial fluorination of the surface and the incorporation of non-polar functionalities. XPS analysis was used to confirm this assumption. Treatment times exceeding 15 s (Fig. 5) seem re-increase the wettability of the surface. This could be attributed to an interaction of the ions generated by the plasma with the substrate, leading to the ablation of the highly fluorinated groups present at the surface (see Table 4).

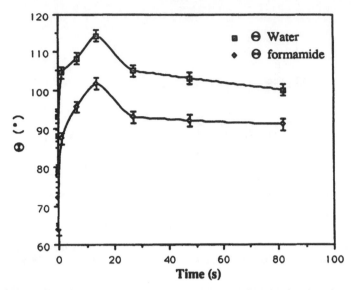

Figure 5. Evolution of the advancing contact angle as a function of the treatment time. $P = 100$ Pa; $Q = 100$ sccm; $P_w = 150$ W; $f = 70$ kHz; nonsymmetrical configuration of electrodes.

Table 2.

Contact angle measurements on treated PE as a function of the treatment time

Treatment time (s)	Θ_{water} (°)	$\Theta_{formamide}$ (°)	γ^p (mJ/m²)	γ^d (mJ/m²)	γ^T (mJ/m²)
0	94	78	4	18	22
0.07	79	64	10	21	30
0.14	88	72	6	20	25
0.2	93	78	5	17	22
1.4	104	87	1	16	18
7	108	95	2	10	12
14	114	102	1	9	10
27	105	93	3	11	13

$P = 100$ Pa; $Q = 100$ sccm; $f = 70$ kHz; nonsymmetrical configuration of electrodes.

Table 3.

Contact angle measurements on PE samples treated in the parallel-plate reactor

Treatment time (s)	Θ_{water} (°)	$\Theta_{formamide}$ (°)	γ^p (mJ/m²)	γ^d (mJ/m²)	γ^T (mJ/m²)
0	94	78	4	18	22
5	106	99	4	6	11
30	110	101	3	6	9
120	117	104	1	8	9

$P = 100$ Pa; $Q = 30$ sccm; $P_w = 80$ W; $f = 13.56$ MHz.

The results obtained in the case of the parallel plate reactor are listed in Table 3. One can clearly see that the values of the contact angles are slightly higher than those obtained with the nonsymmetrical configuration of electrodes and do not decrease at long treatment times. Shorter treatment times could not be used because the experiments were carried out in the static mode (the shortest treatment time, in this case, was 5 s).

3.2.2. XPS analysis of the treated PE samples. Figure 6 displays the XPS C 1*s* spectra of untreated and plasma-treated PE for different treatment times and different reactors. No silicon signal from the anti-blocking agent present in PE was ever observed.

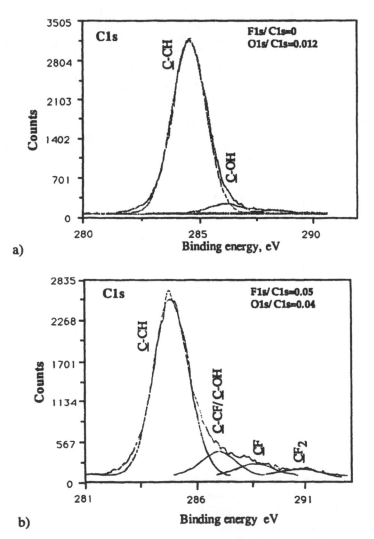

Figure 6. (a) XPS spectrum of untreated PE. (b) XPS spectrum of PE treated in the nonsymmetrical configuration reactor at 70 kHz (treatment time = 0.23 s).

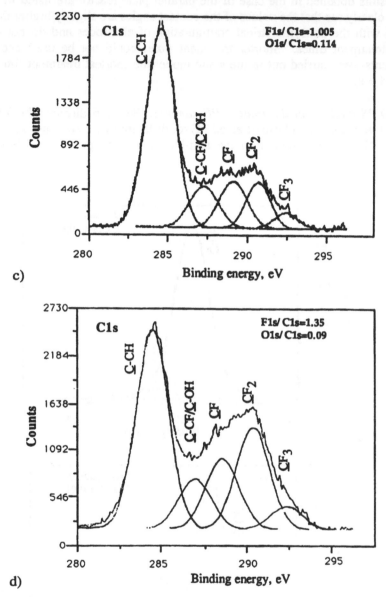

Figure 6. (Continued). (c) XPS spectrum of PE treated in the nonsymmetrical configuration reactor at 70 kHz (treatment time = 27 s). (d) XPS spectrum of PE treated in the parallel-plate reactor at 13.56 MHz (treatment time = 30 s).

Fluorinated functional groups were detected. The percentage contribution of each functional group to the total C 1s spectrum is presented in Table 4. It reveals the incorporation of more highly fluorinated groups with increasing treatment time.

The treatment of polymers with active F atoms generates a number of fluorinated moities, mainly CF and CF_2. We can clearly see that for short treatment times ($t < 0.2$ s), only polar CF groups are present, conferring to the surface its increased

Table 4.
XPS data as a function of the treatment time

Treatment time (s)	F/C[a]	O/C[a]	C–CH (%)[b]	C–CF, C–OH (%)[b]	CF (%)[b]	CF$_2$ (%)[b]	CF$_3$ (%)[b]
0.02	0.05	0.01	95	5	–	–	–
0.14	0.5	0.04	84	12	4	–	–
0.2	0.9	0.04	80	11	6	3	–
0.7	1	0.03	58	11	12	14	5
3	0.9	0.06	55	11	13	16	5
14	1.1	0.07	50	12	14	18	6
27	1	0.11	60	12	13	11	4
48	1	0.14	49	24	13	10	4

[a] F 1s/C 1s and O 1s/C 1s ratios as a function of the treatment time.
[b] Percentage contribution of functional groups to the total C 1s spectrum.
  Plasma conditions: $P = 100$ Pa; $Q = 100$ sccm; $P_w = 150$ W; $f = 70$ kHz; nonsymmetrical configuration of electrodes.

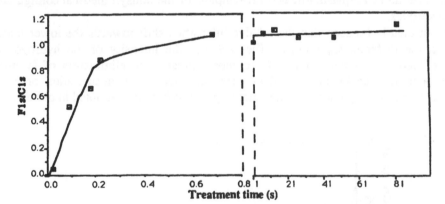

Figure 7. Evolution of the F 1s/C 1s ratio as a function of the treatment time. $P = 100$ Pa; $Q = 100$ sccm; $P_w = 150$ W; $f = 70$ kHz; nonsymmetrical configuration of electrodes.

wettability. For longer times, non-polar groups such as CF$_2$ and CF$_3$ appear. The enhanced wettability of the surface thereby decreases. These results are in good agreement with the contact angle measurements. The F 1s/C 1s ratio (Fig. 7) reaches a plateau at a value approximately equal to 1 after 0.7 s. We can point out that this corresponds to the formation of CF$_3$ groups, as observed in Table 4.

The results obtained with the diode reactor are listed in Table 5. Highly fluorinated peaks (CF$_2$ and CF$_3$) are also observed. Their intensities are always higher then those observed in the nonsymmetrical configuration of electrodes. This is an agreement with the results obtained with the contact angle measurements, which were higher in the diode reactor. The interpretation for the lower fluorine content in the nonsymmetrical configuration of electrodes is the reaction of the fluorinated groups, present through a depth less than or equal to that analyzed by XPS, with the ions generated by the 70 kHz plasma, giving rise to the degradation of the fluorinated layer and a loss of fluorine.

Table 5.
XPS data as a function of the treatment time

Treatment time (s)	F/C[a]	O/C[a]	CH–CH (%)[b]	C–CF (%)[b]	CF/C–OH (%)[b]	CF$_2$ (%)[b]	CF$_3$ (%)[b]
5	1	0.09	55	12	12	17	4
30	1.35	0.09	41	11	18	25	5
120	1.43	0.09	37	11	19	26	7

[a] F 1s/C 1s and O 1s/C 1s ratios as a function of the treatment time.
[b] Percentage contribution of functional groups to the total C 1s spectrum.
Plasma conditions: $P = 100$ Pa; $Q = 30$ sccm; $P_w = 80$ W; $f = 13.56$ MHz; diode reactor.

3.2.3. Static SIMS study of the chemical modifications. Static SIMS analyses were performed in both the positive and the negative mode.

3.2.3.1. Positive SSIMS: The peak intensities were normalized with respect to the overall ion yield recorded in the spectra. Figure 8 shows the results obtained on untreated PE and CF$_4$-plasma-treated PE samples for the nonsymmetrical configuration of electrodes.

It can clearly be seen that the relative intensities shift towards the lower masses within the different C$_n$ groups. The fact that the intensity of the hydrogenated peaks decreases in the different C$_n$ groups indicates the elimination of H atoms. This behavior can be interpreted as either an increase in unsaturation or as the replacement of H by F atoms. The latter hypothesis has been adopted in our case and

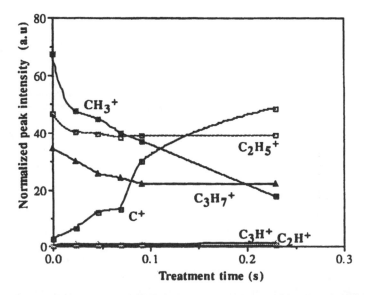

Figure 8. Evolution of the positive SSIMS hydrogen-containing peaks inside the C1 (C$^+$,CH$_3^+$), C2 (C$_2$H$^+$, C$_2$H$_5^+$), and C3 (C$_3$H$^+$,C$_3$H$_7^+$) clusters as a function of the treatment time. $P = 100$ Pa; $Q = 100$ sccm; $P_w = 150$ W; $f = 70$ kHz; nonsymmetrical configuration of electrodes.

was also confirmed by the presence of different fluorinated peaks such as $CF^+(31)$, $CH_2F^+(33)$, $CF_2^+(50)$, and $C_2F_2^+(62)$, the intensities of which increase with the treatment time, and appear significantly after 0.23 s of exposure time.

Figure 9 displays SSIMS spectra for longer treatment times, in the case of the 70 kHz treated PE with a nonsymmetrical configuration of electrodes, and for samples treated in the parallel-plate reactor with an excitation frequency of 13.56 MHz.

Examination of the SSIMS spectra from CF_4-plasma-treated PE leads to the conclusion that the polymer surface has been considerably modified: new peaks can be identified and changes in the relative intensities are also noted for both reactor systems. These new peaks can be compared with those reported in the spectra of fluorinated polymers such as PVDF (33: CH_2F^+; 57: $C_3H_2F^+$) and PTFE (31: CF^+; 69: CF_3^+; 74: $C_3F_2^+$; 93: $C_3F_3^+$; 100: $C_2F_4^+$; 119: $C_2F_5^+$; 131: $C_2F_5^+$) given in the literature [21].

The same shifts towards lower masses are observed in the spectra obtained for samples treated in both reactors (cf. the relative intensity of 12: C^+ as compared with that of 15: CH_3^+). This is a further indication of the incorporation of fluorine, which is confirmed by the increase in the intensities of purely fluorinated groups and fluoro-hydrocarbon groups (Table 6). These intensities were calculated from the average of three different measurements made on different portions of the samples. Typical shifts of $\pm 20\%$ with respect to this average value were observed. Note that in the case of masses 57 and 69, hydrocarbon groups ($C_4H_9^+$, $C_5H_9^+$) entirely account for the values of the peak intensities on the untreated sample and partially account for the values on the treated ones. Note also that some peaks could also be attributed to other species such as oxygen-containing ions. SSIMS in the negative mode will be helpfull in this context.

We can note, in Table 6, the evolution of the peak intensities, in particular for the fully fluorinated peaks (31: CF^+; 50: CF_2^+; 74: $C_3F_2^+$; etc.). There is a difference as a function of the reactor type. If we compare samples C5s with D5s and C27s with D30s, it is clear that in the case of the nonsymmetrical configuration of electrodes, fluorine incorporation is less compared with the diode reactor. This could be attributed to simultaneous competitive processes such as grafting and ablation occurring under ion bombardment of the surface, ablation being more pronounced in the former case. The variation of the peak intensities of hydrofluorinated (33: CH_2F^+; 57: $C_3H_2F^+$; etc.) groups is less evident but still exists.

Table 6.
Peak intensities of fluorinated and hydrofluorinated groups

m	Assignment	Untreated	C5s	C27s	C48s	C82s	D5s	D30s	D120s
31	CF^+	1	29	23	40	52	34	53	83
33	CH_2F^+	0	1	1	1	1	1	1	2
50	CF_2^+	1	9	8	10	12	9	12	12
57	$C_3H_2F^+/C_4H_9^+$	36	47	32	33	31	29	19	20
69	$CF_3^+/C_5H_9^+$	34	48	43	62	51	63	63	60
74	$C_3F_2^+$	1	3	3	4	6	4	7	7
100	$C_2F_4^+$	2	3	3	4	4	5	8	8

C: Nonsymmetrical configuration, followed by the treatment time in seconds.
D: Diode configuration, followed by the treatment time in seconds.

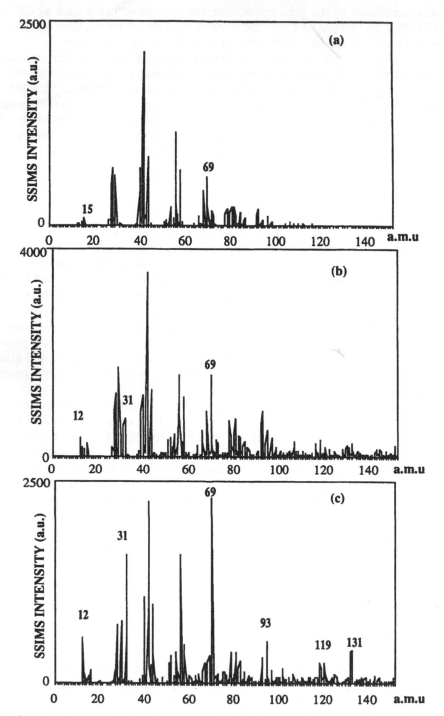

Figure 9. Positive SSIMS spectra of (a) untreated PE; (b) PE treated in the low-frequency nonsymmetrical configuration of electrodes, $t = 27$ s; and (c) PE treated in the parallel-plate reactor at 13.56 MHz, $t = 30$ s.

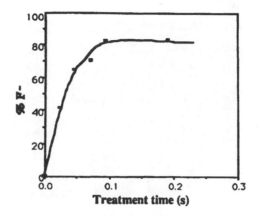

Figure 10. F⁻ intensity as measured by negative SSIMS for the case of the nonsymmetrical configuration of electrodes. $P = 100$ Pa; $Q = 100$ sccm; $P_w = 150$ W.

3.2.3.2. Negative SSIMS: The relative intensity of F⁻ to the total intensity minus the intensity of mass 1 (H⁻), which is highly dependent on the spectrometer settings, was analyzed and the results are reported in Fig. 10. Rapid saturation of the surface with fluorine (80% of the total intensity) within the first 0.1 s of treatment time is observed. The level of fluorine remains constant for longer treatment times. In the case of the parallel-plate reactor, the treatment times studied being longer than 5 s, the intensity of fluorine remained constant at $\approx 85\%$ of the total intensity.

The oxidation level, as measured by negative SSIMS, was the same for almost all of the samples. This is a very important observation because it excludes a variable contribution from the oxidized functionalities in the results obtained with positive SSIMS. In the same way, the possibility that even masses could be attributed to nitrogen-containing groups is also excluded (absence of a significant value for CN⁻ ion).

3.2.4. RBS measurements. The RBS analyses were performed only for the case of the low-frequency nonsymmetrical configuration of electrodes. The results showed an increase in the fluorine concentration with the treatment time. Figure 11 presents

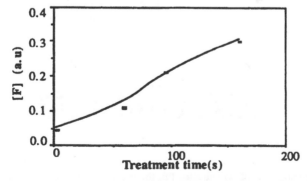

Figure 11. RBS analysis of CF₄ plasma-treated PE as a function of the treatment time. ($P = 100$ Pa; $Q = 100$ sccm; $P_w = 150$ W, $f = 70$ kHz, nonsymmetrical configuration of electrodes).

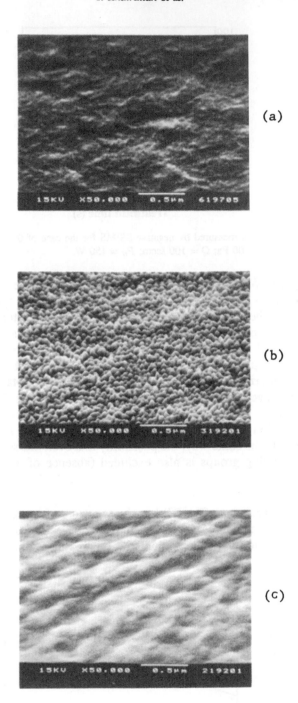

Figure 12. SEM photographs of (a) untreated PE, (b) treated PE (excitation frequency: 70 kHz; non-symmetrical configuration of electrodes; $t = 27$ s), and (c) treated PE (excitation frequency: 13.56 MHz; parallel-plate reactor; $t = 30$ s).

the evolution of the fluorine peak area as a function of the treatment time. The peak broadening was the same for all of the samples studied. This broadening is due to the limitations of the detector and corresponds to an analyzed depth less than 100 nm.

We can only conclude that there is quantitatively more fluorine in this depth as a function of the treatment time. Although XPS and SSIMS analyses showed, due to their lower depth sensitivities, that the surface was highly fluorinated, the RBS results can only be related to fluorine diffusion further into the material.

3.2.5. Analysis of the surface topography by scanning electron microscopy (SEM). SEM was used to study the modification of the surface topography. Figures 12a, 12b, and 12c display the photographs obtained for untreated PE, PE treated by a low-frequency nonsymmetrical configuration of electrodes, and PE treated in the diode reactor at an excitation frequency of 13.56 MHz.

It can be clearly observed, if we compare the topography of both treated samples with that of the untreated PE, that the surface of the PE treated at a frequency of 70 kHz with the nonsymmetrical configuration of electrodes is more altered than the PE treated in the parallel-plate electrodes at a frequency of 13.56 MHz. In the former case, a nodular structure can be observed. This underlines the ablative role of the plasma-generated ions, which is more pronounced in the low-frequency discharge than in the high-frequency discharge. Indeed, at high frequencies, due to the fact that the ion transit time is higher than the half-period of the plasma frequency, the ions are trapped by the applied electric field and, consequently, do not bombard the film surface. The treated samples were coated by a sputtered gold layer ($\approx 15-20$ nm) in order to bury the polar functions and to allow us to take into consideration only the effect of surface roughness on the measured contact angles. The measurements performed on the different gold-coated treated films gave rise to similar contact angle values showing, therefore, that roughness had no significant influence on the surface wettability.

4. CONCLUSION

The emission spectra of CF_4 plasmas obtained by different glow discharges were examined. The excited species were identified and the energetic aspect was studied. The nonsymmetrical configuration of electrodes with an excitation frequency of 70 kHz gave a discharge more energetic than that given by a 13.56 MHz parallel-plate reactor. This was shown by the relative intensities of the excited CF_2 lines compared with that of the CF_2^+ broad continuum, as observed by optical emission spectroscopy. The ratio of N_2^+/N_2 also confirmed that the mean electronic temperature was higher in the case of the nonsymmetrical configuration at 70 kHz. The configuration of the electrodes and the low excitation frequency can account for these differences.

Surface analysis of the treated polyethylene led to the conclusion that the polymer surface was considerably modified in both reactors. XPS and SSIMS showed that fluorinated peaks are present and that their intensities increase with increasing treatment time. Very short treatment times (less than 1 s) were studied only in the low-pressure nonsymmetrical electrodes, which provide the ability to simulate industrial applications. The results obtained in this case showed that the wettability of the treated surface decreased with treatment times exceeding 0.2 s, whereas for shorter treatment times the wettability increased with respect to the untreated reference. The decrease in the wettability was shown by XPS to be caused by the grafting of fluorinated groups such as CF_2 and CF_3. SSIMS analysis showed saturation of the surface with fluorine within 0.1 s. This was shown by a shift of the peak intensities of the hydrocarbon peaks towards lower masses caused by the replacement of hydrogen by fluorine. XPS analysis showed that saturation was obtained after 0.7 s of treatment time. This difference in saturation time can be attributed to the differences in the depths analyzed by the two techniques. RBS measurements indicated the diffusion of fluorine into the polymer matrix with increasing treatment time.

Similar results were obtained in a parallel-plate RF reactor for treatment times above 5 s (the shortest time studied). However, samples treated in this discharge were more hydrophobic than those treated in the nonsymmetrical reactor at 70 kHz and the intensities of the CF_2 and CF_3 groups observed by XPS were higher for the corresponding treatment times. The SSIMS results also exhibit clear differences in the fluorination obtained on samples treated in the two reactors. For identical treatment times, less F is incorporated in the case of the nonsymmetrical configuration of electrodes as compared with the diode reactor. This is attributed to the ablation of the surface under ion bombardment, which is more pronounced in the case of the low-frequency nonsymmetrical electrodes. No effect other than the replacement of H atoms by F atoms was observed by SSIMS, even though the surface topography was not the same as that observed by SEM. The surface was revealed to be smoother for samples treated in a 13.56 MHz diode reactor. The reason for this can also be related to the ablative role of energetic ions in a nonsymmetrical configuration of electrodes at low frequencies.

REFERENCES

1. G. Smolinski and D. L. Flamm, *J. Appl. Phys.* **50**, 4982–4990 (1979).
2. F. Arefi, P. Montazer-Rahmati, V. Andre and J. Amouroux, *J. Appl. Polym. Sci.: Appl. Polym. Symp.* **46**, 33–60 (1990).
3. F. Arefi, V. Andre, P. Montazer-Rahmati and J. Amouroux, *Pure Appl. Chem.* **64**, 715–723 (1992).
4. R. d'Agostino, F. Cramarossa, S. De Benedictis and G. Ferraro, *J. Appl. Phys.* **52**, 1259–1265 (1981).
5. A. Ricard, *Spectroscopy of Glow Discharges*. CNRS Report, LP 231 (1989).
6. J. W. Coburn and M. Chen, *J. Appl. Phys.* **51**, 3134–3136 (1980).
7. R. d'Agostino, F. Cramarossa and S. De Benedictis, *Plasma Chem. Plasma Process.* **2**, 213–231 (1982).
8. M. Shimozuma, G. Tochitani and H. Tagashira, *J. Appl. Phys.* **70**, 645–648 (1991).
9. D. L. Flamm and V. M. Donnelly, *J. Appl. Phys.* **59**, 1052–1062 (1986).
10. M. Anand, R. E. Cohen and R. F. Baddour, *Polymer* **22**, 361–371 (1981).
11. R. d'Agostino, F. Cramarossa and F. Fracassi, in: *Plasma Polymerization of Fluorocarbon in Plasma Deposition, Treatment and Etching of Polymers*, R. d'Agostino (Ed.), pp. 45–61. Academic Press, Boston (1990).
12. T. Yasuda, M. Gazicki and H. Yasuda, *J. Appl. Polym. Sci.: Appl. Polym. Symp.* **38**, 201–214 (1984).
13. A. Brown and J. C. Vickerman, *Surface Interface Anal.* **6**, 1–15 (1984).
14. D. Briggs, *Surface Interface Anal.* **9**, 391–400 (1986).
15. Y. De Puydt, D. Leonard, P. Bertrand, Y. Novis, M. Chtaib and P. Lutgen, *Vacuum* **42**, 811–817 (1991).
16. R. Foerch and D. Johnson, *Surface Interface Anal.* **17**, 847–860 (1991).
17. E. Occhiello, F. Garbassi and M. Morra, *Surface Sci.* **211–212**, 218 (1989).
18. D. T. Clark, W. J. Feast, W. K. R. Musgrave and I. Ritchie, *J. Polym. Sci.* **13**, 857–890 (1975).

19. E. Occhiello, M. Morra, G. Morini, F. Garbassi and P. Humphrey, *J. Appl. Polym. Sci.* **42**, 551–559 (1991).
20. D. Briggs, *Surface Interface Anal.* **15**, 734–745 (1990).
21. W. J. van Ooij and R. H. G. Brinkhuis, *Surface Interface Anal.* **11**, 430–442 (1988).
22. P. Montazer-Rahmati, F. Arefi, R. Borrin, A. Delacroix and J. Amouroux, *Bull. Soc. Chim. Fr.* **5**, 811–816 (1988).
23. D. H. Kaelble, *J. Adhesion* **2**, 66–81 (1970).
24. P. Montazer-Rahmati, F. Arefi, J. Amouroux and A. Ricard, *9th Int. Symp. Plasma Chem.* **2**, 1195–1200 (1989).
25. H. Yasuda, *Plasma Polymerization.* Academic Press, New York (1985).
26. G. Turban, *Interactions Plasmas Froids-Materiaux*, pp. 79–112. Journée d'études 'OLERON 87'. Editeur Scientifique GRECO 57 du CNRS, les Editions de Physique (1987).

Plasma Surface Modification of Polymers, pp. 167–180
M. Strobel, C. Lyons and K. L. Mittal (Eds)
© VSP 1994

Reactivity of a polypropylene surface modified in a nitrogen plasma

FABIENNE PONCIN-EPAILLARD,[1,*] BRUNO CHEVET[1,2] and
JEAN-CLAUDE BROSSE[1]

[1]*Laboratoire de Chimie et Physicochimie Macromoléculaire, Université du Maine,
Avenue Olivier Messiaen, 72017 Le Mans cedex, France*
[2]*Institut Textile de France, Avenue Guy de Collongue, 69132 Ecully cedex, France*

Revised version received 6 December 1993

Abstract—The modification of a polypropylene surface in a nitrogen plasma is studied in terms of degradation, crosslinking, functionalization, and activation (radical site creation.) The most important reactions are functionalization through primary amine formation and radical creation through methyl-group elimination. Degradation and crosslinking are competitive and the latter is more important when the plasma conditions (exposure time, discharge power) are severe. Exomethylenic bonds ($CH_2=C$) are formed preferentially over crosslinking reactions. Chemical titrations are described for radical and amine concentrations on the polypropylene surface.

Keywords: Polypropylene; nitrogen plasma modification; functionalization; activation; radical and amino group titrations.

1. INTRODUCTION

As polymers are increasingly used in composite materials, the compatibility of each component in such heterophase systems must be controlled. Most common polymers are insoluble with each other and their mixture is unstable. Several solutions have been proposed, such as the copolymerization of different monomers or the grafting of a compatible monomer onto each polymer. But these synthetic methods are limited by the inherent problems of macromolecular synthesis (yield and secondary reactions) and sometimes by a loss of mechanical properties of the final composite material. We propose a new class of grafted polymers, namely polymers modified by a cold plasma.

Cold plasma treatment is known to modify only the surface of a polymer substrate. Therefore, the mechanical properties of the polymer will not be affected by such a treatment, and the post-grafting reaction of a comonomer will give the appropriate solubility properties. In this paper, we describe the modification of polypropylene (PP) in a nitrogen microwave plasma as a preactivation for a post-grafting reaction. The grafting of acrylic acid on such a modified PP will be described in another paper.

The selection of the type of plasma depends on two factors:

---

*To whom correspondence should be addressed.

(1) For a post-grafting reaction through a radical mechanism, the modified surface will be the starting material for the growth of grafted layer; therefore, it must present efficient radicals. Radical efficiency is defined as the radical fraction which can initiate the grafting or polymerization reaction.

(2) The plasma treatment must not seriously degrade the polymer surface as radical efficiency towards grafting depends on the degradation yield of the PP surface.

A balance between these two factors has been found with nitrogen plasma [1].

2. EXPERIMENTAL

2.1. Plasma equipment

The reactor (a 433 MHz microwave plasma apparatus) and polymer treatment have already been described in [1]. The conditions for plasma treatment were as follows: incident power, $P_i = 60$ W; reflected power, $P_r < 2 \times 10^{-2}$ W; nitrogen flow, $Q = 20$ sccm; nitrogen purity, 99.985%; distance between sample and excitator, $d = 10$ cm; duration, 3 min; pressure during plasma, 0.3 mbar; system ultimate pressure, 10^{-5} mbar; volume of the reactor, 2.7 l; and volume of the plasma, 1 l. When a study was carried out with respect to one of these parameters, the other parameters remained constant.

Before treatment, the following conditions were applied: a primary pumping stage to 10^{-2} mbar, followed by a secondary pumping stage at 10^{-5} mbar for 30 min. N_2 introduction was run over 5 min; then the discharge power was switched on for 5 min and the plasma purity was checked by optical emission spectroscopy. The discharge power was switched off, the slide valve was closed, and the sample was introduced into the reactor chamber. A secondary pumping stage was run for 7.5 min. After the introduction of nitrogen, the treatment was carried out.

By optical emission spectroscopy, we analyzed the light emitted by the plasma. We used a Jobin-Yvon spectrometer with a focal length of 320 mm. The photons were detected with a Hamazu R928 photomultiplier whose spectral response ranged from 185 to 930 nm. The output signal was linearly amplified by a Spectralink (Jobin-Yvon) connected to a chart recorder. Two gratings (1200 groves/mm) with spectral responses in the ranges 200–400 nm and 400–700 nm were used.

2.2. Material

The polypropylene [PP; $\overline{M_n} = 50\,000$, $\overline{DP_n} = 1200$, $T_m = 160°$, crystallinity yield = 62% (through differential thermal analysis) and 52% (through X-ray diffraction)], supplied by ITF Lyon, was synthesized by Ziegler-Natta catalysis in a heterogeneous phase. The PP was a film of 100 ± 10 μm thickness without additives. Each sample had a surface area of 15 cm^2 and a mass of about 100 mg. The samples were washed with pure acetone before the plasma treatment and the treated samples were kept under reduced nitrogen pressure before the analyses.

2.3. Surface analyses

SEM micrographs ($\times 500$) were taken on a Hitachi model S 2300 in the 'Laboratoire de Physique des Matériaux' (URA 807) at the Université du Maine in Le Mans, France.

Static secondary ion mass spectra (SIMS; 4 keV xenon ions for the primary beam; take-off angle 75°; total ion dose 10^{13} ions/cm^2) were run on Riber Q 156 equipment in the 'Unité de Physique et de Physicochimie des Matériaux' at the Université de Louvain-La-Neuve, Belgium. Treated samples stored under reduced nitrogen pressure (10 mbar) were analyzed 4 or 5 days after the plasma treatment.

X-ray photoelectron spectroscopy (XPS) was performed on Leybold LHS 12 equipment in the 'Laboratoire de Physique des Couches Minces' at the Université de Nantes, France. The X-ray source was monochromatic MgK$_\alpha$ (1.2536 keV) with a 90° electron take-off angle with respect to the surface. Spectra were referenced with respect to C−CH at 284.6 eV. Samples stored under reduced nitrogen pressure were analyzed by ESCA on the same day as the plasma treatment.

As described in [1] and [2], the surface energy was calculated from the Dupré equation. The surface energies of the different liquids used, which are not solvents for PP, and their dispersive (d) and nondispersive (nd) components are $\gamma = 72.8$ mJ m^{-2}, $\gamma^d = 21.8$ mJ m^{-2}, and $\gamma^{nd} = 51.0$ mJ m^{-2} for distilled water; and $\gamma = 50.8$ mJ m^{-2}, $\gamma^d = 49.5$ mJ m^{-2}, and $\gamma^{nd} = 1.3$ mJ m^{-2} for diiodomethane (Aldrich, spectrophotometer grade). The reported liquid surface energies are literature values. The estimated error in the surface energy values is 0.4 mJ m^{-2}.

2.4. Titration of amines or radicals

Surface amines and radicals were titrated separately using Ponceau 2R (3-hydroxy-4-[2,4-methyl-phenylazo]-2,7-naphthalenedisulfonic acid, disodium salt) acidified with 0.1 N HCl solution (Fluka) and diphenylpicrylhydrazyl (DPPH; Aldrich, 99%), respectively. The titration for the amines proceeds by the following reactions:

$$R-SO_3^- \ Na^+ \quad \xrightarrow{HCl} \quad R-SO_3H \quad \xrightarrow[70°C/12\,h]{RNH_2} \quad R-SO_3^- \ RNH_3^+$$

Immediately after treatment and without contact with the air atmosphere, the PP sample was dipped into a 10 ml solution of reagent (DPPH solution concentration in benzene: 1.2×10^{-4} mol l^{-1}; Ponceau 2R concentration in water: 2.0×10^{-6} mol l^{-1}). The polymer and the analysis set were degassed before polymer immersion, and the solution was degassed just after. Then the solution was heated to 70°C for 12 h while stirring. A back-titration of the solution, giving a high sensitivity, was carried out on a UV spectrometer (Varian DMS 100). Any temperature effect should be negligible because these titrations were run below the α-phase transition of PP (80°C); treated surfaces were not altered by exposure to water and weak acidic solutions. A blank titration was run. The amino group concentration on the virgin PP was found to be negligible, while the radical concentration of treated samples was calculated from the difference between the experimental values for the treated sample and the virgin PP because virgin samples can have some radicals present. Radicals on the virgin sample represent 1.13×10^{-8} mol eq cm^2. The purity, stability, and activity of each reagent solution were checked before each titration. The errors associated with the titrations were evaluated by numerous tests made on samples treated under the same conditions. The titrations appeared to be very reproducible, and the error in these values was less than 1%.

2.5. Other analyses

Nuclear magnetic resonance (NMR) and Fourier transform infrared (FTIR) spectroscopy were run, respectively, in deuterated chloroform on a 90 MHz Varian apparatus and on a 1750 Perkin-Elmer apparatus. Size exclusion chromatography (SEC) was run on a Waters apparatus with five columns (μStyragel, 100 Å, 3×500 Å, 1000 Å), using THF as the solvent.

2.6. Extraction on amorphous phase

The amorphous phase was extracted in a Soxhlet extractor with freshly distilled dichloromethane (500 ml) for 44 h. The crystalline phase remained insoluble.

3. RESULTS AND DISCUSSION

The modification of polymer surfaces with a cold plasma can lead to the following four effects: degradation, crosslinking, functionalization, and activation.

Each effect is present in every plasma treatment and the PP modification will be described in terms of these effects.

3.1. Plasma phase composition

The nitrogen plasma was characterized by optical emission spectroscopy with or without PP on the sample holder. Between 100 and 700 nm, several transitions of excited states of N_2^+ or N_2 molecules were noticed:

(a) N_2: $B^3\Pi_g \rightarrow A^3\Sigma_u^+$, the first positive system ($500 \leqslant \lambda \leqslant 670$ nm);

(b) N_2: $C^3\Pi_u \rightarrow B^3\Pi_g$, the second positive system ($260 \leqslant \lambda \leqslant 470$ nm); and

(c) N_2^+: $B^2\Sigma_u^+ \rightarrow X^2\Sigma_g^+$, the first negative system ($390 \leqslant \lambda \leqslant 430$ nm).

A system with a very weak intensity was also detected in the range 200–280 nm, coming from traces of NO γ ($A^2\Sigma^+ - X^2$), an impurity.

When a PP sheet was introduced into the plasma, no new band or peak corresponding to a new volatile product, such as atomic oxygen, hydroxyl, hydrogen, carbon, or a carbon oxide, adsorbed either on the PP sheet or on the reactor walls, was detected. Therefore, if degradation of PP takes place, is should be a minor reaction. But the emission of N_2^+ (391.4 nm) and NO (236.3 nm) depends on the presence of PP (Fig. 1), whereas N_2^* molecule emission (394.3 nm) is constant. Only the excited ionized state of the nitrogen molecule is consumed during PP treatment and the disappearance of NO* could lead to surface oxidation. Concerning Fig. 1, futher experiments are in progress. We use the actinometry method with Ar lines as an actinometer. The first results have confirmed those described in this paper.

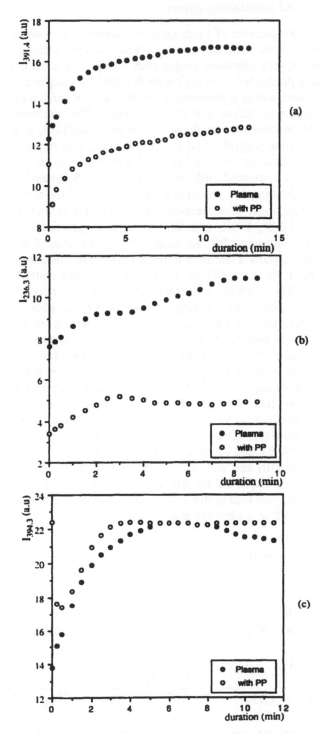

Figure 1. Dependence of the intensity of N_2^+ (391.4 nm) (a), NO^* (236.3 nm) (b), and N_2^* (394.3 nm) (c) on the duration of polypropylene plasma modification.

3.2. Degradation and crosslinking effects

SEM analysis shows pictures of homogeneous surfaces for both blank and treated samples. However, some particles were observed when the plasma conditions (time and power) were severe (duration longer than 1 h), as noticed before [3, 4]. No significant volatile products (> 0.4 mg) were detected by weight-loss measurements. Weight-loss measurement is a gravimetric method and is carried out by a differential weight measurement before and after treatment. The precision of this method is 0.2 mg. The PP surface topography seems to be unmodified, as predicted by Yasuda for semicrystalline polymers that are treated in H_2, He, Ar, and N_2 plasmas [3].

The amorphous phase, as extracted in dichloromethane, is in a lower proportion in the treated PP as compared with the blank PP (4.9% instead of 6%; the error is less than 1%), showing the presence of an insoluble three-dimensional network and crosslinking reactions. The chemical structure of the treated amorphous phase as determined by NMR, FTIR, and SEC analysis is practically the same as the amorphous phase of untreated PP. No oxidation of this extracted phase was noticed, but the NMR spectra show a decrease in the methyl peak intensity of about 5% and an increase of about 5% in the CH and CH_2 peak intensities, indicating the presence of $-CH_2-C^•H-CH_2-$ radical or exomethylene-bond formation.

SIMS analysis of treated PP surfaces (Fig. 2) also gives some indication of the PP degradation and crosslinking during the nitrogen plasma treatment. The SIMS spectra show different groups of peaks corresponding to linear or aromatic C_nH_m fragments. The peak corresponding to $n = 2$ is chosen as a reference [5–10]. Plasma treatment leads to an increase in the C_n/C_2 ratio for $n = 3$ and a decrease for $n > 3$, which could be explained by chain scission. The appearance of high-mass peaks (> 100 amu) indicates the possibility of branching and crosslinking [5]. The decrease of aromatic fragments (at 77, 91, 105, 115, 128, and 129 amu) could be interpreted as plasma modification through chain scission rather than graphitization. A decrease of the peaks at 55, 69, 83, and 97 amu, first assigned to (di)methylcyclopropylium [9] then to allyl cations [10] and derivatives (addition of methyl group: +14 amu) is observed and could be explained by scission of the CH_3 groups on the PP chain.

These different analyses lead to three major conclusions in terms of the degradation and crosslinking of PP in a nitrogen plasma:

 (i) slight degradation;

 (ii) a decrease in the methyl groups; and

(iii) surface crosslinking.

These results are opposite to those obtained by O_2 or CO_2 plasma treatment [1, 11, 12], where degradation is the most important reaction.

The decrease of methyl groups suggests the following reaction:

$$-CH_2-\underset{\displaystyle |}{\overset{\displaystyle CH_3}{CH}}-CH_2- \quad \xrightarrow{\quad -^•CH_3 \quad} \quad -CH_2-\overset{\displaystyle •}{CH}-CH_2-$$

After elimination of CH_3, the monomer unit having this radical is similar to a polyethylene unit, which could lead to branching under irradiation [13–15].

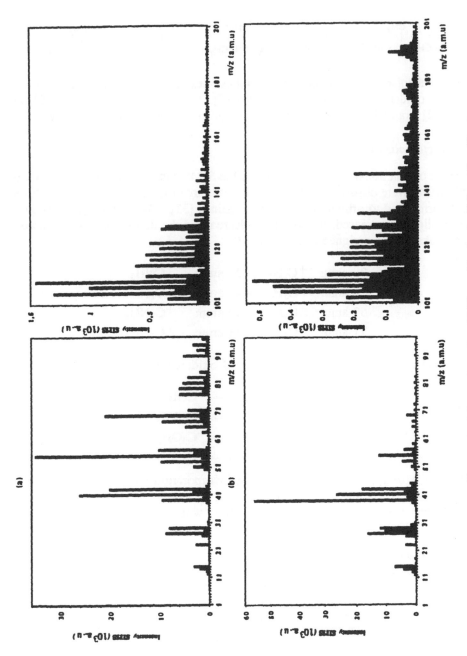

Figure 2. Positive SIMS spectra of blank (a) and treated (b) polypropylene. $P_i = 24$ W; $t = 60$ min; $d = 5.5$ cm; $Q = 23$ sccm.

Exomethylene double bonds have already been observed in Ar or O_2 plasma treatment [8]. The proposed mechanism of plasma treatment of polyolefins [16, 17] is a crosslinking mechanism where double bonds seem to initiate the reaction.

$$-CH_2-\overset{\underset{\displaystyle CH_3}{|}}{CH}-CH_2- \quad \xrightarrow{\cdot H^\cdot} \quad -CH_2-\overset{\underset{\displaystyle \dot{C}H_2}{|}}{CH}-CH_2- \quad \xrightarrow{\cdot H^\cdot} \quad -CH_2-\overset{\underset{\displaystyle CH_2}{\|}}{C}-CH_2-$$

These mechanisms can be contrasted with those described for thermal oxidation [18], chemical oxidation [19], ozonization [20–22], UV [23–25], ion beam [26], and γ [27–32] irradiation of PP, where tertiary hydrogen abstraction is proposed.

3.3. Functionalization

Evidence of functionalization was found by the SIMS and ESCA analyses. New clusters in SIMS were observed with a possible structure of $C_{n-1}NH_{2n+1}$. These clusters could correspond to primary or secondary amines [31]. The C_nH_{2n-1}/C_nH_{2n+1} ratio was slightly decreased. Rather than a decrease in the double-bond concentration, these peaks can be assigned as follows [33]:

$$
\begin{aligned}
29\ \text{amu:} & \quad ^+C_2H_5 \rightarrow 31\ \text{amu:} \quad ^+CNH_5 \\
43\ \text{amu:} & \quad ^+C_3H_7 \rightarrow 45\ \text{amu:} \quad ^+C_2NH_7 \\
57\ \text{amu:} & \quad ^+C_4H_9 \rightarrow 59\ \text{amu:} \quad ^+C_3NH_9 \\
71\ \text{amu:} & \quad ^+C_5H_{11} \rightarrow 73\ \text{amu:} \quad ^+C_4NH_{11}
\end{aligned}
$$

These peaks can also indicate the presence of oxidized groups:

$$
\begin{aligned}
29\ \text{amu:} & \quad ^+C_2H_5 \rightarrow ^+COH \\
43\ \text{amu:} & \quad ^+C_3H_7 \rightarrow ^+C_2OH_3 \\
57\ \text{amu:} & \quad ^+C_4H_9 \rightarrow ^+C_3OH_5 \\
71\ \text{amu:} & \quad ^+C_5H_{11} \rightarrow ^+C_4OH_7
\end{aligned}
$$

The ESCA data (Table 1) show nitrogen incorporation with a plateau around 18 atomic % and oxidation with a plateau around 11 atomic %. The nitrogen concentration plateau is somewhat low in comparison with literature values [34–37] but because the ESCA analysis was not run *in situ*, an aging reaction could be possible

Table 1.

Dependence of the atomic concentration (ESCA values) of PP treated in a nitrogen plasma on the plasma exposure time ($P_i = 24$ W; $d = 5.5$ cm; $Q = 23$ sccm)

Treatment time (min)	Carbon (%)	Oxygen (%)	Nitrogen (%)
0	95	5	0
7	75	10	15
9	73	11	16
60	71	11	18

through the hydrolysis of imine groups [38] or as an *in situ* oxidation as the excited NO* decreased. The high resolution of the C_{1s}, N_{1s}, and O_{1s} peaks shows that the amino groups are only in a reduced form, as primary amines, and that the oxide groups are carbonyl (aldehyde, ketone) and C—O (alcohol and/or peroxide).

The chemical titration gives more evidence of amino groups. Ponceau 2R (Aldrich) is an aromatic disulfonate that can react with basic-character functional groups after acidification. This molecule contains two sulfonic groups that are able to react. A three-dimensional simulation allows us to calculate the molarity of such a compound (from Chem 3D software):

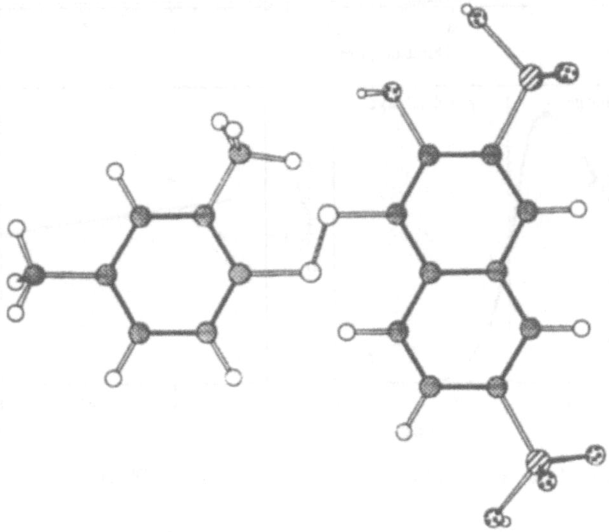

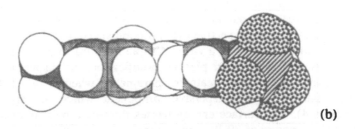

(a)

rotation 90° :

(b)

When one sulfonic site has reacted (a), the second one cannot reach another amino group on an ideal (perfectly smooth, rigid, and impermeable) surface. Thus, the molarity of Ponceau 2R should be considered as one. From representation (b), steric

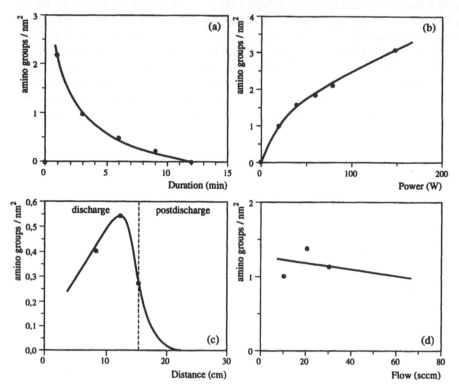

Figure 3. Dependence of the amino group concentration of treated PP surfaces on different plasma conditions.

hindrance was estimated and led to the maximum theoretical concentration of Ponceau 2R that could be fixed on an ideal surface. The concentration is $1.7 \, mol \, eq \, nm^{-2}$. This estimate does not take into account the possibility of diffusion effects or roughness effects on a molecular scale.

The dependence of the amino group concentration on the plasma treatment parameters (time, power, distance, and flow) was also studied (Fig. 3). Plasma treatment conditions generating a higher concentration of amino groups are short durations, powers greater than 50 W, a sample position close to the visible part of the plasma, and a low nitrogen flow, even if this final parameter does not have a large influence. The 'lower' the plasma treatment parameters are, the greater the amine concentration is.

The surface energy and its dispersive (d) and nondispersive (nd) components were determined. The surface energy of a virgin sample is $33 \, mJ \, m^{-2}$ and the nondispersive component is near zero. After plasma modification, the dispersive component is constant whatever the plasma treatment parameter, whereas the surface energy variations are related to the nondispersive component, which corresponds to functionalization (Fig. 4). The surface energy passes through a maximum for a plasma duration of 2 min, a power of 30 W, and a sample position at the end of the visible part of the plasma.

These surface energy values agree with those obtained after N_2 RF plasma [39] or N_2 corona discharge [40] treatments.

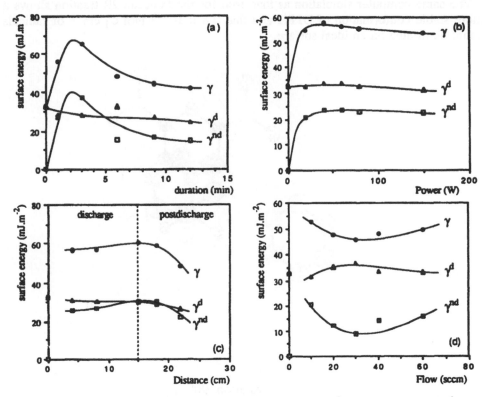

Figure 4. Dependence of the surface energy (γ) and the dispersive (γ^d) and nondispersive (γ^{nd}) components of treated PP surfaces on different plasma conditions.

The nitrogen and oxygen concentrations obtained from the ESCA data reach a plateau after a short plasma exposure, while the titration and the surface energy decrease after 2 min of treatment. This could be explained by crosslinking reactions, which limit polar group mobility, affect the non-dispersive energy, and limit the diffusion of Ponceau 2R, which is then unable to react with the amino groups. The same explanation can be given for the discharge power effect.

3.4. Activation

Activation can be defined as the creation of reactive sites, such as radicals, on the PP surface. Evidence of radicals ($R^{\bullet}$) was obtained by colorimetric chemical titration with diphenylpicrylhydrazyl radical (DPPH) as the reagent.

The same computer simulation as that used for the Ponceau 2R titration shows a more stable conformation of DPPH and that 2.8×10^{-10} mol eq cm^{-2} of radicals could be titrated on an ideal surface.

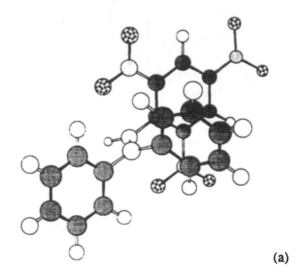

(a)

rotation 90° :

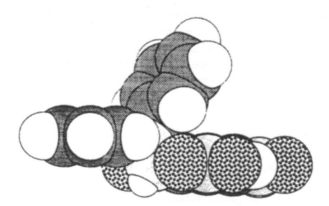

The dependence of the radical concentration on different treatment parameters (duration, power, flow, and distance) is shown in Fig. 5. A high radical concentration results from exposure times longer than 5 min, powers higher than 100 W, sample substrate in the plasma phase, and a low nitrogen flow. The more severe the plasma treatment parameters are, the greater the radical concentration is, but these conditions also lead to competitive reactions such as degradation or crosslinking.

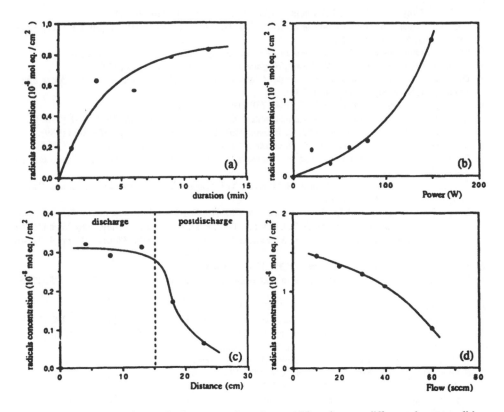

Figure 5. Dependence of the radical concentration of treated PP surfaces on different plasma conditions.

4. CONCLUSION

The interaction between a nitrogen plasma and PP leads to two main effects: functionalization, and activation through primary amines, oxidized groups, and radicals. Two competitive reactions also exist: degradation and crosslinking; the latter is more important when the plasma parameters are severe and could lead to new surface properties of PP such as selective permeability or higher solvent inertness. The mechanism of nitrogen plasma modification of PP is described by methyl group degradation, or exomethylenic bond formation.

Acknowledgements

We are greatly indebted to Professor P. Bertrand (Université Catholique de Louvain-la-Neuve, Belgium) and his collaborators for the many fruitful discussions on the SIMS technique.

REFERENCES

1. F. Poncin-Epaillard, B. Chevet and J. C. Brosse, *Eur. Polym. J.* **26**, 333–339 (1990).
2. H. Gleich, R. M. Criens, H. G. Mosle and U. Leute, *Int. J. Adhesion Adhesives* **9**, 88–94 (1989).
3. H. Yasuda, *J. Macromol. Sci.-Chem.* **A10**, 383–420 (1976).
4. P. W. Kramer, Y.-S. Yeh and H. Yasuda, *J. Membr. Sci.* **46**, 1–28 (1989).

5. W. J. van Ooij and R. H. G. Brinkhuis, *Surface Interf. Anal.* **11**, 430–440 (1988).
6. J. C. Vickerman, in: *Spectroscopy of Surface*, R. J. H. Clark and R. E. Hester (Eds), pp. 155–214. John Wiley, New York (1988).
7. W. J. van Ooij and R. S. Michael, *Mater. Res. Soc. Symp. Proc.* **119**, 285–290 (1988).
8. E. Occhiello, F. Garbassi and M. Morra, *Surface Sci.* **211/212**, 218–226 (1989).
9. D. Briggs, A. Brown and J. C. Vickerman (Eds), *Handbook of Static Secondary Ion Mass Spectrometry*. John Wiley, New York (1989).
10. D. Briggs, *Surface Interface Anal.* **15**, 734–738 (1990).
11. A. J. Muller, J. L. Feijoo, C. A. Villamizar and P. E. Vasquez, *Met. Finish.* **84**, 57–61 (1986).
12. F. Poncin-Epaillard, B. Chevet and J. C. Brosse, *Vide Couches Minces* **246**, 207–209 (1989).
13. A. K. Mukherjee, P. K. Tyayi and B. D. Gupta, *Angew. Makromol. Chem.* **161**, 77–87 (1988).
14. Q. Zhu, F. Horii, R. Kitamaru and H. Yamaoka, *J. Appl. Polym. Sci.: Polym. Chem. Ed.* **28**, 2741–2751 (1990).
15. T. Seguchi, N. Hayakawa, N. Tamura, N. Hayashi, Y. Hayashi, Y. Katsumura and Y. Tabuta, *Radiat. Phys. Chem.* **33**, 114–119 (1989).
16. D. T. Clark, A. Dilks and D. Schuttleworth, in: *The Application of Plasma to the Synthesis and Surface Modification of Polymers*, D. T. Clark and W. J. Feast (Eds). John Wiley, New York (1986).
17. F. Clouet and F. Poncin-Epaillard, *Journ. Formation Colloque Int. Procédés Plasmas 91*, pp. 407–461 (1991).
18. F. Tüdos and M. Iring, *Acta Polimerica* **39**, 19–25 (1988).
19. J. De Andres-Llopis, A. Aguilar-Navarro and J. Domenech, *J. Macromol. Sci.-Chem.* **A25**, 1575–1585 (1988).
20. R. Oueslati and B. Catoire, *Eur. Polym. J.* **27**, 331–340 (1991).
21. J. F. Rabek, J. Lucki, B. Ranby, Y. Watanabe and B. J. Qu, *ACS Symp. Ser.* **364**, 187–200 (1988).
22. J. Yamauchi, A. Yamaoka, K. Ikemoto and T. Matsui, *J. Appl. Polym. Sci.* **43**, 1197–1203 (1991).
23. Z. P. Yao and B. Ranby, *J. Appl. Polym. Sci.* **41**, 1469–1478 (1990).
24. W. Kesting, D. Knittel and E. Schollmeyer, *Angew. Makromol. Chem.* **182**, 177–186 (1990).
25. F. Gugumus, *Angew. Makromol. Chem.* **176/177**, 27–42 (1990).
26. U. W. Gedde, K. Pellfolk, M. Braun and C. Rodehed, *J. Appl. Polym. Sci.* **39**, 477–482 (1990).
27. I. Y. Prazdnikova, R. R. Shifrina, S. A. Pavlov, M. A. Bruck and E. N. Teleshov, *Polym. Sci. USSR* **31**, 1789–1794 (1989).
28. I. Kaur, R. Barsola and B. N. Misra, *Polym. Prepr.* **32**, 650–652 (1991).
29. I. Kaur and B. N. Misra, *Desalination* **64**, 271–284 (1987).
30. I. K. Mehta, D. S. Sood and B. N. Misra, *J. Polym. Sci.: Polym. Chem.* **27**, 53–62 (1989).
31. D. J. Carlsson, R. Brousseau, C. Zhang and D. M. Wiles, *ACS Symp. Ser.* **36**, 376–389 (1988).
32. I. Kaur and R. Barsola, *J. Appl. Polym. Sci.* **41**, 2067–2076 (1990).
33. W. J. van Ooij and R. S. Michael, *ACS Symp. Ser.* **440**, 60–87 (1990).
34. J. E. Klemberg-Sapieha, L. Martinu, E. Sacher and M. R. Wertheimer, *Proc. 10th Int. Symp. Plasma Chem.* **2**, 40–44 (1991).
35. R. Foerch, G. Beamson and D. Briggs, *Proc. 10th Int. Symp. Plasma Chem.* **2**, 2–6 (1991).
36. R. Foerch, G. Beamson and D. Briggs, *Surface Interface Anal.* **17**, 842–846 (1991).
37. R. Foerch and D. H. Hunter, *J. Polym. Sci.: Polym. Chem.* **30**, 279–286 (1992).
38. L. J. Gerenser, *J. Adhesion Sci. Technol.* **1**, 303–318 (1987).
39. R. E. Marchant, C. J. Chou and C. Khoo, *J. Appl. Polym. Sci.: Appl. Polym. Symp.* **42**, 125–138 (1988).
40. V. André, F. Tchoubineh, F. Arefi, J. Amouroux, Y. De Puydt, P. Bertrand and M. Goldman, *Proc. 9th Int. Symp. Plasma Chem.* **3**, 1601–1603 (1989).

3
Practical Applications
of Plasma-treated Surfaces

Plasma Surface Modification of Polymers, pp. 183–195
M. Strobel, C. Lyons and K. L. Mittal (Eds)
© VSP 1994

Chemical reactions on plasma-treated polyethylene surfaces

M. MORRA, E. OCCHIELLO and F. GARBASSI*
Istituto G. Donegani S.p.A., V. Fauser 4, 28100 Novara, Italy

Revised version received 22 April 1993

Abstract—Oxygen plasma treatment as a surface functionalization technique is discussed. Oxygen-containing functionalities were introduced on the surface of high- (HDPE) and low-density polyethylene (LDPE) by glow discharge. The number of surface hydroxyl groups was increased by a post-discharge wet treatment in a reducing solution. The effects of the substrate nature, the discharge parameters, and the post-discharge wet treatment on the surface functional groups are discussed, and the effectiveness of functionalized surfaces on the yield of coupling reactions is shown.

Keywords: Plasma treatment; surface functionalization; surface coupling; surface analysis.

1. INTRODUCTION

Many technological applications of polymers, from adhesion [1] to biocompatibility [2], require optimization of the surface composition and properties of the material, which can be accomplished by several surface modification techniques [1–3]. The coverage of the substrate by a suitable overcoat can be obtained by purely 'physical' surface-modification techniques, which do not involve the formation of covalent chemical bonds between the substrate and the overcoat. Among them, the coating technology is probably the most widespread and commercially exploited [4]. Conversely, the establishement of a true chemical bond is often desired, especially when a high degree of adhesion between the substrate and the overcoat is required for a particular application. Unfortunately, most synthetic polymers do not have surface reactive groups, and chemical surface-modification techniques require necessarily an activation step to create reactive species in the surface region (surface functionalization). Due to the very stable nature of the bonds in organic polymers, these treatments involve either highly reactive reagents [1] or high-energy-density techniques [1, 5].

Polyethylene (PE) is a typical example of a polymer which requires an energetic surface-activation step. Carboxyl, carbonyl, and hydroxyl groups can be introduced on the surface of PE by treatment in strong oxidizing solutions [6–13]. Whitesides and co-workers have accurately described the surface functionalization of PE by oxidizing treatments [14–18].

From an environmental standpoint, the exploitation of wet oxidizing treatments is even more hindered by the highly pollutant nature of the chemicals required. The surface activation by plasma treatment is, among its other interesting features [19], an environmentally friendly alternative. However, several difficulties limit the industrial application of this technique.

---

*To whom correspondence should be addressed.

Surface functionalization by low-temperature plasmas is often achieved by the deposition of thin polymeric films from oxygen- or nitrogen-containing monomers [20, 21]. This technique offers the advantage of a closer control of the surface chemistry of the deposited coating (even if plasma-deposited polymeric films are definitely multifunctional [22]), which, of course, is a key issue in surface functionalization. However, it is rather time-consuming, since the rate of deposition is generally low. As discussed by Yasuda [20], the increase of the deposition rate often involves a greater fragmentation of the parent monomer, and the consequent loss of oxygen or nitrogen.

The plasma treatment of a substrate with non-polymerizable gases is a very quick and economical process that can be performed with inexpensive gases such as air, N, or O_2 [1]. With respect to chemical surface functionalization, the main drawback of this approach is the lack of chemical selectivity: a wide range of groups are introduced on the substrate surface upon plasma treatment with gas, and optimization of the surface functionalities is practically impossible by plasma treatment alone. However, Nuzzo and Smolinsky, [23] demonstrated that the coupling of plasma and post-treatment by wet-chemical techniques could be successfully used for the preparation of functionalized low-density PE (LDPE) surfaces.

Our interest in the preparation of functionalized PE surfaces, on which further coupling of selected molecules or chains can be performed, by a quick, economic, and clean process, prompted us to investigate further the combined plasma–wet treatment. In order to move the process from the laboratory to industry, several fundamental issues must be answered. For instance, different kinds of PE, LDPE, or HDPE are known to respond in different ways to plasma treatments [1]. This, in turn, can affect the yield of the wet post-treatment reaction and that of subsequent coupling reactions. Also, it is important to know the kinetics of reactions involving a surface and a liquid phase and, more generally, how the overall reactivity is affected by the reduction of dimensionality when a reaction occurs at an interface [24, 25].

In this paper we discuss some of the above points. HDPE and LDPE were subjected to oxidizing treatments by an O_2 plasma and to a wet reducing treatment in aqueous $NaBH_4$ solutions. Yu and Marchant [26] have shown the effectiveness of this wet treatment in the reduction of carbonyl to hydroxyl groups in plasma-polymerized *N*-vinyl-2-pyrrolidone. The effect of these treatments on the surface composition and the surface functional groups was evaluated by X-ray photoelectron spectroscopy (XPS) and derivatization reactions. The effectiveness of the functionalization treatment in coupling reactions with polyethylene glycol is evaluated.

2. EXPERIMENTAL

2.1. Materials and reagents

Samples of HDPE (Eraclene H, EniChem Polimeri) plates ($10 \times 10 \times 1$ mm) and blow-molded LDPE films ($10 \times 10 \times 0.1$ mm, EniChem) were used. The amount of additives and slip agents in these samples was less than 0.1%. Samples were cleaned by extraction in boiling CH_3OH for several hours and then vacuum-dried. No oxygen signal was observed by XPS analysis on the surface of the samples after this purification routine. The crystallinity, as measured by a Siemens D-500 X-ray diffractometer, was *ca.* 58% for HDPE and 35% for LDPE.

Oxygen (Carlo Erba) was used for the plasma treatments. NaBH$_4$, trifluoroacetic anhydride (TFAA), metallic Na, thionyl chloride (SOCl$_2$), dimethylformamide (DMFA), and poly(ethylene glycol) monoethyl ether, molecular weight 350 (PEG$_{350}$) were purchased from Aldrich and used without further purification. Deionized water was used for rinsing and wet treatments.

2.2. Plasma treatment

Plasma treatments were performed in a capacitively coupled parallel-plate reactor, with the samples located on the water-cooled grounded electrode. Both the reactor and the electrodes were made of aluminum. The volume of the reactor was about 20 dm^3, and the distance between the electrodes was 15 cm. The diameter of the lower, grounded electrode was 35 cm, while the diameter of the upper electrode was 15 cm. The base pressure of the reactor was 1×10^{-4} Pa. The plasma parameters were as follows: excitation frequency, 13.56 MHz; flow rate, 16 sccm; pressure, 2 Pa. The gas flow was controlled by MKS mass flow meters and flow controllers. The treatment time ranged from 5 to 60 s, and power from 10 to 80 W, as discussed below.

2.3. Chemical modification

Treated samples were reduced in a 0.2 M aqueous solution of NaBH$_4$ at room temperature. Unless otherwise indicated, the treatment time was 30 min. In another set of experiments, the effect of the immersion time in the reducing solution was assessed. In this case, the treatment time ranged from 10 s to 24 h.

2.4. Derivatization with TFAA

Derivitization with TFAA was used to identify surface hydroxyl groups in XPS analysis [2, 27]. HDPE or LDPE samples were placed on the top of glass test-tubes containing liquid TFAA and then exposed to TFAA vapors. The distance from the liquid to the surface of the sample was 10 cm. Based on XPS measurements, an exposure time of 30 min was used. The surface fluorine concentration was not increased by longer exposure times.

2.5. Coupling reaction

The sodium alcoholate of PEG$_{350}$ (PEG-Na) was prepared by stirring PEG$_{350}$ and excess Na, overnight, at room temperature, under a dry nitrogen purge. Unreacted Na was mechanically removed before the coupling reaction.

Alkyl chloride and acyl chloride were produced on the sample surface by immersion in SOCl$_2$ containing 0.1 DMFA [15] for 4 h, at room temperature. The reaction vessel was sealed with an anhydrous CaCl$_2$ cap. At the end of the reaction, the samples were dried under flowing dry N$_2$ and then exposed to the PEG–Na solution for 20 h under dry N$_2$. After that, the samples were extracted overnight in refluxing, deionized water and vacuum-dried overnight.

2.6. Surface composition and surface morphology

XPS analysis was performed with a Perkin Elmer PHI 5500 ESCA system. The instrument was equipped with a monochromatic X-ray source (Al $K\alpha$ anode) operating at 14 kV and 250 W. The diameter of the analyzed spot was 400 μm. The base pressure was 10^{-8} Pa. Peak deconvolution and quantification of the elements were accomplished using the software and the sensitivity factors supplied by the manufacturer. In fixed-angle measurements the electron take-off angle (i.e. the angle between the electron analyzer and the sample surface) was maintained at 45°, while in angle-resolved measurements it ranged from 10° to 70°.

The surface compositions reported in this paper are the mean of three measured values. In no case was the standard deviation greater than ± 1.1%.

In high-resolution spectra, all binding energies were referenced by setting the CH_x peak maximum in the resolved $C\,1s$ spectra to 285.0 eV.

The surface morphology of the plasma-treated samples and the controls was observed using a Cambridge Stereoscan 360 scanning electron microscope equipped with a LaB$_6$ filament.

3. RESULTS

3.1. Effect of the plasma treatment

The surface morphology of the samples was unaffected by the treatment, within the resolution of SEM observation. XPS analysis of the O$_2$-plasma-treated samples showed the well-known time–power dependence of the surface concentration of oxygen [1], which on LDPE ranged from 5.5%, in the case of a 5 s treatment at 10 W, to 18.8% for a 60 s treatment at 80 W. (The values for HDPE were slightly lower, as discussed in the following.) A decrease in the oxygen content was observed after rinsing with deionized water, and a further decrease after the 30 min reduction in NaBH$_4$. These results are shown in Fig. 1, which also highlights the treatment-time dependence of the surface concentration of oxygen. No further reduction of the oxygen surface concentration was observed when the storage time in water was increased above 30 min. The most likely explanation of this behavior is the well-documented solubilization of the outermost layers of plasma-treated PE [1]. An alternative hypothesis, i.e. the building up of a contaminant layer at the water–PE interface, is ruled out by angle-resolved analysis (Table 1), which shows a decreasing trend of the O/C ratio as the sampling depth increases.

The same general trend is observed in the case of HDPE, with a slight shift towards lower amounts of oxygen (Table 2).

3.2. Surface functionalities and effect of the post-discharge wet treatment

Some clue on the nature of the oxygen-containing surface functionalities on plasma-treated PE can be obtained by derivatization of the treated samples by TFAA. Table 2 shows the surface atomic ratios on LDPE, detected by fixed-angle XPS, before and after derivatization, for an 80 W discharge treatment. Of course, no fluorine was detected when untreated HDPE or LDPE was subjected to the derivatization reaction. In the case of as-treated and rinsed samples, it is

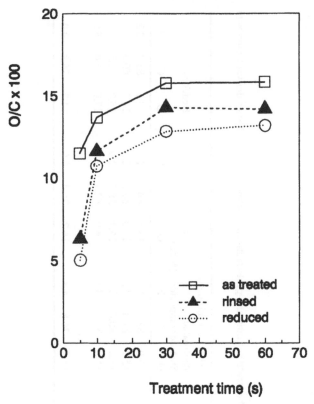

Fig. 1. Effect of the treatment time and of the post-treatment routine on the O/C ratio for LDPE treated with a 40 W oxygen discharge.

Table 1.
Angle-resolved XPS analysis of treated, rinsed, and reduced LDPE

(a) 10 W, 5 s

| Electron take-off angle (°) | Surface composition (% at.) | | | | | |
| | As-treated | | Rinsed | | Reduced | |
	O	C	O	C	O	C
10	5.2	94.8	4.1	95.9	3.1	96.9
45	4.9	95.1	3.7	96.3	2.7	97.3
70	3.7	96.3	2.9	97.1	2.2	97.8

(b) 80 W, 60 s

| Electron take-off angle (°) | Surface composition (% at.) | | | | | |
| | As-treated | | Rinsed | | Reduced | |
	O	C	O	C	O	C
10	18.7	81.3	15.4	84.6	15.2	84.8
45	17.2	82.8	14.1	85.9	14.1	85.9
70	16.5	83.5	13.9	86.1	13.5	86.5

Table 2.
Surface composition of LDPE O_2 plasma-treated at 80 W as a function of the treatment time and post-treatment routine, before and after derivatization with TFAA

Treatment time (s)	Before derivatization						After derivatization								
	As-treated Surface composition (% at.)		Rinsed Surface composition (% at.)		Reduced		As-treated Surface composition (% at.)			Rinsed Surface composition (% at.)			Reduced		
	O	C	O	C	O	C	O	C	F	O	C	F	O	C	F
5	13.7	86.3	12.6	87.4	11.9	88.1	15.0	83.1	1.9	13.7	84.2	2.1	13.1	83.7	3.2
10	14.6	85.4	12.8	87.2	12.7	87.3	15.9	81.8	2.3	16.0	81.8	2.2	14.1	80.3	5.6
30	18.2	81.8	14.1	85.9	14.0	86.0	16.0	80.4	3.6	16.4	80.7	2.9	15.2	78.5	6.3
60	18.7	81.3	15.9	84.1	14.7	85.3	15.8	81.1	3.1	14.9	82.2	2.9	17.0	75.0	8.0
HDPE 60	17.4	82.6	14.9	83.1	13.8	86.2	18.4	76.8	4.8	16.7	76.4	6.9	14.2	76.8	9.0

interesting to note that despite the 13–19% surface oxygen concentration, only a few percent of fluorine was observed on the TFAA-derivatized samples. Note that the reaction between TFAA and hydroxyl groups [2, 27] introduces three fluorine atoms for each hydroxyl. A definite increase of the surface amount of fluorine was observed after reduction with $NaBH_4$, even if, also in this case, most of the oxygen-containing functionalities remained unreacted. A larger amount of fluorine, especially in the case of the as-treated and rinsed samples, was observed on HDPE. As Chilkoti and Ratner have shown that TFAA also reacts with surface epoxy groups [27], so it is important to evaluate the contribution of epoxy functionalities in the derivatization reactions. As hydroxyl groups readily react with the oxirane ring and no coupling was observed when PEG was reacted with plasma-treated PE over a wide range of time and temperature, we believe that the concentration of epoxy functionalities is negligible.

To evaluate the kinetic aspects of the reduction of oxygen-containing surface groups by aqueous $NaBH_4$, a set of experiments was conducted with different reduction times. The fixed-angle XPS results are shown in Fig. 2. It can be seen

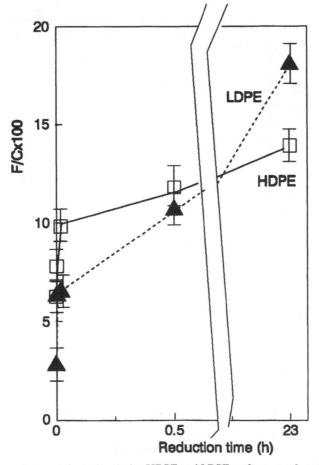

Figure 2. Time dependence of the F/C ratio for HDPE and LDPE surfaces as a function of the time of reduction in 0.3 M $NaBH_4$ solution. Samples were treated for 60 s at 80 W and derivatized with TFAA after reduction.

that for a low reduction time, the surface amount of fluorine is greater on HDPE than on LDPE, while the reverse is observed after prolonged reduction. To gain more information, angle-resolved XPS was performed on these samples. The results are shown in Fig. 3: in the case of LDPE, it is interesting to observe that at the lowest sampling depth, the saturation value is already reached after 10 s reduction. As the reduction time increases, only the data at greater sampling depth are affected, probably reflecting the diffusion and the reaction of $NaBH_4$ in the modified layer.

A completely different behavior is observed on HDPE surfaces: a generally increasing trend of the F/C ratio with reduction time is clearly detected; but, contrary to LDPE, the reaction does not seem so sharply controlled by diffusion. In Fig. 3b, the highest value of the F/C ratio is generally observed at a 45° take-off angle, confirming that the reduction reaction is less surface-specific than in the case of LDPE.

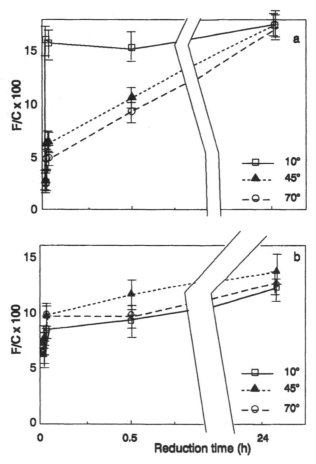

Figure 3. Time dependence of the F/C ratio as a function of the time of reduction in 0.3 M $NaBH_4$ solution and of the electron take-off angle. Samples were treated for 60 s at 80 W and derivatized with TFAA after reduction: (a) LDPE; (b) HDPE.

3.3. Effect of plasma–wet treatment on the coupling reaction

Treated surfaces were exposed to thionyl chloride and subsequently to PEG–Na. The effect of the coupling reaction can be evaluated by the relative ratio of the 285.0 and 286.5 eV components of the C 1s XPS peak, as shown in Fig. 4 for HDPE (similar results were observed on LDPE). The former is due to C- and H-bonded carbon, and is the only component in clean, untreated PE (Fig. 4a). After plasma treatment, several other components are introduced on the sample (Fig. 4b) with an approximately 1.5 eV shift for every carbon-to-oxygen bond [1, 2]. The 286.5 eV component is therefore due to hydroxyl or ether functionalities, and it is the only peak observed in the C 1s peak of pure PEG. After the coupling reaction on plasma-treated HDPE (Fig. 4c), a strong increase of the C—O component is observed, but the C—C feature is still clearly observable. Angle-resolved XPS shows that the sample is vertically homogeneous; thus, the peak shown in Fig. 4c indicates that the surface coverage by PEG is incomplete. On the other hand, when the coupling reaction is performed after reduction with aqueous $NaBH_4$ (Fig. 4d), the 286.5 eV component overwhelms the other peaks, suggesting an increased density of coupled PEG chains. Of course, no coupling was observed when PEG–Na was reacted with untreated HDPE or LDPE.

4. DISCUSSION

The first point to consider is that O_2-plasma treatment alone is a rather poor technique for the preparation of hydroxyl-functionalized PE surfaces. Derivatization of as-treated samples shows that only a minor fraction of oxygen-containing surface groups, especially in the case of LDPE, react readily with TFAA. On the one hand, this means that the great majority of oxygen-containing functionalities should be either carboxyl, carbonyl, or ether; while, on the other hand, this finding underlines one of the reasons why the deposition of polymeric films from plasma is so often preferred over the plain plasma treatment as a surface-functionalization technique.

The coupling of wet treatment and plasma discharge deserves some discussion. First of all, even if the surface concentration of fluorine in the reduced samples increases, the yield of the reaction seems rather low. In the most favorable case—the reduction of LDPE treated for 60 s at 80 W—it can be calculated from the XPS data at a 10° take-off angle (Table 1 and Fig. 3a) that no more than 30% of the surface oxygen atoms react with TFAA after reduction with $NaBH_4$. Since $NaBH_4$ can easily reduce carbonyl groups in solution [28], these results could suggest a lower reactivity in the reaction involving a solid interface. In this respect, more information and higher yields could be obtained using a stronger reducing agent, for instance $LiAlH_4$. In the present case, our choice was mainly dictated by the need for an inexpensive chemical which could be used in aqueous solutions.

A striking result is the different effects of the reduction time on LDPE and HDPE. In the former case, the reaction seems rather straightforward, at least within the previously discussed limits, and the rate-determining step is clearly the diffusion of the aqueous solution into the modified layer. An interesting consequence is that the functionalized layer can be maintained as very thin using a short reduction time (note that using the commonly accepted escape depth of C 1s photoelectrons [2], the depth sampled at a 10° take-off angle is about

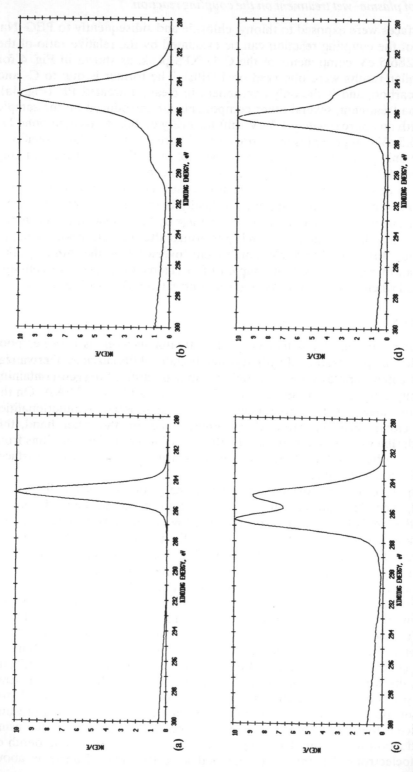

Figure 4. High resolution C 1*s* peak of (a) untreated HDPE; (b) HDPE O₂-plasma-treated at 80 W for 60 s; (c) HDPE O₂-plasma-treated at 80 W for 60 s, reacted with SOCl₂, and coupled with PEG–Na for 20 h; (d) HDPE O₂-plasma-treated at 80 W for 60 s, reduced with NaBH₄ for 24 h, reacted with SOCl₂, and coupled with PEG–Na for 20 h.

1.6 nm). From a practical point of view, this means that very short treatment times are required to produce functionalized surfaces by plasma treatment–wet reaction: besides the discharge time (maximum 1 min), the chemical structure of the outermost layer is practically completed after 10 s reduction.

In the case of HDPE, the picture is less clear. The reaction is not so sharply diffusion-limited as in the case of LDPE. The data shown in Fig. 3b suggests that the reduction itself requires a longer time. In this respect, it is interesting to note that Whitesides and co-workers [15] observed species of different reactivies on PE functionalized by wet oxidizing treatment, possibly reflecting the attachment of surface functional groups to amorphous or crystalline regions. Thus, it is tempting to attribute the greater reactivity of the surface groups of the plasma-treated LDPE to their greater rotational freedom granted by the higher percentage of amorphous domains. On the other hand, it is important to bear in mind that neither the true percentage of crystallinity of the surface layers [2] involved in the reduction reaction nor the effect on the surface crystallinity of the plasma treatment [1] is known. Another possible explanation for the different reactivities of surface LDPE and HDPE groups arises from a well-known side-effect of plasma treatment, i.e. crosslinking: HDPE is one of the most readily crosslinked polymers, whereas LDPE can only be moderately crosslinked [1]. A high degree of crosslinking can lead to a low mobility of functional groups, and hence to reduced reactivity [15].

An interesting parallel can be drawn between the reactivity and wettability data: in fact, both HDPE and LDPE are made hydrophilic by O_2 plasma; but while the surfaces of O_2-plasma-treated HDPE show a very small amount of hydrophobic recovery when stored in air at room temperature, and remain definitely hydrophilic as the time from treatment elapses [29], the contact angle of O_2-plasma-treated LDPE is not so stable and reverts to hydrophobic values [30], as shown in Fig. 5. (For details of the contact angle measurement, see ref. 29.) Since hydrophobic recovery in plasma-treated olefins is mainly due to short-range reorientation of side-groups [31], this is indirect confirmation of the different mobilities and possibly reactivities of surface functional groups.

Finally, it must be noted that the results observed on HDPE could be artificially induced by an undulating surface topography, with slopes exceeding 10°. When working at very low take-off angles, this surface topography results in indistinct depth profiles. No definite proof of the contribution of this feature to the results could be obtained by electron microscopy, since the LDPE and HDPE surfaces looked very similar when observed by SEM.

The most important aspect from a practical point of view is the effectiveness of the surface-functionalization routine in the yield of coupling reactions. Figure 4 shows that plasma–wet treated samples are effective coupling substrates, even in the worst case, i.e. HDPE. Clearly, despite the incomplete conversion of oxygen-containing functionalities to hydroxyl groups, the number of reactive groups per unit area is enough to produce a high fractional coverage by PEG chains. It is important to note that the lack of a homogeneous surface coverage, as shown in Fig. 4c, is often observed in coupling reactions of PEG chains on functionalized surfaces [21, 32]. (In the case of PEG of higher molecular weights, this behavior can be due to steric constraints. With lower molecular weights, as in the present case, the main reason is probably the heterogeneity of the distribution of surface

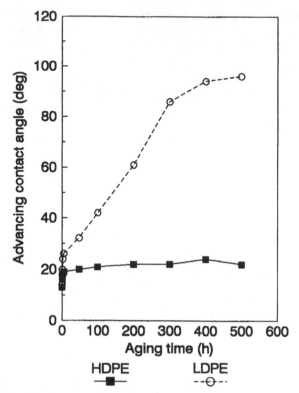

Figure 5. Effect of the ageing time at room temperature in air on the advancing contact angle of water on HDPE and LDPE, both O_2-plasma-treated at 80 W for 60 s. The advancing contact angles of the untreated samples were $93 \pm 3°$ for HDPE and $94 \pm 3°$ for LDPE.

functional groups.) In the present case, the shape of the C 1s peak shown in Fig. 4d indicates a greatly increased fractional coverage, confirming the effectiveness of the functionalization routine. Studies on the effect of time and PEG molecular weight on this coupling reaction are currently in progress.

5. CONCLUSIONS

The surface chemistry of O_2-plasma-treated PE can be refined by post-discharge treatments with aqueous $NaBH_4$. The coupling of plasma and wet treatments is an effective method of surface functionalization.

Substrate effects are clearly observed in the yield of the wet reaction. In the case of LDPE, the reduction is diffusion-controlled, and 100% conversion in the outermost layers is reached after a few seconds. Surface functionalities on HDPE seem less reactive, possibly because of reduced mobility. The latter can be due to the higher percentage of crystalline domains or to the more extensive crosslinking induced by the discharge.

PEG_{350} was successfully coupled onto functionalized surfaces. The comparison between plasma-treated and plasma–wet treated samples shows that the yield of the coupling reaction is much higher in the latter case, indicating that the density of surface-reactive groups is greatly increased by the aqueous $NaBH_4$ treatment.

Acknowledgments

We wish to thank Mr. L. Pozzi and Mr. L. Torelli for experimental assistance.

REFERENCES

1. S. Wu, *Polymer Interfaces and Adhesion*, Ch. 9. Marcel Dekker, New York (1982).
2. J. D. Andrade (Ed.), *Surface and Interfacial Aspects of Biomedical Polymers*, Ch. 1. Plenum Press, New York (1985).
3. W. J. Ward and T. J. McCarthy, in: *Encyclopedia of Polymer Science and Engineering*, Supplement Volume, pp. 674–689. John Wiley, New York (1986).
4. B. N. Chapman and J. C. Anderson (Eds), *Science and Technology of Surface Coating*. Academic Press, London (1974).
5. B. D. Ratner and A. S. Hoffman, in: *Hydrogels for Medical and Related Applications*, J. D. Andrade (Ed.), ACS Symposium Series No. 31, pp. 1–36. American Chemical Society, Washington, DC (1976).
6. D. Briggs, D. M. Brewis and M. B. Konieczko, *J. Mater. Sci.* **11**, 1270–1277 (1976).
7. D. Briggs, V. J. I. Zichy, D. M. Brewis, J. Comyn, R. H. Dahm, M. A. Green and M. B. Konieczko, *Surface Interface Anal.* **2**, 107–112 (1980).
8. A. Baszkin and L. Ter-Minassian-Saraga, *J. Polym. Sci.* **C34**, 243–246 (1971).
9. A. Baszkin and L. Ter-Minassian-Saraga, *J. Colloid Interface Sci.* **43**, 190–202 (1973).
10. A. Baszkin and L. Ter-Minassian-Saraga, *Polymer* **15**, 759–760 (1974).
11. A. Baszkin, M. Nishino and L. Ter-Minassian-Saraga, *J. Colloid Interface Sci.* **54**, 317–328 (1976).
12. B. Catoire, P. Bouriot, A. Baszkin, L. Ter-Minassian-Saraga and M. M. Boissonnade, *J. Colloid Interface Sci.* **79**, 143–150 (1981).
13. J. C. Eriksson, C. G. Golander, A. Baszkin and L. Ter-Minassian-Saraga, *J. Colloid Interface Sci.* **100**, 381–392 (1984).
14. J. R. Rasmussen, E. R. Stedronsky and G. M. Whitesides, *J. Am. Chem. Soc.* **99**, 4736–4745 (1977).
15. J. R. Rasmussen, D. E. Bergbreiter and G. M. Whitesides, *J. Am. Chem. Soc.* **99**, 4746–4756 (1977).
16. S. R. Holmes-Farley, R. H. Reamey, T. J. McCarthy, J. Deutch and G. M. Whitesides, *Langmuir* **1**, 725–740 (1985).
17. S. R. Holmes-Farley and G. M. Whitesides, *Langmuir* **2**, 266–281 (1986).
18. S. R. Holmes-Farley and G. M. Whitesides, *Langmuir* **3**, 62–76 (1987),
19. B. D. Ratner, A. Chilkoti and G. P. Lopez, in: *Plasma Deposition, Treatment and Etching of Polymers*, R. d'Agostino (Ed.), pp. 463–516. Academic Press, San Diego (1990).
20. H. Yasuda, *Plasma Polymerization*, Ch. 8. Academic Press, Orlando, FL (1985).
21. W. R. Gombotz, W. Guanghui and A. S. Hoffman, *J. Appl. Polym. Sci.* **37**, 91–107 (1989).
22. A. Chilkoti, B. D. Ratner and D. Briggs, *Chem. Mater.* **3**, 51–61 (1991).
23. R. G. Nuzzo and G. Smolinsky, *Macromolecules* **17**, 1013–1019 (1984).
24. P. Pfeifer and D. Avnir, *J. Chem. Phys.* **79**, 3558–3565 (1983).
25. R. D. Astumian and Z. A. Schelly, *J. Am. Chem. Soc.* **106**, 304–308 (1984).
26. D. Yu and R. E. Marchant, *Macromolecules* **22**, 2957–2961 (1989).
27. A. Chilkoti and B. D. Ratner, *Surface Interface Anal.* **17**, 567–574 (1991).
28. J. March, *Advanced Organic Chemistry*, pp. 809–813. John Wiley, New York (1985).
29. M. Morra, E. Occhiello, L. Gila and F. Garbassi, *J. Adhesion* **33**, 77–83 (1990).
30. M. Morra, E. Occhiello and F. Garbassi, *J. Appl. Polym. Sc.*, submitted.
31. M. Morra, E. Occhiello and F. Garbassi, in: *Polymer–Solid Interfaces*, J. J. Pireaux, P. Bertrand and J. L. Bredas (Eds), pp. 407–428. Institute of Physics Publishing, Bristol (1992).
32. N. P. Desai and J. A. Hubbell, *J. Biomed. Mater. Res.* **25**, 829–843 (1991).

Plasma Surface Modification of Polymers, pp. 197–208
M. Strobel, C. Lyons and K. L. Mittal (Eds)
© VSP 1994

Immobilization of polyethylene oxide surfactants for non-fouling biomaterial surfaces using an argon glow discharge treatment

M.-S. SHEU,[1,]* A. S. HOFFMAN,[1,]† B. D. RATNER,[1] J. FEIJEN[2] and J. M. HARRIS[3]

[1] *Center for Bioengineering, FL-20, University of Washington, Seattle, WA 98195, USA*
[2] *Department of Chemical Engineering, University of Twente, P.O. Box 217, 7500 AE, Enschede, The Netherlands*
[3] *Department of Chemistry, University of Alabama, Huntsville, AL 35899, USA*

Revised version received 26 April 1993

Abstract—A non-fouling (protein-resistant) polymer surface is achieved by the covalent immobilization of polyethylene oxide (PEO) surfactants using an inert gas discharge treatment. Treated surfaces have been characterized using electron spectroscopy for chemical analysis (ESCA), static secondary ion mass spectrometry (SSIMS), water contact angle measurement, fibrinogen adsorption, and platelet adhesion. This paper is intended to review our recent work in using this simple surface modification process to obtain wettable polymer surfaces in general, and non-fouling biomaterial surfaces in particular.

Keywords: Surface modification; glow discharge treatment; non-fouling surfaces; wettable polymer surfaces; poly(ethylene oxide) surfactants.

1. INTRODUCTION

Surfaces modified with polyethylene oxide (PEO) exhibit resistance to fouling by protein adsorption and platelet adhesion [1, 2], mainly due to the non-ionic hydrophilic characteristics (high water content) and the large excluded volume of the PEO molecule [3] as well as its high chain mobility in water [4]. Such non-fouling surfaces are important to most biotechnological and medical applications, such as diagnostic assays, drug-delivery systems, biosensors, bio-separations, and implants and medical devices. As a result, a wide variety of surface treatments for generating PEO-containing surfaces have been investigated, such as physical adsorption [5–8], surface entrapment [9, 10], chemical immobilization [11, 12], surface grafting [4, 13], and plasma (or glow discharge) polymerization [14], shown schematically in Fig. 1 [15]. Bulk polymers containing PEO have also been prepared as block copolymers [16–19] and self-crosslinked hydrogels [20] (see also Fig. 1). Many of these surface modification methods have limitations, such as physical or chemical instability, lack of functional groups, low surface coverage, undesirable changes of the bulk properties of the substrate, multiple or costly process steps, or extreme reaction conditions.

---

*Present address: Advanced Surface Technology, Inc., 9 Linnell Circle, Billerica, MA 01821, USA.
†To whom correspondence should be addressed.

Surface Modifications to Obtain PEO Surfaces

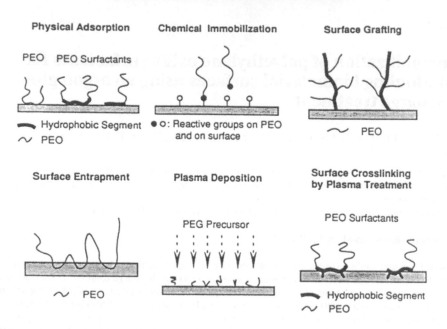

Bulk Modifications to Obtain PEO Surfaces

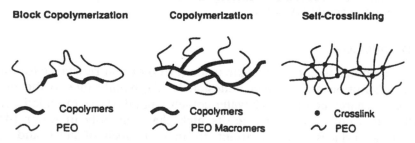

Figure 1. Schematic diagram of the surface and bulk modifications of polymers used to yield PEO surfaces [15].

Glow discharge processes have been widely used for the surface modification of polymers, mainly due to their localized surface treatment without changing the bulk properties of the polymer. Recently, the plasma polymerization (deposition) process has shown great potential to modify the surface composition of polymers by forming a deposited polymer layer or an organic thin film. In this process, the polymerizable gases are used and can be introduced to the plasma reactor during or after the glow discharge treatment. In principle, the surface chemistry of the modified substrate can be tailored by selecting the proper monomers or organic precursors in the glow discharge process. However, due to the complexity of the plasma reactions, it is difficult to obtain a specific or desired surface chemistry in the plasma-deposited layers. Also, the treatment is usually limited to using small volatile molecules in the plasma.

A modification of the plasma polymerization process was developed to graft/ polymerize non-volatile monomers using plasma treatment [21, 22]. In this process, monomers, mainly acrylates, were first adsorbed or coated on the substrate from solutions. An inert gas plasma treatment was then applied to initiate the grafting polymerization on the precoated substrate. The rate of polymerization or grafting was also enhanced by adding conventional free-radical initiators [23], suggesting a free-radical mechanism for the graft polymerization. However, if the monomers used are incompatible with the substrate and form an uneven precoated layer before plasma treatment, then an uneven surface coating on the treated substrate will result.

In the mid-1960s, Schonhorn and co-workers improved the adhesion strength of polymers using a plasma treatment with inert gases [24, 25]. Their treatment was based on the premise that the adhesion strength of polymers is mainly limited by the presence of 'weak boundary layers' on surfaces, e.g. low-molecular-weight polymers and impurities. When treated with inert gas plasmas, these weak boundary layers can be crosslinked to the larger molecules in the surface, and this enhances the adhesion strength of the treated surface. This process was called the 'CASING' technique (Crosslinking by Activated Species of INert Gases) [24, 25]. Such a plasma treatment process is a simple and convenient way to effect covalent crosslinking within a thin layer at a substrate surface.

In our laboratory, we have developed a process which extends the CASING technique to the covalent immobilization of surface-active compounds on hydrophobic polymer surfaces. This is shown schematically in Fig. 2 [15]. In this process, surface-active compounds (e.g. PEO surfactants) are first absorbed or deposited from aqueous or organic solutions onto polymer films. These pre-coated surfactants, acting as the 'weak boundary layer', are then crosslinked to the surfaces and to each other by an inert gas discharge treatment (e.g. argon). This glow discharge immobilization process shows unique advantages: a fast and simple two-step process; covalent immobilization; no change in the bulk properties of the polymer substrate; low dependency on the surface composition of the polymer substrate; high and uniform surface coverage; and no need or requirement to use volatile organic vapors.

A non-fouling polymer surface was prepared using an inert gas discharge treatment of a low-density polyethylene (LDPE) surface which had been precoated with an oleyl PEO surfactant (Brij99) and polyethylene oxide–polypropylene oxide–polyethylene oxide (PEO–PPO–PEO) tri-block copolymer surfactants (Pluronic) [15, 26]. Electron spectroscopy for chemical analysis (ESCA) was used to estimate the retention of the PEO surfactants on the treated surfaces. The enhanced wettability of the modified surfaces was characterized using water contact angle measurements. The non-fouling properties of the treated surfaces were examined by adsorption of ^{125}I-labeled baboon fibrinogen and *in vitro* adhesion of ^{111}In-labeled baboon platelets. Static secondary ion mass spectrometry (SSIMS) was used as a complementary method to ESCA and water contact angle goniometry to characterize the glow-discharge-treated surfaces, particularly to correlate the surface structure of the treated surfactants,

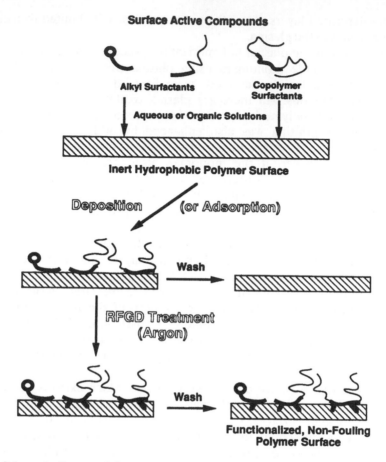

Figure 2. Schematic diagram of the argon glow discharge treatment used for the immobilization of surface-active compounds [15].

especially PEO chains, to the protein/platelet adsorption results. This paper is a review of our recent work in immobilizing PEO surfactants to generate a permanent non-fouling surface using an argon glow discharge treatment.

2. EXPERIMENTAL

2.1. Materials

Low-density polyethylene (LDPE) films (Cadillac Plastics, Seattle, WA) with an average thickness of about 0.3 mm were precleaned by sequential extraction with methylene chloride, acetone, and water for 15 min each in a sonicator. Brij99, an oleyl PEO (Sigma Chemical Co.), and Pluronic surfactants (gifts from BASF Corp.) were used as received. PEO (Sigma Chemical Co.) and PPO (Scientific Polymer Products, Inc.) homopolymers were selected as controls for comparing with the Brij99 and Pluronic surfactants. The chemical properties of the PEO, PPO, and PEO surfactants used are listed in Table 1.

Fibrinogen and platelets were purified from fresh baboon blood (from the Regional Primate Research Center, Seattle, WA) [27] and radiolabelled with [125]I

Table 1.
Properties of the PEO, PPO polymers, and PEO surfactants

Sample code	Average molecular weight	EO/PO/EO[a] (repeat units)	Theoretical O/C atomic ratio
Brij99	1100	[b]	0.36
PEO1K	1000	23 (EO)	0.50
PEO10K	10 000	227 (EO)	0.50
PPO4K	4000	0/69/0	0.33
Pluronic121	4400	6/67/6	0.36
Pluronic122	5000	13/67/13	0.37
Pluronic127	11 500	98/67/98	0.44

[a] EO: ethylene oxide unit; PO: propylene oxide unit.
[b] An oleyl ether with average 20 repeat units in PEO segments.

and ^{111}In, respectively, according to previously published protocols [28, 29]. Deionized and distilled water was used in the contact angle measurements.

2.2. Glow discharge immobilization of PEO surfactants

The glow discharge immobilization of PEO surfactants on LDPE is a two-step process: surfactant deposition on polymer substrates followed by surface treatment with argon plasmas, as shown schematically in Fig. 2. Detailed procedures for the surfactant deposition, the glow discharge treatments, and the washing protocol have been described in previous publications and are briefly presented here [15]. PEO surfactants were physically deposited onto LDPE using a simple dip-coating method. LDPE films were dipped in 1% (w/v) chloroform solutions of the surfactants for 30 s. After drying overnight, the films were then treated with an argon glow discharge.

A capacitive radio-frequency glow discharge (RFGD) at 13.56 MHz (HF-300, ENI Power Systems Inc., Rochester, NY) was used to treat the LDPE or PEO surfactant/LDPE surfaces in a glass-cylinder reactor (11.5 cm inside diameter × 80 cm long). After evacuating the chamber three times to a base pressure of 5–7 mTorr, a static argon (Air Products and Chemicals Inc., Allentown, PA; pre-pure grade: >99.95%) gas discharge without flow was generated at a reactor pressure of 25 mTorr, a low power of ≤5 W, and ambient temperature. The RFGD treatment time was varied from 0 to 300 s.

After the treatment, the surfaces were washed in chloroform twice for 30 min each and then soaked in fresh chloroform overnight. The samples were then dried in air and stored in a laminar flow hood before further surface characterization. This washing protocol was performed to completely remove all of the physically precoated PEO surfactants from the treated surface.

2.3. Surface characterization

The RFGD-treated surfaces were characterized by ESCA, SSIMS, and water contact angle measurements. ESCA measurements were done on an SSX-100 spectrometer (Surface Science Instruments, Mountain View, CA) using a mono-chromatic Al K_a X-ray source with a 5 eV floodgun. The X-ray spot size (analyzing area) on the sample surfaces was about 1000 μm in diameter. A standard 55° take-off angle (the angle between the surface normal and the axis of the analyzer lens) was used for all measurements. Surface oxygen to carbon ratios (O/C) from survey scans and the other carbon peak (286.4 eV) in high-resolution C $1s$ spectra were used to detect the presence of PEO surfactants on the treated surfaces. A Ramé-Hart goniometer (A-100) was used to measure the advancing water contact angles on the treated films at room temperature in air.

SSIMS analysis was performed on the SSX-100 surface analysis system equipped with a static SIMS add-on (SubMonolayer System, Mountain View, CA). The primary ion source was a 3.5 keV, 1.5 nA Xe$^+$ beam. Positive-ion SSIMS spectra for the treated films were recorded from $m/z = 0$ to 100 and three samples for each surface were measured. In order to observe the relative amount of PEO to hydrocarbon on the treated surface, a PEO index is defined as the ratio of intensities from the sum of two SSIMS peaks, each pair of peaks being characteristic of either the PEO ($m/z = 45$ and 89) or the LDPE ($m/z = 41$ and 55) [30]:

$$\text{PEO index} = \frac{m/z(45) + m/z(89)}{m/z(41) + m/z(55)}.$$

2.4. Fibrinogen adsorption and in vitro platelet adhesion

Polymer films (15 mm × 11 mm) were prehydrated citrated phosphate-buffered saline with 0.02% sodium azide and 0.01 M sodium iodide (CPBSzi) at 37°C for 4 h. Then an ^{125}I-labeled baboon fibrinogen solution was added to reach a 0.2 mg/ml total protein concentration (average counts = 2×10^6 cpm/mg fibrinogen) and the films were incubated at 37°C for 2 h. After incubation, the protein-adsorbed films were washed with 100 ml of fresh CPBSzi and then counted in a gamma counter. The amount of protein adsorbed per unit area was calculated from the specific activity of the fibrinogen and the planar surface area of films.

In the *in vitro* platelet-adhesion experiments, after prehydration in CPBSzi, the samples were immersed in 1×10^8 ^{111}In-labeled platelets per ml at 37°C for 2 h. Detailed protocols for protein adsorption and platelet adhesion have been described previously [15, 29].

3. RESULTS AND DISCUSSION

Figures 3 and 4 show the results of ESCA and advancing water contact angle measurements on the RFGD-treated/CHCl$_3$-washed surfaces, respectively [26]. Without RFGD treatment, i.e. RFGD treatment time at 0 s, the physically deposited surfactants are completely removed from the CHCl$_3$-washed surfaces; these washed surfaces are similar to the untreated LDPE control. With RFGD treatment, on the other hand, significant increases in the surface O/C and the surface wettability (water contact angle <30°) at all treatment times (15–300 s) were found. These reveal that the PEO surfactants are retained on the RFGD-treated surfaces. Even after extensive washing in chloroform for 4 days, the surfaces remained the same, i.e. high O/C ratios (0.29) and low water contact angles (<30°). This strong retention of the surfactants on the treated surfaces

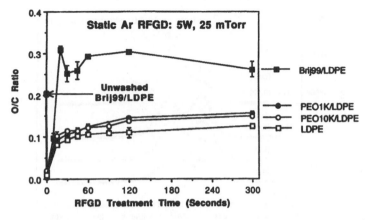

Figure 3. ESCA O/C atomic ratios for LDPE, PEO/LDPE, and Brij99/LDPE surfaces after Ar RFGD treatment and washing in chloroform [26]. Number of samples, *n*, equals 3.

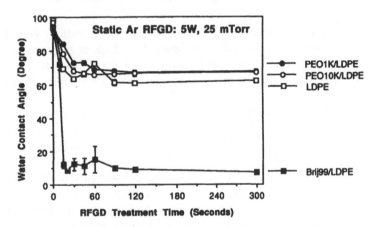

Figure 4. Advancing water contact angles on LDPE, PEO/LDPE, and Brij99/LDPE surfaces after Ar RFGD treatment and washing in chloroform [26]. Number of samples, *n*, equals 5.

suggests that the immobilization of PEO surfactants is probably not due to physical adsorption and/or surface entrapment.

In addition, the ether carbons in the high-resolution C $1s$ spectra on the RFGD-treated/CHCl$_3$-washed Pluronic/LDPE surfaces increase with increasing chain length of PEO in the surfactants, as shown in Fig. 5 [15]. This is another indication of the presence of the surfactants on the treated/washed surfaces. In contrast, only minor increases in surface oxidation and wettability were observed on the RFGD-treated CHCl$_3$-washed LDPE and PEO/LDPE control surfaces. The latter clearly demonstrates that PEO homopolymers cannot be immobilized by this method. This inefficient immobilization of PEO homopolymers suggests that the glow discharge immobilization of the PEO surfactants on the LDPE films may be through the hydrophobic segments rather than the PEO segments.

Both static (without argon gas flow) and dynamic (with argon gas flow) RFGD have also been compared under the same treatment power and reactor pressure.

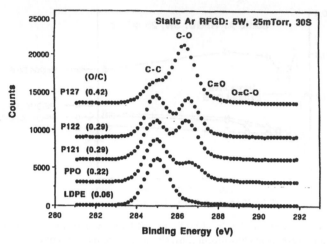

Figure 5. ESCA high-resolution C 1*s* spectra for LDPE, PPO/LDPE, and Pluronic/LDPE surfaces after Ar RFGD treatment for 30 s and washing in chloroform [15]. P denotes Pluronic.

However, no significant differences in the surface O/C ratios were found for either the LDPE or the Brij99/LDPE surfaces. This lack of influence of gas flow in the surfactant immobilization is due to the inert properties of the argon gas, which is not consumed by the surfactant or the LDPE film during RFGD treatment.

The mechanisms responsible for the surfactant immobilization were also investigated [31]. A study of surfactants on gold substrates was designed to examine self-crosslinking in the glow–discharge–treated surfactants

An argon gas discharge treatment at 2.5 W and 25 mTorr was applied to the Brij99/Au surface for 30 s. The treated surfaces were then washed in chloroform for various soaking times. ESCA results indicated that the treated Brij99 could be removed from the Au surface only when an overnight soak in chloroform was used, while the untreated surfactant was completely removed in a 2 h wash. These results suggest that self-crosslinking does occur in the treated Brij99 and causes a reduction of its solubility in chloroform. However, because this self-crosslinked Brij99 can eventually be removed by soaking overnight, we conclude that self-crosslinking may not be the major mechanism for the glow discharge immobilization of the surfactant. Crosslinking between the treated (self-crosslinked) surfactant and the polymer substrate may occur and thus permanently immobilize the PEO. In addition, this study also indicates that the established washing protocol is effective in removing both the deposited Brij99 and the crosslinked surfactant.

Protein adsorption on RFGD-treated/CHCl$_3$-washed LDPE and Brij99/LDPE surfaces was studied using ^{125}I-labeled baboon fibrinogen. The results are shown in Fig. 6 [26]. Fibrinogen adsorption on the RFGD-treated/CHCl$_3$-washed control LDPE films increased as both the treatment time and the treatment power were increased, probably as a result of the increasing surface oxidation suggested by the ESCA results. On the other hand, the RFGD-treated/CHCl$_3$-washed Brij99/LDPE surfaces exhibited a significant reduction in fibrinogen adsorption when a short treatment was used (less than 30 s). Similar observations were also

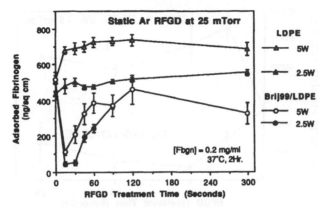

Figure 6. Fibrinogen adsorption on LDPE and Brij99/LDPE surfaces after Ar RFGD treatment and washing in chloroform [26]. Number of samples, *n*, equals 3.

noted for the RFGD-treated/CHCl₃-washed Pluronic127/LDPE surfaces, as shown in Fig. 7 [15]. Also, as expected, the non-fouling properties of the treated surfaces are enhanced when longer PEO chains in the surfactants are used (see Fig. 7).

The results of the protein adsorption studies reveal that the PEO segments in the RFGD-immobilized surfactants exhibit non-fouling properties. Taken together with the previously described ESCA and water contact angle results, the protein adsorption studies support the proposed crosslinking of the PEO surfactants to the LDPE surface molecules via the alkyl segment of Brij99 surfactants or via the PPO segment in the Pluronic surfactants.

However, when the RFGD treatment time is prolonged, fibrinogen adsorption increases on the treated surfaces. A similar trend is also observed in platelet adhesion to the treated surfaces (see Fig. 8). SSIMS was used to investigate the

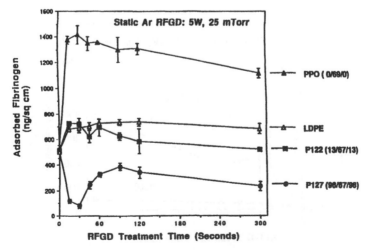

Figure 7. Fibrinogen adsorption on LDPE and Pluronic/LDPE surfaces after Ar RFGD treatment and washing in chloroform [15]. Number of samples, *n*, equals 3.

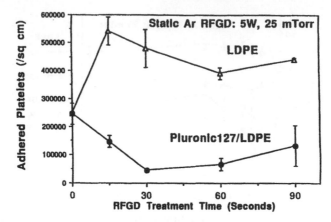

Figure 8. Platelet adhesion to LDPE and Pluronic/LDPE surfaces after Ar RFGD treatment and washing in chloroform. Number of samples, n, equals 3.

possible structure changes of the PEO in the RFGD-treated surfactants [30]. Figure 9 shows the PEO index of the treated/washed surfaces as a function of the RFGD treatment time. On the treated Brij99/LDPE, a maximum in the PEO index at 30 s treatment is seen after $CHCl_3$ washing. Interestingly, this curve is a mirror image of the protein adsorption and platelet adhesion curves on the treated/washed Brij99/LDPE surfaces. When the treatment time was prolonged to 120 s, the PEO index of the treated/washed Brij99/LDPE decreased. This revealed that the relative amount of PEO chains to hydrophobic tails on the surface decreased and suggested that the PEO chains in the treated surfactant were degrading. However, when treated for 300 s, the PEO index slightly increases again, which may have been due to the RFGD oxidation of the alkyl tails or the LDPE. The results from the SSIMS study suggest that the increases in protein adsorption and platelet adhesion at longer plasma treatment times are mainly due to the argon-plasma-induced degradation and oxidation of the PEO chains.

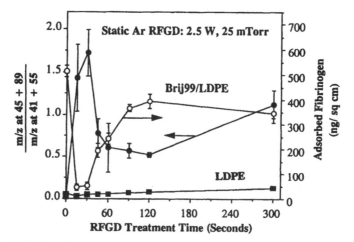

Figure 9. Static SIMS and fibrinogen adsorption for the RFGD-treated Brij99/LDPE surfaces after washing in chloroform [31]. Number of samples, n, equals 3.

4. CONCLUSION

Polyethylene with improved surface wettability and non-fouling (protein- and platelet-resistant) properties is obtained using a short, low power Ar RFGD treatment on the surface that has been precoated with a PEO surfactant. This glow discharge immobilization process has also been applied to functionalize polymer surfaces with sulfate groups and primary amines, when sodium dodecyl sulfate and decyl amine hydrochloride were used as the precoatings, respectively [32, 33]. The simple glow discharge process developed in this study may have wide applicability for modifying polymer surfaces in general and biomaterial surfaces in particular.

Acknowledgements

We thank NIH (GM 40111-2, 3, and 4) and the Washington Technology Centers for their financial support for this study. We also thank NESAC/BIO at the University of Washington (NIH, RR 01296) for the use of the ESCA and static SIMS facilities, and the Regional Primate Center at the University of Washington (NIH, RR 00166) for providing fresh baboon blood for the fibrinogen adsorption and platelet adhesion experiments.

REFERENCES

1. J. D. Andrade, S. Nagaoka, S. Cooper, T. Okano and S. W. Kim, *Trans. ASAIO* **33**, 75–84 (1987).
2. D. E. Gregonis, D. E. Buerger, R. A. Van Wagenen, S. K. Hunter and J. D. Andrade, *Trans. 2nd World Congr. on Biomater.* Washington, DC, pp. 266–267 (1984).
3. J. Hermans, *J. Chem. Phys.* **77**, 2193–2203 (1982).
4. Y. Mori, S. Nagaoka, H. Takiuchi, T. Kikuchi, N. Noguchi, H. Tanzawa and Y. Noishiki, *Trans. ASAIO* **28**, 459–463 (1982).
5. D. H. Randerson and J. A. Taylor, *Plasmapheresis, New Trends in Therapeutic Applications.* ISAO Press, Cleveland, OH (1983).
6. J. E. O'Mullane, C. J. Davison, K. Petrak and E. Tomlinson, *Biomaterials* **9**, 203–204 (1988).
7. J. H. Lee, J. Kopecek and J. D. Andrade, *J. Biomed. Mater. Res.* **23**, 351–368 (1989).
8. M. J. Bridgett, M. C. Davies and S. P. Denyer, *Biomaterials* **10**, 411–416 (1989).
9. N. P. Desai and J. A. Hubbell, *Biomaterials* **12**, 144–153 (1991).
10. N. P. Desai, S. F. A. Hossainy and J. A. Hubbell, *Biomaterials* **13**, 417–420 (1992).
11. W. R. Gombotz, G. H. Wang and A. S. Hoffman, *J. Appl. Polym. Sci.* **37**, 91–107 (1989).
12. S. W. Kim, H. Jacobs, J. Y. Lin, C. Nojori and T. Okano, *Ann. N.Y. Acad. Sci.* **516**, 116–130 (1988).
13. Y. H. Sun, W. R. Gombotz and A. S. Hoffman, *J. Bioact. Compat. Polym.* **1**, 316–334 (1986).
14. G. P. Lopez, B. D. Ratner, C. D. Tidwell, C. L. Haycox, R. J. Rapoza and T. A. Horbett, *J. Biomed. Mater. Res.* **26**, 415–439 (1992).
15. M.-S. Sheu, A. S. Hoffman and J. Feijen, *J. Adhesion Sci. Technol.* **6**, 995–1009 (1992).
16. A. Z. Okkema, T. G. Grasel, R. J. Zdrahala, D. D. Solomon and S. L. Cooper, *J. Biomater. Sci. Polym. Ed.* **1**, 43–62 (1989).
17. D. W. Grainger, S. W. Kim and J. Feijen, *J. Biomed. Mater. Res.* **22**, 231–249 (1988).
18. S. G. Wang, C. F. Chen, Z. F. Li, X. F. Li and H. Q. Gu, *J. Macromol. Sci. Chem.* **A26**, 505–518 (1989).
19. D. W. Grainger, C. Nojiri, T. Okano and S. W. Kim, *J. Biomed. Mater. Res.* **23**, 979–1005 (1989).
20. K. A. Dennison, Ph.D. dissertation, Massachusetts Institute of Technology, Cambridge, MA (1986).
21. F. Epaillard, J. C. Brosse and G. Legeay, *J. Appl. Polym. Sci.* **38**, 887–898 (1990).

22. C. I. Simionescu, F. Denes, M. M. Macoveanu and I. Negulescu, *Makromol. Chem. (Suppl.)* **8**, 17–36 (1984).
23. J. C. Brosse, F. Epaillard and G. Legeay, *Eur. Polym. J.* **19**, 743–747 (1983).
24. L. H. Sharpe and H. Schonhorn, *Adv. Chem. Ser.* **43**, 189–201 (1964).
25. R. H. Hansen and H. Schonhorn, *J. Polym. Sci. B, Polym. Lett. Ed.* **4**, 203–209 (1966).
26. M.-S. Sheu, A. S. Hoffman, J. G. A. Terlingen and J. Feijen, *Clin. Mater.* (in press).
27. R. J. Rapoza and T. A. Horbett, *J. Biomed. Mater. Res.* **24**, 1263–1287 (1990).
28. T. A. Horbett, *J. Biomed. Mater. Res.* **15**, 673–695 (1981).
29. D. X. Kiaei, A. S. Hoffman and S. R. Hanson, *J. Biomed. Mater. Res.* **26**, 357–372 (1992).
30. M.-S. Sheu, A. S. Hoffman, J. Feijen and J. M. Harries, in preparation.
31. M.-S. Sheu, A. S. Hoffman, B. D. Ratner and J. Feijen, in preparation.
32. J. G. A. Terlingen, J. Feijen and A. S. Hoffman, *J. Colloid Interface Sci.* **155**, 55–65 (1993).
33. J. G. A. Terlingen, L. M. Brenneisen, H. T. J. Super, A. P. Pijpers, A. S. Hoffman and J. Feijen, *J. Biomater. Sci.* (submitted).

Plasma Surface Modification of Polymers, pp. 209–217
M. Strobel, C. Lyons and K. L. Mittal (Eds)
© VSP 1994

Plasma surface treatment of poly(p-phenylene benzobisthiozol) fibers

Y. QIU*, S. DEFLON and P. SCHWARTZ†

Fiber Science Program, Department of Textiles and Apparel, Cornell University, Ithaca, NY 14853-4401, USA

Revised version received 24 March 1993

Abstract—Poly(p-phenylene benzobisthiozol) (PBZT) fibers were subjected to radio-frequency (RF)-induced, glow-discharge plasma treatments using argon and carbon dioxide gases in order to modify the adhesion of the fibers to bisphenol-A epoxy. The interfacial shear strength (IFSS) was used as a measure of the adhesion and was determined using the microbond technique. Scanning electron photomicrographs revealed no visible surface etching at magnifications of up to 10 000 ×. Slight, but statistically significant, improvements in IFSS were noted with the CO_2 plasma-treated fibers as compared with control fibers, but Ar plasma-treated fibers showed no improvement.

Keywords: PBZT; surface modification; plasma; glow discharge; microbond; interfacial shear strength; adhesion.

1. INTRODUCTION

The use of cold gas plasmas to modify the surface of polymers represents a valuable tool for the improvement of the fiber–matrix interface in composite materials. Depending on the plasma-characteristics, various surface reactions may be achieved and the interface may be tailored to maximize selected composite properties. Inert gases can enhance crosslinking [1] or cause oxidation on a fiber surface; this latter effect is the result of interactions between long-lived free radicals on the surface and the ambience. Surface oxidation leads to improved wetting of the surface by the resin. Oxidizing gases provide for etching or direct surface oxidation [2]. Etching of the fibers can improve adhesion by providing better mechanical interlocking between the fiber and the resin.

Poly(p-phenylene benzobisthiozol) (PBZT), whose structure is represented in Fig. 1, is one of the newer high-performance fibers considered for use in composite materials, especially for aircraft applications. While plasma surface

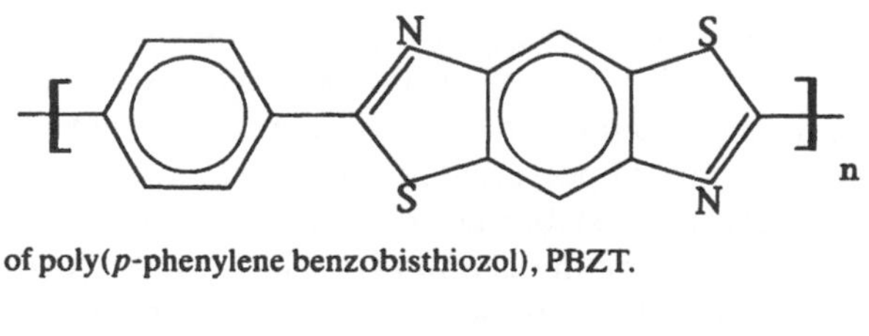

Figure 1. Structure of poly(p-phenylene benzobisthiozol), PBZT.

---

*Present address: Department of Mechanical Engineering, Massachusetts Institute of Technology, Cambridge, MA 02139, USA.
†To whom correspondence should be addressed.

modifications of other high-performance polymeric fibers, such as poly(p-phenylene terephthalamide) (PPTA) and co-poly(p-phenylene/3,4'-diphenyl-ether terephthalamide) (CPDT), have been reported in the literature [3], because of the relative newness of PBZT, very little has appeared to date regarding surface modifications of this polymer. Considerable work, however, has been done to characterize the adhesion between PBZT and epoxies. For example, Vratsanos *et al.* [4] reported temperature effects on the adhesion behavior of PBZT/epoxy composites, while others [5, 6], using single fiber pull-out results, have reported adhesion strengths in the 30–40 MPa range.

In this study we investigated the effects of inert and oxidizing gas plasmas on the interfacial shear strengths (IFSS) of PBZT/epoxy composites. We used argon, an inert gas, and carbon dioxide, an oxidizing gas, in our work. Using ESCA to study the surface of polyethylene (PE) fibers treated with argon and carbon dioxide plasmas, Hild and Schwartz [7] found an increase in oxygen on the surface (8.3% on the control vs. ~ 19% after Ar or CO_2 treatment), with a corresponding broadening of the C $1s$ peak relative to that of untreated polyethylene. After deconvolution, the presence of carbonyl and carboxyl groups was determined. Also, small amounts of nitrogen were found on fibers treated with carbon dioxide, and this was thought to be the result of residual air in the chamber.

The IFSS was measured using the microbond test [8]. As described by Herrera-Franco and Drzal [9], the microbond test is becoming the accepted characterization method for IFSS. In this test, a variation of the fiber pull-out test, a small droplet of resin is used rather than a large, thin coupon. The small size of the droplet reduces the number of tests lost due to fiber breakage prior to pull out.

2. EXPERIMENTAL

2.1. Materials

The PBZT fibers used in this work were produced at Wright-Patterson AFB, and were received in multifilament yarn form, each filament ~ 20 μm in diameter. The matrix resin was DER 331 (diglycidal ether of bisphenol-A, DGEBA) with 14.57 phr (parts per 100 parts) of a curing agent, DEH 26 (tetraethylenepentamine, TEPA), both manufactured by Dow Chemical Company.

2.2. Plasma treatment

An inductively coupled, radio-frequency plasma cleaner (13.56 MHz Model PDC-32G, Harrick Scientific Corporation) was used to treat the fibers. The cylindrical quartz plasma chamber had a volume of 630 cm³. The PBZT yarns were wrapped on a cylindrical polyethylene mesh frame and placed at the center of the chamber, leaving at least 0.5 cm clearance between the sample and the chamber wall. After evacuating the chamber to the ultimate level achievable with our system (≤ 100 mTorr), it was flooded with the desired gas until a pressure of 5 Torr was achieved, the latter an attempt to remove any remaining atmospheric gases. The gas flow rate was then adjusted to 200 cm³/min, using a flow meter, to provide a constant pressure of 250 mTorr before the RF-generator was switched on. Each group was treated for 5 min at a power level of 100 W. The maximum power level obtainable with our system was 100 W, and 5 min exposure was used as it provided good results in earlier treatments on aramid fibers [3] and polyethylene [7]. At the end of the treatment, the gas flow was stopped and the chamber was vented to the atmosphere.

2.3. Microbond test

Single fibers were selected at random from various locations on the frame in an attempt to reduce sampling bias. The diameters of the fibers were individually measured by means of a vibroscope (ASTM D1577-79) prior to applying the epoxy beads. The fibers were attached to a rigid frame with eight positions as described in ref. 10, and a bead of epoxy was applied to the test sample by bringing an epoxy-coated fiber into light contact, allowing resin transfer. Figure 2 is a photomicrograph of a typical microbond specimen showing a typical bead of epoxy on a PBZT fiber. On each of eight positions, three fibers from three groups, i.e. control, Ar-treated, and CO_2-treated, were mounted. The sequence of applying the beads on the three fibers were random so as to ensure a randomized complete block design (RBCD) [11] using positions as the blocks so that the potential variation of the IFSS created by the increase of resin viscosity over sample preparation time could be eliminated in the statistical analyses of the results. After the beads were applied, the specimens were cured at 80°C for 3 h. Using a light microscope, the beads were then inspected and measured to determine the embedment lengths.

The microbond tests were carried out at 21°C and 65% relative humidity using

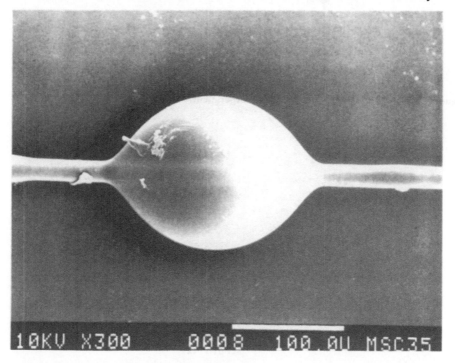

Figure 2. Typical microbond specimen prior to testing.

an Instron Model 1122. The cross head speed was 1 mm/min. A schematic diagram of the test procedure is shown in Fig. 3. In Fig. 4 we show a typical load–displacement trace for a microbond test. The IFSS, τ, is calculated as

$$\tau = \frac{F_{max}}{2\pi rl},$$

Where F_{max} is the peak load required to break the bond between the fiber and the bead, r is the fiber radius (10 μm), and l is the length of the bead.

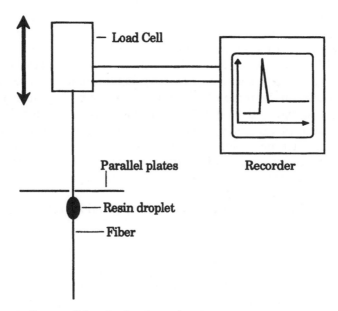

Figure 3. Schematic diagram of the microbond experiment.

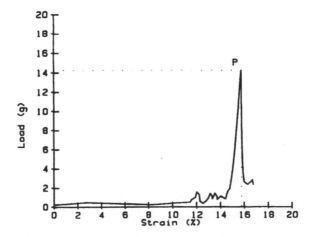

Figure 4. Typical trace in a microbond experiment.

3. RESULTS AND DISCUSSION

Scanning electron photomicrographs at 4000 × magnification (Figs 5–7) show no visible differences in the surfaces of treated and untreated fibers. Based on our previous work with PPTA [3], we would expect to see no differences in the fibers after either treatment. The plasma treatment with CO_2 was not harsh enough to produce any noticeable etching even when viewed at magnifications of up to 10 000 ×. If etching were taking place, it would seem to be on a submicrometer scale. On the left-hand side of Fig. 7, a kink band on the PBZT fiber can be seen. Rigid rod fibers such as PBZT and PPTA are especially vulnerable to buckling under small compressive loads and readily form kink bands. These affect the overall fiber strength and may affect the measured adhesion strength in the test when a bead encapsulates one or more kink bands. But their role in adhesion strength as measured by the microbond test or in a real composite is not clear. Gaur *et al.* [12] have presented wettability evidence using graphite fibers in epoxy to indicate possible 'submicroscopic cohesive failure (less than 100 nm thick) at the interface during fiber pull-out', and one might speculate that this effect may be present where a kink band occurs. Because we did not inspect the specimens for kink bands prior to testing, our results can provide no data about this phenomenon.

To detect the differences among the control and two treatments as regards IFSS, covariance analysis of the peak load, with the embedment area as covariate and the positions as blocks, was employed. As described above, the block design was chosen so as to account for the possible effect of the viscosity difference of

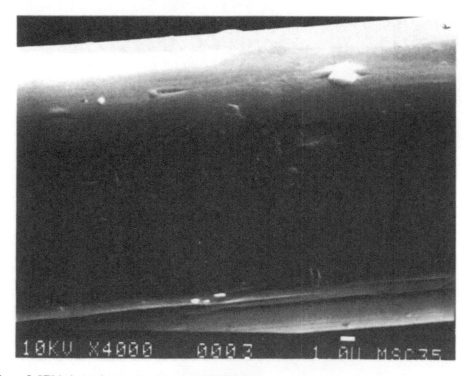

Figure 5. SEM photomicrograph of a control PBZT fiber.

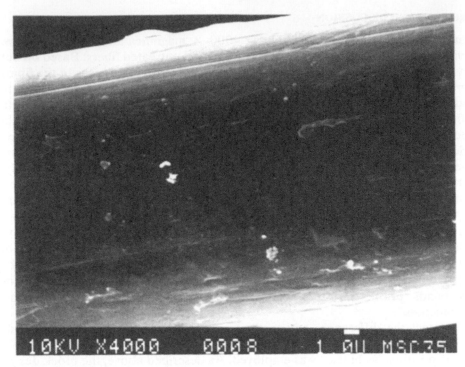

Figure 6. SEM photomicrograph of an argon-treated PBZT fiber.

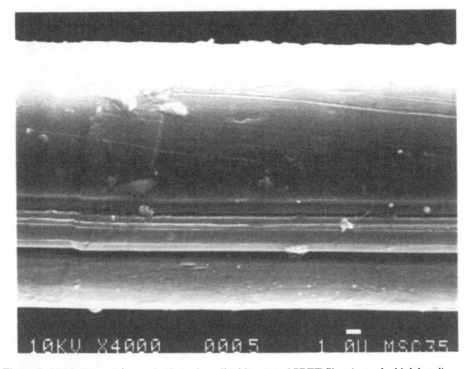

Figure 7. SEM photomicrograph of a carbon dioxide-treated PBZT fiber (note the kink band).

the epoxy when the beads were applied on the fibers on different positions; the fibers on one position were taken as one block. From this analysis we found that the time of application of the resin (total epoxy pot time ~ 0.5 h) *did* affect the peak load ($p < 0.05$) given the *same embedment area*, indicating an effect due to the increase in resin viscosity over time. When this is accounted for, however, no differences ($p = 0.39$) in average peak load were seen among the control and plasma-treated fibers, as can be seen in Table 1. In Fig. 8, where we plot the relationship between the measured peak load and the embedment area for all of the samples, we see (i) a positive correlation between the peak load and the embedment area and (ii) that all of the data are well intermingled, as we would expect if there were no differences in peak load across the treatments.

Dividing the peak load by the embedment area to get τ, we find that the mean IFSS (Table 2) for the CO_2-plasma-treated specimens (21.1 MPa) was higher ($p < 0.01$) than those of the control (18.3 MPa) and Ar-treated (17.8 MPa) samples, with no differences seen between the latter two ($p < 0.0001$). As there were no visible signs of etching at magnifications up to 10 000 ×, and as CO_2 treatment of PE produced an increase in carbonyl and carboxyl groups on the surface [7], we postulate that these groups were present on the surface of PBZT

Table 1.
Mean values for the peak load

Treatment	μm	Mean peak load (mN)	Standard deviation (mN)
Control	26	182	6.8
Ar plasma	16	180	11
CO_2 plasma	22	177	6.2

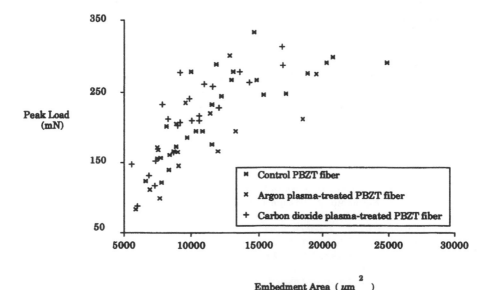

Figure 8. Relationship between the peak load and the embedment area.

Table 2.
Mean values for the interfacial shear strength

Treatment	μm	Mean IFSS (MPa)	Stables (MPa)
Control	26	18.3	3.6
Ar plasma	16	17.8	4.3
CO_2 plasma	22	21.1	4.0

and were responsible for the improvement in mean IFSS. In the same work, Hild reported that there was approximately the same oxygen increase in PE fibers exposed to Ar, and we could thus expect to see a similar increase in IFSS for Ar plasma-treated PBZT as well. Also, in our work with PPTA and CPDT [3], we obtained similar results with Ar and CO_2. We believe that due to its structure, there are fewer free radicals formed on PBZT exposed to the inert Ar as opposed to the oxidizing CO_2. PPTA and CPDT, also with ring structures, have an accessible amide link ($-CO-NH-$) which is missing in PBZT, while in PE there are many more C$-$H groups on the surface.

4. CONCLUSIONS

For the treatment conditions above:
(a) argon and carbon dioxide plasma treatments at 100 W for 5 min did not noticeably etch the surface of PBZT fibers when viewed at magnifications up to 10 000 ×;
(b) argon plasma-treated fibers did not show a significant difference ($p < 0.0001$) in mean IFSS over that of control fibers; and
(c) carbon dioxide plasma-treated fibers showed a significant increase ($p < 0.01$) in mean IFSS over that of control fibers, although the surface was not noticeably changed.

Acknowledgements

We would like to thank Professor Satish Kumar, School of Fiber and Textile Engineering, Georgia Institute of Technology, for providing the fibers used in this study. The resin and hardener were supplied by Dow Chemical Company. This research was funded in part through grant No. NYS-329424, Cornell University Agricultural Experiment Station.

REFERENCES

1. R. H. Hansen and H. Schonhorn, *J. Polym Sci. Polym. Lett. Ed.* **4**, 203 (1966).
2. R. Yosomiya, K. Morimoto, A. Nakajima, Y. Ikada and T. Suzuki, *Adhesion and Bonding in Composites*. Marcel Dekker, New York (1990).
3. K. Küpper and P. Schwartz, *J. Adhesion Sci. Technol.* **5**, 165 (1991).
4. M. S. Vratsanos, E. L. Thomas and R. J. Farris, in: *Composite Interfaces*, H. Ishida and J. L. Koenig (Eds), p. 151. Elsevier, Amsterdam (1986).
5. J. F. Mammone and W. C. Uy, U.S. Air Force Technical Report AFWAL-TR-82-4154 (1984).
6. E. C. Chenevy, U.S. Air Force Technical Report AFWAL-TR-82-4194 (1985).
7. D. N. Hild and P. Schwartz, *J. Adhesion Sci. Technol.* **6**, 879 (1992).
8. B. Miller, P. Muri and L. Rebenfeld, *Compos. Sci. Technol.* **28**, 17 (1987).

9. P. J. Herrera-Franco and L. T. Drzal, *Composites* **23**, 2 (1992).
10. Y. Qiu and P. Schwartz, *J. Adhesion Sci. Technol.* **5**, 741 (1991).
11. G. W. Snedecor and W. G. Cochran, *Statistical Methods*, 7th edn. Iowa State University Press, Ames, IA (1980).
12. U. Gaur, C. T. Chou and B. Miller, *Compos. Sci. Technol.* (in press).

Plasma Surface Modification of Polymers, pp. 219–230
M. Strobel, C. Lyons and K. L. Mittal (Eds)
© VSP 1994

CO$_2$ plasma modification of high-modulus carbon fibers and their adhesion to epoxy resins

RONALD E. ALLRED[1],[*] and WARREN C. SCHIMPF[2]

[1]*Adherent Technologies, 9621 Camino del Sol NE, Albuquerque, NM 87111, USA*
[2]*Hercules Research Center, Wilmington, DE 19894, USA*

Revised version received 18 October 1993

Abstract—Interfacial bond strength is often a performance-limiting factor of carbon-fiber-reinforced composites. This limitation is most prevalent when higher-modulus fibers or relatively unreactive matrix resins, such as engineering thermoplastics or high-temperature thermoset resin systems, are used. Radio-frequency (RF) glow discharge plasmas are an effective means of modifying carbon-fiber surface chemical characteristics to promote adhesion. It has been previously shown that oxidizing plasmas are especially effective compared with electro-oxidative treatments for treating carbon fiber surfaces as revealed by titrations, electron spectroscopy, wetting, and inverse gas chromatography measurements. This study evaluated the effectiveness of CO$_2$ plasmas on two experimental high-modulus carbon/graphite fibers and correlated the plasma surface modification with interfacial adhesion in an epoxy matrix composite system. The results show that CO$_2$ plasma treatment increased the surface oxygen content by nearly a factor of 2 over typical electro-oxidation treatments. The increased oxygen is mainly in the form of hydroxyl, ketone, and carboxyl-like moieties. Unidirectional composites were prepared from as-received and plasma-modified versions of each type of experimental fiber. The composites containing plasma-modified filaments exhibited 1.5–3.0 times the strength of composites fabricated with untreated or electro-oxidized filaments in transverse-flexural tests. Short-beam shear strength increased by two times over those with as-produced filaments and is equivalent to that of composites containing electro-oxidized filaments.

Keywords: Adhesion; carbon fibers; plasma treatments; carbon dioxide; interface.

1. INTRODUCTION

Approaches for improving the interfacial bonding between carbon fibers and polymer matrices can be based on three fundamental adhesion mechanisms or combinations thereof: (1) mechanical interlocking, (2) acid–base interactions, or (3) chemical-bond formation through reactive intermediates.

With the exception of fabricating filaments with complex shapes, mechanical interlocking cannot be used to any great extent to improve adhesion without extensively degrading the desirable bulk properties of the filament. Acid/base bonding can be an effective means of promoting interfacial adhesion [1–3], but may be adversely affected by environmental moisture. Chemical-bond formation is more difficult to achieve and, when possible, is usually based on a particular polymer chemistry.

---

*To whom correspondence should be addressed.

Surface treatments for carbon and graphite fibers are generally based on electrolytic oxidation processes in acidic or basic solutions that oxidize the fiber surface and remove the weakly bonded material formed during the carbonization cycle [4, 5]. These treatments have been optimized for bonding to epoxy resins, and, for lower-modulus carbon fibers, display adequate adhesion with most epoxy-based systems. Electro-oxidative treatments form a wide range of acidic and basic oxygen-containing moieties on carbon surfaces, including ethers, hydroxyls, lactones, carbonyls, carboxylic acids, and carbonates [6]. The electro-oxidative process cannot be easily controlled to the extent that only a particular surface species is emphasized. For this reason, electro-oxidative processes cannot be tailored for forming acid/base or chemical bonds with matrix resins.

RF glow discharge plasmas offer a surface modification technique that can overcome the inherent limitations of the electro-oxidative process. There is a large volume of literature on the plasma modification of solid surfaces; [7] and [8] offer a compendium of background information on plasma technology and plasma–surface interactions. Recently, a number of articles on the plasma treatment of carbon have appeared in the literature.

Early references to the plasma treatment of carbon fibers are found in the patent literature [9–13]. These patents entail the use of oxidizing gases (NO_2, O_2, SO_2, CO_2, NO, H_2O, and air) or ammonia. In the literature, a number of studies on the plasma modification of carbon fibers began to appear in the mid-1980s, continuing to the present time [14–33]. These studies, in general, show that plasmas can effectively oxidize or aminate carbon surfaces, and that oxidizing plasmas are more aggressive and lead to greater improvements in adhesion than ammonia or nitrogen plasmas.

In the plasma process, far-ultraviolet radiation and ion impacts volatilize surface species from the solid substrate, creating surface-active sites. Weak boundary layers and contaminants are also removed by these mechanisms. The gas phase contains numerous reactive species including ions, electrons, free radicals, and molecules with a variety of electronically excited states. By the proper choice of gas and plasma operating parameters, particular moieties may be emphasized when functionalizing a surface. Plasma processes thus offer a technique to tailor carbon-fiber surface chemistry to interact with matrix resins without sacrificing fiber bulk properties.

The objectives of this study were to demonstrate the potential for improving adhesion in high-modulus carbon and graphite/epoxy composites using plasma surface modification techniques and to characterize the modified surface. CO_2 plasmas were examined for their effects on two Hercules Research Center experimental carbon fibers. The surface chemistry of the plasma-treated fibers was characterized using the electron spectroscopy for chemical analysis (ESCA) technique. Unidirectional composite laminates were fabricated and tested in transverse flexure and short-beam shear.

Our results show that plasmas are an effective technique for the modification of high-modulus carbon fibers and may be substituted for conventional treatments to effectively tailor filament surface chemistry and remove weakly bound material on as-produced fibers.

2. EXPERIMENTAL

2.1. Materials

Two types of carbon fiber supplied by the Hercules Research Center were examined: (1) an unsized polyacrylonitrile (PAN)-based HM-type (Young's modulus, $E = 379$ GPa; density, $\rho = 1.837$ g/cm^3; 12k tow), and (2) a nonround, pitch-based ultrahigh-modulus carbon fiber (UHM-type) sized with ethylene glycol ($E = 689$ GPa, $\rho = 2.116$ g/cm^3). Neither fiber was surface-treated, but some of Hercules' HM-type with their proprietary surface treatment were also included in the test matrix to a limited extent. The surface-treated fibers were designated HMS-type. The ethylene glycol size on the UHM-type fibers was removed with an acetone wash before the plasma treatments.

Epoxy resin components were obtained from Shell Chemical Co. as Epon 826 (diglycidyl ether of bisphenol A) and Curing Agent Z (eutectic blend of m-phenylene diamine and methylene dianiline). Laboratory chemicals and gases were standard research grade.

2.2. Plasma treatments

Carbon dioxide-based plasma treatments were selected in an attempt to form carboxyl and hydroxyl groups on the fiber surface. Plasma treatments were conducted using a research reactor designed to continuously treat fiber tows and rovings. The reactor consists of three separately controlled reaction zones 1 m long and 4 cm in diameter constructed of Pyrex tubing. The 12k carbon tows are introduced into the plasma chambers through a series of wire drawing dies that allow the reduced pressure necessary to generate a glow discharge plasma. Power from a Tegal Scientific 300 W RF generator operating at 13.56 MHz is capacitively coupled to copper-foil electrodes through an impedance-matching antenna tuner. Gas flow is maintained by MKS model 2159B mass flow controllers, thermocouple vacuum gages, and a butterfly valve on each of the three vacuum pumps. A detailed description of the plasma research reactor is given in [34].

With the carbon yarn in place, the reactor chamber was initially pumped down to a pressure of 40 mTorr before the introduction of CO$_2$ gas. The CO$_2$ plasma was run at 15–20 W RF power per zone at a pressure of 500 mTorr. At a pass-through rate of 2.5 cm/s, the total exposure time to the plasma was 120 s. The operating parameters were selected from earlier results on other types of carbon fibers [25, 30]. Unidirectional laminae impregnated with epoxy resin (prepreg), were fabricated by dipping the carbon yarns into an epoxy bath and winding on a 10 cm diameter and 30 cm wide drum. Prepregs containing untreated carbon fibers were fabricated in the same manner, but with no plasma in the reactor chamber.

2.3. Fiber characterization

As received, surface-treated (HMS-type), and CO$_2$-plasma-treated HM-type fibers were analyzed by ESCA as fiber bundles. Approximately 1 week elapsed between the plasma treatment and the ESCA analysis. Survey spectra were taken to examine the as-received surface chemical composition. High-resolution spectra were taken of the C_{1s} peaks to determine possible surface bonding states. The analyses were run at Rocky Mountain Laboratories, Inc. (Golden, Colorado) using a Surface Sciences SSX-100 spectrometer with a monochromatic Al K_α source. C_{1s} high-resolution peaks were averaged over 10–15 scans using a spot size of 600 μm^2 with a 1.0 eV flood gun. The spectra were not adjusted for any effects of surface charging.

A Gaussian curve-fitting routine was used to resolve the high-resolution photoionization peaks into components based on binding-energy references from model hydrocarbon compounds [35]. The approach used in the curve fitting was to assign beginning peak locations to binding energies for carbon-carbon bonds and various states of carbon bound to oxygen. Once those peaks were established, a peak for β-, γ-, and δ-shifted carbon atoms (referred to as β-shifted hereafter) was added to complete the curve fit.

2.4. Prepreg and laminate fabrication

Prepregs with all fiber types were fabricated by tow impregnation using the continuous plasma-impregnation facility [34]. Because the resin bath is in-line with the plasma reactor, only a few seconds elapsed between the plasma treatment and resin impregnation. Composite laminates were fabricated by hot-press molding in a U-shaped mold with interior dimensions of 10.2 × 15.2 cm. The mold was designed with bolt-on sides and ends to restrain the composite during pressing and for easy removal of the consolidated part. Restraining the epoxy-resin flow was necessary to control the volume fraction and void content. A thermocouple inserted in the laminate was used to control the thermal cure cycle.

Contact pressure was initially applied at room temperature to improve heat transfer and to remove air from the laminate stack. The Epon 826/Z matrix laminate was heated to 80°C at a rate 1°C/min. A pressure of 689 kPa was applied after the laminate reached 80°C; it was then held for 8 h before cooling to room temperature in 30 min. The laminates were post-cured in an air circulating oven at 180°C for 4 h.

Samples cut from each laminate were cross-sectioned normal to the fiber direction, polished, and then examined with optical microscopy for void content and fiber distribution. Only laminates with void contents less than 1% and a uniform fiber distribution were mechanically tested.

2.5. Mechanical testing

Two interface-sensitive tests were performed on the unidirectional plates to evaluate their relative mechanical strengths: transverse flexure and short-beam shear (SBS). Both tests were conducted on a screw-driven tension-compression machine with a 8800 N capacity load cell.

The transverse (90°) three-point flexure test was run with an aspect ratio (L/d) of 10, which was large enough to induce a tensile failure rather than a shear failure. Specimens were nominally 1.27 cm wide and spanned 1.5 cm between supports. A compression load was applied at a constant rate of 0.25 cm/min until failure occurred. The SBS test was run using as aspect ratio of 4. Other parameters were similar to the transverse-flexure tests.

3. RESULTS AND DISCUSSION

The objectives of this program were (1) to determine the effectiveness of the plasma treatments for modifying the surface chemistry of two experimental high-modulus carbon fibers, and (2) to determine how plasma modification affects the interfacial bond strength with an aromatic-amine-cured epoxy-resin system.

3.1. ESCA analysis of plasma-modified HM carbon fibers

To gain insight into the plasma interactions with the HM-type fiber surface, the fiber was analyzed for surface chemical composition in three conditions: (1) untreated as-received, (2) Hercules surface treated (HMS-type), and (3) CO_2 plasma treated. Relative elemental surface compositions derived from survey spectra are given in Table 1. The data given in Table 1 have a relative error of $\pm 10\%$.

The three fibers show carbon and oxygen as the primary surface components. There is no detectable nitrogen in any of the fibers, which is surprising since a small amount of residual nitrogen from PAN-based HM carbon fibers is normally present. A trace of Cl is seen on both the untreated and the plasma-treated surfaces. The plasma-treated surface also shows a trace amount of Na, and the fibers subjected to the Hercules' surface treatment show a trace of Ca. These are likely contaminants from handling or processing. The main effect of the surface treatments is to increase the surface oxygen content. The electro-oxidative treatment increases surface oxygen by three times over that of the as-produced fiber. The CO_2 plasma increases surface oxygen by five times over that of the as-produced fiber.

A potential advantage of using RF glow discharge plasma treatments is that surface species can be weighted towards acids, hydroxyls, etc. by proper selection of the gas and the plasma operational parameters. To compare the binding-energy states of the surface oxygen species, high-resolution C_{1s} photoionization peaks taken from the three fiber surfaces are shown in Figs 1–3. As seen in the nondescript high-binding-energy tail of the C_{1s} peaks, the surface chemistries of carbon fibers are quite complex with no well-defined peaks. Rather, there is a continuum of binding-energy states arising from a multitude of molecular structures and resonance states that diffuse the electron density throughout the surface species. Because there is no clear peak definition, it is difficult to obtain quantification of the types of surface moiety present.

The approach used here is to fit peaks to groups of known species having well-characterized binding-energy shifts using the same methodology for each spectrum so that relative comparisons can be made between fiber treatments. These peak

Table 1.
Surface chemical composition (in %) of carbon fibers

Fiber (unsized)	C	O	Na	Cl	Ca
HM-type (untreated)	97.7	2.3	n.d.[a]	< 0.1	n.d.
HMS-type (treated)	92.0	7.3	n.d.	n.d.	0.7
HM-type (CO_2 plasma treated)	87.5	12.0	0.4	0.1	n.d.

[a] n.d. = not detected.

assignments to the Gaussian curve-fitting routine are given in Table 2 along with
the component peak values. The results of resolving the spectra are also shown as
dashed lines in Figs 1–3. It can be seen that the complex electron fields surrounding
the surface carbon atoms lead to considerable broadening in the fitted peaks.

Broadening is due to perturbation of the electron field around carbon atoms in the
proximity of heteroatoms compared to the relatively unperturbed graphitic carbon
that displays a narrow peak. Such behavior is commonly seen for oxidized species;
for example, note the reference spectra for Si and SiO_2 given in [36]. Figures 1–3
also show a deviation from the peak fit in the low-binding-energy tail of the hydrocarbon peaks. That deviation has been shown to be due to the aromatic character of
graphite [37].

Examination of the data given in Table 2 shows that β-shifted peak values do not
agree with the sum of the fitted peaks at 286.4, 287.6, and 288.7 eV. The lack of
agreement is due to two sources. There is more error in fitting the β-shifted peaks
because of their close proximity to the main hydrocarbon peak, and some species
likely to be present on the surface do not have β carbons. Sample groups that do
not have β carbons are methoxy and methyl esters.

The C_{1s} component peaks given in Table 2 show that the increased oxygen im-
parted to the fiber surface by the plasma treatment is primarily in the form of hydroxyl, ketone, and carboxyl-like functionalities. Other groups that could also be
present on the fiber surface include epoxides, peroxy, and hydroperoxides. While
the electro-oxidative treatment also increases these moieties, it is not as effective as
the CO_2 plasma. CO_2 plasma treatment increases the surface oxygen concentration
by augmenting the surface oxygen species present at crystallite edges and defects
and possibly by attacking the basal planes of the graphite crystallites. The increased
oxygen resulting from plasma treatment has also been shown to correspond to high-
energy sites [25, 30]. Those high-energy sites are believed to play a primary role in
the adhesion between carbon fibers and polymer matrix resins.

Table 2.
Components of the C_{1s} high-resolution photoionization peaks

Equivalent shift groups	Binding energy (± 0.3 eV)	Untreated unsized HM-type	Treated unsized HMS-type	CO_2-plasma-treated unsized HM-type
		Percent of C_{1s} photoionization peak		
Primary C—C bond	284.6	72	69	66
β-shifted carbon from C—O	285.3	15	14	13
Hydroxyl, ether, ester (single bond)	286.4	6	6	9
Ketone, amide	287.6	2	2	4
Carboxyl, ester (double bond)	288.7	2	4	5
Aromatic shake-up, carbonate	290.7	4	5	4

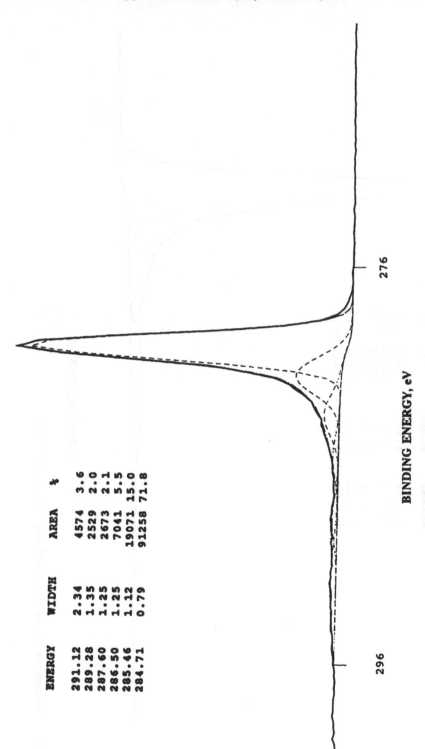

ENERGY	WIDTH	AREA	%
291.12	2.34	4574	3.6
289.28	1.35	2529	2.0
287.60	1.25	2673	2.1
286.50	1.25	7041	5.5
285.46	1.12	19071	15.0
284.71	0.79	91258	71.8

BINDING ENERGY, eV

Figure 1. High-resolution C_{1s} scan of the as-received HM-type carbon fiber.

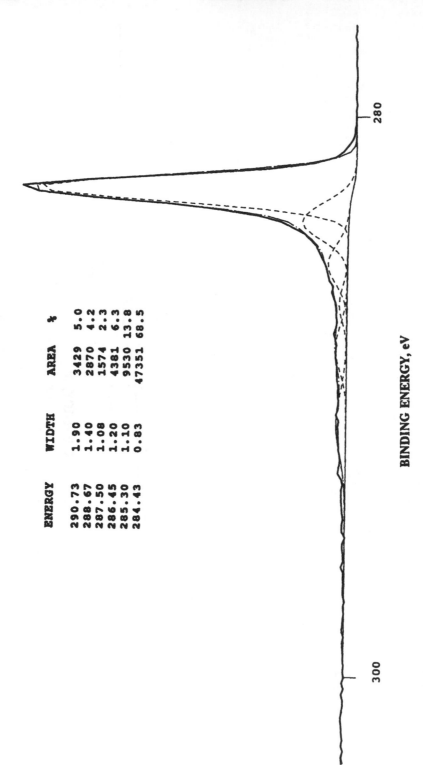

ENERGY	WIDTH	AREA	%
290.73	1.90	3429	5.0
288.67	1.40	2870	4.2
287.50	1.08	1574	2.3
286.45	1.20	4381	6.3
285.30	1.10	9530	13.8
284.43	0.83	47351	68.5

BINDING ENERGY, eV

Figure 2. High-resolution C_{1s} scan of the HMS-type (electro-oxidized) fiber.

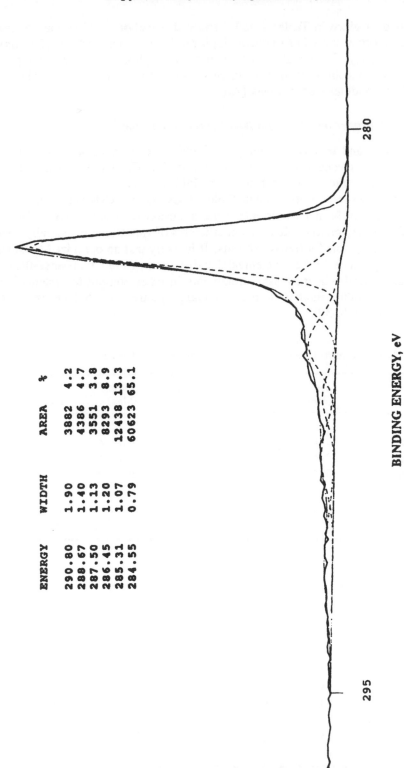

ENERGY	WIDTH	AREA	%
290.80	1.90	3882	4.2
288.67	1.40	4386	4.7
287.50	1.13	3551	3.8
286.45	1.20	8293	8.9
285.31	1.07	12438	13.3
284.55	0.79	60623	65.1

BINDING ENERGY, eV

Figure 3. High-resolution C_{1s} scan of the CO_2-plasma-treated HM-type carbon fiber.

The ESCA data given in Tables 1 and 2 show that carbon dioxide plasma treatments are an effective means of oxidizing high-modulus carbon surfaces. Carboxyl, ketone, and hydroxyl functionalities are increased by plasma treatment compared with conventional electro-oxidative treatments. Similar results have been obtained on intermediate-modulus carbon fibers [25].

3.2. Mechanical properties of carbon fiber/epoxy composites

Short-beam shear and transverse three-point bend (flexure) tests were selected as interfacial-sensitive mechanical tests for evaluating the effectiveness of the plasma treatments. The test results are summarized in Tables 3 and 4.

The transverse-flexural results given in Table 3 show that the CO_2 plasma treatment produces at least a 50% increase in interfacial adhesion with both carbon fibers in an Epon 826/Z epoxy matrix. Measured coefficients of variation are generally about 10%, which is acceptable for transverse tests. It is likely that an optimized treatment process would produce further incremental gains in measured transverse-flexural strength with reduced scatter. Plasma treatments in other composite systems have been observed to produce more uniform interfacial properties that produce less scatter in mechanical data [38, 39].

Table 3.
Transverse-flexural strengths of unidirectional carbon fiber/epoxy laminates

Fiber	S_{22} (MPa)	SD	Coefficient of variation (%)	Range (MPa)	No. of tests
Controls					
HM-type	21.0	2.2	10	17.3–24.7	12
HMS-type	29.6	5.6	19	20.8–36.7	8
UHM-type	30.6	3.1	10	26.8–37.4	10
CO_2-plasma-treated					
HM-type	46.4	4.9	11	38.6–57.6	10
UHM-type	47.6	4.4	9	43.2–52.6	6

Table 4.
Short-beam shear strengths of unidirectional carbon fiber/epoxy laminates

Fiber	S_{22} (MPa)	SD	Coefficient of variation (%)	Range (MPa)	No. of tests
Controls					
HM-type	20.0	0.4	2	19.6–20.5	8
HMS-type	64.0	1.9	3	61.5–66.9	11
UHM-type	42.0	5.7	14	34.6–50.4	5
CO_2-plasma-treated					
HM-type	67.1	1.2	2	64.8–68.8	12
UHM-type	66.6	4.6	7	62.3–72.6	6

The short-beam shear results given in Table 4 show similar, but less pronounced, trends to the transverse-flexural data. The CO_2 plasma treatment results in a significant increase in interlaminar-shear strength compared with the untreated controls; however, the plasma treatment is not statistically different from the electro-oxidative treatment (HMS-type) in short-beam shear strength. Low coefficients of variation were observed in all the data except the as-produced high-modulus fiber (UHM-type). The untreated HM-type fiber exhibited an extremely low short-beam shear strength of 20 MPa. It is likely that the shear strength is dominated by loosely bound material found on the as-produced fiber surfaces. Weak surface layers are commonly found on as-produced carbon fiber surfaces and require removal by some type of surface treatment before optimum composite properties can be attained [4, 5]. Both types of surface treatment examined here appear to be effective in removing this weak material on the fiber surface and promoting interfacial adhesion. It appears that after the weak boundary layer is removed, failure may be matrix-dominated in shear for this resin system or dominated by the shear strength of the filament skin. Either of these failure mechanisms would reduce the effects produced by altering interfacial adhesion and make both surface treatments appear to have similar effectiveness.

In summary, CO_2 plasma treatments have been shown to be effective for increasing the interfacial bond strength between two experimental high-modulus carbon fibers and an aromatic-amine-cured epoxy resin. CO_2-plasma surface chemical modification increases total oxygen content. The increased oxygen moieties appear to be in the form of hydroxyl and carboxyl groups on the filament surface that may form chemical bonds with the epoxy. It is expected that covalent bonds at the interface would be much more resistant to fatigue and environmental moisture.

4. CONCLUSIONS

The results of this study show that plasmas are an effective technique for the modification of high-modulus carbon fibers. Composites fabricated with CO_2-treated UHM-type fibers in an Epon 826/Z epoxy matrix showed a 56% increase in transverse-flexural strength and a 59% increase in short-beam shear strength compared with similar composites containing untreated fibers. Plasma treatment of HM-type fibers resulted in a five-fold increase in surface oxygen content as shown by ESCA. The majority of the surface oxygen corresponds to binding energies expected for hydroxyl, ketone, and carboxyl-like functionalities. Composites fabricated with CO_2-plasma-treated HM-type fibers displayed 120% higher transverse-flexural strengths and 236% higher short-beam shear strengths than those containing as-produced fibers. A Hercules' electro-oxidative surface treatment produced 41% and 220% gains, respectively. These results show that plasma surface treatments may be substituted for conventional treatments to effectively tailor filament surface chemistry and remove weakly bound material on as-produced fibers.

Acknowledgements

We would like to thank Rik M. Salas and Brent W. Gordon for preparation and testing of the composite specimens. Craig Butler, Rocky Mountain Laboratories, provided considerable assistance in the interpretation of the ESCA data.

REFERENCES

1. F. M. Fowkes, in: *Physicochemical Aspects of Polymer Surfaces*, K. L. Mittal (Ed.), pp. 583–603. Plenum Press, New York (1983).
2. L.-H. Lee, *J. Adhesion* **36**, 39–54 (1991).
3. K. L. Mittal and H. R. Anderson, Jr (Eds), *Acid-Base Interactions: Relevance to Adhesion, Science and Technology*. VSP, Zeist, The Netherlands (1991).
4. D. M. Riggs, R. J. Shuford and R. W. Lewis, in: *Handbook of Composites*, G. Lubin (Ed.), Ch. 11. Van Nostrand Reinhold, New York (1982).
5. E. Fitzer and R. Weiss, *Carbon* **25**, 455–467 (1987).
6. A. Ishitani, *Carbon* **19**, 269–275 (1981).
7. J. R. Hollahan and A. T. Bell (Eds), *Techniques and Applications of Plasma Chemistry*, pp. 113–147 and refs cited therein. John Wiley, New York (1974).
8. H. V. Boenig, *Plasma Science and Technology*. Cornell University Press, Ithaca, NY (1982).
9. J. C. Goan, US Patent 3 634 220 (1972).
10. K. C. Hou, US Patent 3 745 104 (1973).
11. J. C. Goan, US Patent 3 776 829 (1973).
12. H. Fujimaki *et al.*, US Patent 4 009 305 (1977).
13. S. Ueno and H. Kamata, US Patent 4 487 880 (1984).
14. H. H. Madden and R. E. Allred, *J. Vacuum Sci. Technol* **A4**, 1705–1707 (1986).
15. A. Benatar and T. G. Gutowski, *Polym. Composites* **7**, 84–90 (1986).
16. J. B. Donnet, M. Brendle, T. L. Dhami and O. P. Bahl, *Carbon* **24**, 757–770 (1986).
17. I.-H. Loh, R. E. Cohen and R. F. Baddour, *J. Mater. Sci.* **22**, 2937–2947 (1987).
18. J. B. Donnet, T. L. Dhami, S. Dong and M. Brendle, *J. Phys. D: Appl. Phys.* **20**, 269–275 (1987).
19. Y. Da, D. Wang, M. Sun, C. Chen and J. Yue, *Composite Sci. Technol.* **30**, 119–126 (1987).
20. B. Z. Jang, H. Das, L. R. Hwang and T. C. Chang, in: *Interfaces in Polymer, Ceramic and Metal Matrix Composites*, H. Ishida (Ed.), pp. 319–333. Elsevier, New York (1988).
21. J. B. Donnet, S. Dong, G. Guilpain and M. Brendle, in: *Interfaces in Polymer, Ceramic and Metal Matrix Composites*, H. Ishida (Ed.), pp. 35–42. Elsevier, New York (1988).
22. J. Su, X. Tao, Y. Wei, Z. Zhang and L. Liu, in: *Interfaces in Polymer, Ceramic and Metal Matrix Composites*, H. Ishida (Ed.), pp. 269–277. Elsevier, New York (1988).
23. M. Sun, B. Hu, Y. Wu, Y. Tang, W. Huang and Y. Da, *Composite Sci. Technol.* **34**, 353–364 (1989).
24. Y. Xie and P. M. A. Sherwood, *Appl. Spectrosc.* **43**, 1153–1158 (1989).
25. S. P. Wesson and R. E. Allred, in: *Inverse Gas Chromatography*, D. R. Lloyd, T. C. Ward, H. P. Schreiber and C. C. Pizana (Eds), ACS Symp. Series No. 391, Ch. 15. American Chemical Society, Washington, DC (1989).
26. R. E. Allred and L. A. Harrah, in: *Proc. 34th Int. SAMPE Symp. Exhibition*, pp. 2559–2568. Society for the Advancement of Material and Process Engineering, Covina, CA (May 1989).
27. H. Kawamura, T. Okubo, K. Kusakabe and S. Morooka, *J. Mater. Sci. Lett.* **9**, 1033–1035 (1990).
28. C. Jones and E. Sammann, *Carbon* **28**, 509–514 (1990).
29. C. Jones and E. Sammann, *Carbon* **28**, 515–519 (1990).
30. S. P. Wesson and R. E. Allred, *J. Adhesion Sci. Technol.* **4**, 277–301 (1990).
31. Y. Xie and P. M. A. Sherwood, *Appl. Spectrosc.* **44**, 797–803 (1990).
32. W. D. Bascom and W.-J. Chen, *J. Adhesion* **34**, 99–119 (1991).
33. P. Commercon and J. P. Wightman, *J. Adhesion* **38**, 55–78 (1992).
34. R. E. Allred, L. A. Harrah, R. M. Salas and B. W. Gordon, *US Army Laboratory Command Materials Technology Laboratory Report No. MTL TR 90–60*. Available from Defense Technical Information Center (1990).
35. D. T. Clark and H. R. Thomas, *J. Polym. Sci.* **16**, 791–820 (1978).
36. C. D. Wagner, W. M. Riggs, L. E. Davis, J. F. Moulder and G. E. Muilenberg, *Handbook of X-ray Photoelectron Spectroscopy*, p. 52. Perkin-Elmer Corp., Eden Prairie, MN (1979).
37. G. Barth, R. Linder and C. Bryson, *Surface Interface Anal.* **11**, 307–311 (1988).
38. R. E. Allred, E. W. Merrill and D. K. Roylance, in: *Molecular Characterization of Composite Interfaces*, H. Ishida and G. Kumar (Eds), pp. 333–375. Plenum Press, New York (1985).
39. R. E. Allred, in: *Proc. 29th Natl SAMPE Symp. Exhibition*, pp. 947–957. Society for the Advancement of Material and Process Engineering, Covina, CA (April 1984).

Plasma Surface Modification of Polymers, pp. 231–253
M. Strobel, C. Lyons and K. L. Mittal (Eds)
© VSP 1994

Oxygen plasma modification of polyimide webs: effect of ion bombardment on metal adhesion

F. D. EGITTO,* L. J. MATIENZO, K. J. BLACKWELL and A. R. KNOLL

IBM Corporation, IBM Microelectronics, Endicott, NY 13760, USA

Revised version received 14 January 1994

Abstract—Webs of Kapton 200-H and Upilex-S polyimide films were treated using oxygen plasma prior to sequential sputter deposition of chromium and copper in a roll metallization system. Two plasma system configurations were employed for treatment. In one configuration, the sample traveled downstream from a microwave plasma; in the other, the web moved through a DC-generated glow discharge. For the DC-glow treatment, the potential difference between the plasma and the web, ϕ_f, and relative ion densities, n_+, were measured at various values of chamber pressure and DC power using a Langmuir probe. Although samples treated downstream from the microwave plasma were not subjected to bombardment by energetic ions, ϕ_f for the DC-glow operating conditions was between 5 and 13 eV. For both films, advancing DI water contact angles of less than 20° were achieved using both modes of treatment. Contact angles for untreated films were greater than 60°. However, 90° peel tests yielded values of 15 to 20 g/mm for microwave plasma treatments and 40 to 60 g/mm for DC-glow treatment. Peel values for untreated Kapton and Upilex films were about 25 g/mm. High-resolution X-ray photoelectron spectroscopy in the $C1s$ region for Kapton film surfaces treated downstream from the microwave plasma showed increases in carbonyl groups, with concentrations inversely proportional to web speed. In contrast, DC-glow modification was due mainly to formation of carboxylates with a small increase in carbonyl component. It is proposed that treatment downstream from the microwave plasma results in formation of a weak boundary layer at the polyimide surface. Ion bombardment occurring in the DC-plasma configuration results in relatively more crosslinking at the polymer surface. Furthermore, adhesion between the sputter-deposited chromium and the DC-glow modified polyimide improved with increasing values of $\phi_f n_+$.

Keywords: Polyimide; chromium; microwave plasma; DC-glow discharge; Langmuir probe; crosslinking; XPS; contact angle; adhesion.

1. INTRODUCTION

Flexible circuits are now in common use as low-cost integrated circuit chip carriers. One strategy for reducing cost is to fabricate these carriers in a roll format. Polyimide films are commonly the preferred dielectric for this application because of their high thermal stability and good mechanical properties. One such polyimide is Kapton®-H film formed from pyromellitic dianhydride (PMDA) and oxydianiline (ODA) precursors. Another candidate material is Upilex®-S film (formed from

---

*To whom correspondence should be addressed.

®Kapton is a registered trademark of E. I. du Pont de Nemours & Co., Wilmington, DE.
 Upilex is a registered trademark of UBE Industries, Ltd., Tokyo, Japan.

biphenyl tetracarboxylic dianhydride/phenylene diamine, BPDA-PDA). This polyimide offers advantages in terms of lower moisture absorption and a coefficient of thermal expansion more closely matched to copper.

Properties of polyimide films depend on the fabrication method selected and commercial films like Kapton and Upilex polyimides may differ from films prepared using their precursors in a laboratory. In general, molecular segregation, molecular orientation, thermal degradation and crosslink density are properties that can vary from the surface to the bulk of the film. For example, opposite sides of a drum-cast polyimide film can differ in properties that affect film characteristics, e.g. surface roughness and solvent diffusion [1]. In addition, surface impurities, treatments employed during film making, and moisture content must be considered during metallization of the web. Surface impurities can result from low molecular weight silicones introduced by transfer of thermally decomposed silicone rubber rollers in curing ovens [2]. Kapton-HN, although a PMDA-ODA film, contains calcium hydrogen phosphate as a slip additive [3].

Metallization of polymer webs can be accomplished by a seed-and-plate operation, or by lamination of copper foil to the polyimide using an adhesive interlayer. These composites are known as two-layer and three-layer structures, respectively. Two-layer processing is particularly advantageous when metallization is required on both sides of the dielectric. Metal on one side is used as a signal layer while metal on the other side is commonly used as a ground plane to minimize electrical noise. Formation of through holes and vias for electrical connections between the two metal layers of the three-layer structure requires removal of the adhesive layer. Difficulties in removing conventional adhesives with environmentally sound solvents makes the three-layer composite less attractive for two-sided products. The seed layer required for electroplating of the two-layer structure can be deposited by a wet-seed technique or by using a sputter-deposited seed layer, such as chromium (as an adhesion promoter) followed by copper. The metal-coated roll can then be pattern-plated to form circuitry on one side and a ground plane on the other. Good line adhesion is particularly important for higher density packages that utilize fine-line circuitry since chemical attack during wet chemical processing can induce degradation of the metal-polymer interface at line edges. To achieve high seed-to-polymer bond strengths, it is common to treat polymer surfaces with an oxygen plasma [4]. This typically yields superior adhesion performance over conventional wet seeding techniques [5]. Oxygen plasmas have often been employed to improve wetting properties of polymer surfaces [6–11]. This has been proposed to result from formation of various polar groups such as $-C=O$, $-OH$ and $-COOH$ [6].

Among the components of oxygen plasmas that can interact with polymer surfaces are electrons, reactive neutral oxygen atoms, photons (including those in the highly energetic vacuum ultraviolet region), and energetic positive ions. All surfaces in the plasma system are at a lower electrical potential than the plasma. The voltage drop between the plasma and the solid surfaces occurs across a space charge sheath. Positive ions from the plasma are accelerated across the sheath toward the solid surface. If that surface is electrically isolated, the voltage drop is $\phi_f = (V_p - V_f)$, where V_p is the plasma potential and V_f is the floating potential. Typically, plasma-induced surface modification results from interaction with a combination of these components. However, each constituent can act alone to modify polymer surfaces.

For example, energetic ions have been shown to be sufficient to improve metal-polyimide adhesion. Pappas *et al.* [12] used 200 eV ion beams to modify surfaces of both PMDA-ODA and BPDA-PDA polyimides prior to sequential metallization with chromium and copper. With argon ions, PMDA–ODA films gave 90° peel strengths that were about twice the values measured for untreated polymer films. On the other hand, very little adhesion improvement was observed for BPDA-PDA films having the same treatment. For PMDA-ODA films treated with argon ions and subsequently by oxygen ions, or using only oxygen ions, peel values improved by a factor of two relative to the untreated film. These treatments resulted in a factor of four increase in peel values for the BPDA-PDA films. Hence, chemical structure of the polyimide must play a role in the modification and resulting adhesion.

Ultraviolet and vacuum ultraviolet radiation, abundant in most plasmas, is also sufficient to modify the surface of most polymers [13–15]. The effect of ultraviolet radiation on polyimide surfaces was demonstrated using an excimer laser operating at 248 nm by Kokai *et al.* [16]. They used X-ray photoelectron spectroscopy (XPS) to observe an increase of C−O groups, attributed to formation of polymer free radicals and subsequent interaction with air. The observed absence of nitrogen on these modified surfaces was indicative of volatile product formation. Heitz *et al.* [17] improved the adhesion of evaporated cobalt alloys to poly(ethylene terephthalate) using laser irradiation at 248 nm and 308 nm.

Oxygen-plasma treatment is commonly employed to enhance adhesion of vapor-deposited metal films to polyimide [18–22]. Katnani *et al.* [18] reported an increase with treatment time of the intensity of the C1s XPS peak attributed to −C=O on surfaces of thermally-cured PMDA-ODA treated with radiofrequency (RF) plasma in O_2. This work showed a correlation between the increase in this peak, surface wetting, and the adhesion of subsequently deposited chromium films.

Acid-base interactions have been shown to play an important role in adhesion [23, 24]. In a Lewis sense, polymers can be classified as acidic, basic, or neutral. Incorporation of carboxylic acid groups (−COOH) into a polymer imparts an acidic character to the surface whereas incorporation of amino groups (−NHR_2 or −NH_2R) or carbonyls (−C=O) imparts a basic character. Reactive metals, such as chromium, deposited in industrial systems that typically operate at 10^{-6} Torr or above are likely to oxidize in the gas phase [25, 26]. Chromium oxides are more acidic than chromium metal [27]. Therefore, electron donors such as carbonyl groups can interact more strongly with the oxide. In addition, the presence of oxygen-containing groups at the polymer surface leads to formation of metal-oxides. Burkstrand has shown that the formation of metal-oxygen-polymer complexes formed on oxygen plasma-treated polymer surfaces correlates with adhesion of the metal film [28]. Metals were observed to interact with hydroxyl (−OH), carbonyl (−C=O) and ester (−COOR) groups (typically more basic in nature than the untreated polymer, with respect to the metal) on the plasma-modified surface.

Polymer modification is sometimes performed in a configuration with films residing downstream and remote from the plasma region, such that substrates are not exposed to bombardment by energetic particles (ions, electrons, photons). Tead *et al.* [29] have proposed that for modification of polystyrene (olefinic backbone with aromatic side groups) in such a configuration, the relative amount of chain scission events to crosslinking events occurring during treatment is much greater than for modification performed in the presence of ion bombardment, as occurs during reactive ion etching and reactive ion beam etching. Crosslinking promotes improved cohesive strength in polymers and increased resistance to solvents and moisture diffusion. On the other hand, low molecular weight fragments resulting from excessive chain scission can result in a weak boundary layer leading to poor practical adhesion. Hence, treatments that improve the wetting properties of polymer surfaces are not always conducive to reliable adhesion of metal films, especially when the treatment results in the formation of a weak boundary layer. Burger and Gerenser [30] proposed that overtreatment of polymers in oxygen plasmas produced a greater amount of chain scission leading to a surface that was rich in low molecular weight material. This resulted in a weak boundary layer that was detrimental to metal/polymer adhesion. Hence, characteristics of surfaces well prepared for vapor deposition of metals include an abundance of favorable (in terms of acid-base type interactions with deposited-metal atoms) functional groups on the surface, with no treatment-induced degradation of the structural/mechanical integrity (e.g. cohesive strength) of the polymer surface.

Chin and Wightman [31] proposed that in addition to creating a more hydrophilic, polar surface, O_2 RF-plasma treatment of LARC-TPI® polyimide (Fig. 1) caused chain scission (at the C—N bond), leading to formation of a weak boundary layer that inhibited adhesion, measured by 180° peel tests with pressure sensitive adhesive. In this study, samples rested on an electrically insulated substrate [32] in a

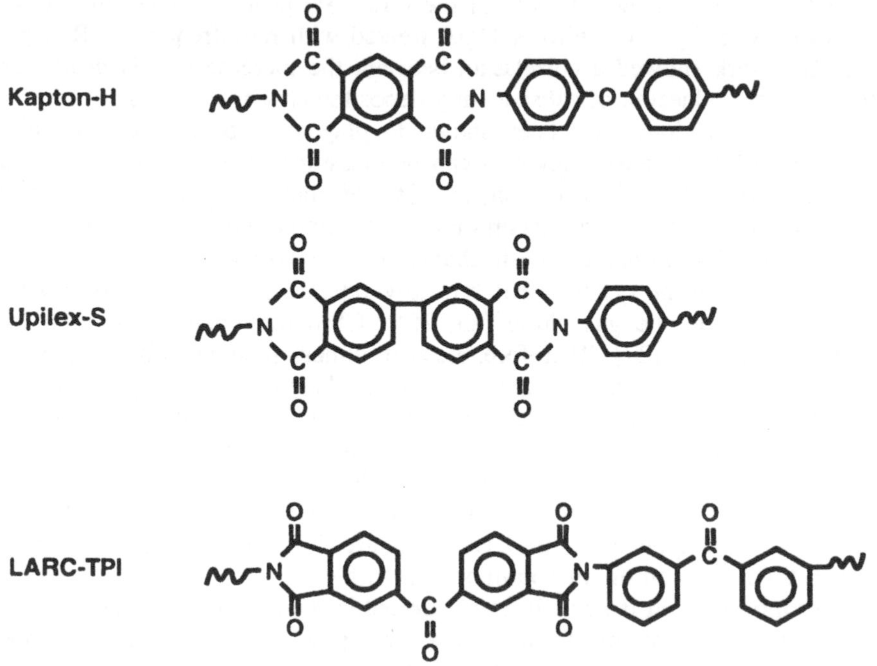

Figure 1. Chemical structures for the monomeric units of several polyimide films.

®LARC-TPI is a registered trademark of NASA-LaRC, Hampton, VA.

plasma operating at high pressures and relatively low power density. This configuration is conducive to relatively low levels of ion bombardment, i.e. low energy and number densities of ions striking the polymer surface. The presence of a weak boundary layer was consistent with the observation that washing of plasma-treated samples with methanol resulted in peel values greater than those observed for both untreated films and as-treated films. For comparison, high density polyethylene, a polyolefin, exhibited improved adhesion with plasma treatment. This was related to the propensity of polyolefin materials to undergo surface crosslinking upon exposure to the high energy radiation present in plasmas.

Matienzo and Egitto [33] observed that advancing DI water contact angles on surfaces of polyimides and poly(etheretherketone) (PEEK), that had been treated downstream from oxygen microwave plasmas, were increased by subsequent water rinsing. For PEEK, also an aromatic polymer, this reversion to less hydrophilic surfaces was accompanied by reappearance of the $\pi \rightarrow \pi^*$ shake-up satellite in high-resolution C1s XPS spectra (an indicator of multiple bonding) that had disappeared as a result of plasma treatment. This suggests that the weak boundary layer formed by the treatment has less aromatic character than both the untreated and washed film surfaces. The depth of modification was quite shallow, on the order of 3 nm or less.

DC-glow treatment is an alternative to RF and remote-microwave plasma processing techniques. A large negative bias is applied to an electrode in a vacuum chamber. Electrons ejected from this electrode collide with gas particles leading to ionization and dissociation into reactive neutrals. Reactive neutrals, ions, and photons all interact with the web surface to induce modification.

In roll metallization systems, plasma treatment is performed on a moving web of polymer. Web speed determines the residence time of material in the plasma treatment zone, and affects web heating rates. Although it has been shown that polymer film temperature can affect the rate of surface modification [33], there is no evidence that temperature affects surface composition once a steady state is achieved (fully-treated surfaces). Rate of treatment increases with increasing oxygen atom concentrations at the film surface. Means of increasing O atom concentration include increasing the amount of power absorbed by the plasma, increasing the linear flow velocity from the plasma to the sample [34], and addition of dopant gases, such as nitrogen [33, 35] to the O_2 gas feed. For both microwave (downstream) and DC-plasma configurations, power supplied to the plasma, gas pressure, gas flow rate, and gas composition (for mixtures) determine the number density of gaseous reactive species at the polymer surface. In addition, for DC plasmas these parameters determine the number density and energies of ions striking the web.

In the present investigation, treatment of Kapton-200H and Upilex-S films in two plasma reactor configurations, downstream from oxygen microwave plasmas and in DC-glow discharges, was performed with various plasma conditions (pressure, gas feed composition, and power). Treatment was evaluated using XPS, DI water contact angle measurements, and adhesion tests. Results of these analyses were correlated with measurements of oxygen atom concentration in the plasma. For the DC-glow configuration, adhesion values were correlated with measurements of floating potentials and relative ion densities in the plasma.

2. EXPERIMENTAL

2.1. Web materials

Rolls of Kapton-200H and Upilex-S films, 30.5 cm wide, were obtained from E. I. du Pont de Nemours & Co. and ICI Americas, respectively. Chemical structures of these polyimide films are shown in Fig. 1. All films were 50 μm thick.

2.2. Plasma treatment and roll metallization

Plasma treatment and in-line sputter deposition were performed in an Ulvac Corporation model SPW-030S roll metallization system shown schematically in Fig. 2a. Differentially pumped sections prevent oxygen from migrating from the plasma chamber to the sputtering chamber. Microwave and DC plasma configurations of the treatment station are shown in Figs 2b and 2c, respectively.

In one configuration, the web passed through a region downstream from an oxygen microwave plasma operating at 2.45 GHz. Microwaves were transmitted to the applicator through a rectangular waveguide; impedance matching was performed manually with a triple stub tuner and a sliding short. A three-port circulator and dummy load protected the generator from reflected power. Oxygen was admitted through a mass flow controller into the base of the microwave applicator. Effluent from the plasma region, including long-lived reactive neutrals, flowed through a quartz tube and into a manifold located in the treatment chamber. The manifold helped disperse this effluent across the web. Separation between the manifold and the web was maintained at 7.5 cm.

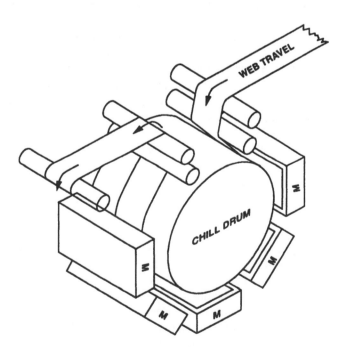

Figure 2a. Schematic diagram of the roll metallization system. *M* indicates a sputtering cathode.

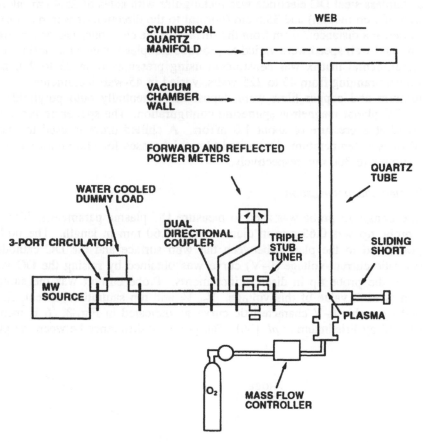

Figure 2b. Schematic diagram of the microwave plasma station. Web travel is in a direction normal to the plane of the page.

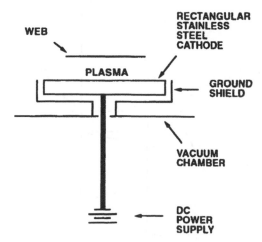

Figure 2c. Schematic diagram of the DC-glow station. Web travel is in a direction normal to the plane of the page.

The stainless steel DC electrode was rectangular with sides of 25.4 cm (along the direction of web motion) and 35.6 cm (normal to the direction of web motion). The web passed at a distance 7.6 cm from the surface of the electrode. DC power, oxygen pressure, and web speed were varied to produce samples for metal adhesion testing. The experimental matrix was constructed using pressures from 75 to 250 mTorr, with powers ranging from 45 to 225 watts, varied in 45-watt increments.

Chromium and copper films were deposited sequentially onto polyimide webs using a DC planar magnetron sputtering configuration. The sputtering process was performed at a pressure of about 1.0 mTorr. A chilled drum is used to maintain control of web temperature. Deposited-film thicknesses for chromium and copper were 20 nm and 300 nm, respectively.

2.3. Plasma characterization

A single Langmuir probe was used to measure DC plasma parameters. The cylindrical probe tip was 0.65 mm in diameter and 1.60 mm in length. The probe tip was positioned in the plasma between the web surface and the DC cathode. A characteristic current-voltage (I–V) curve was obtained by raising the DC voltage applied to the probe tip in discrete increments. Probe current was measured and recorded at each value of this voltage. V_p, V_f and ion saturation current, I_i, were obtained from this I–V characteristic curve as indicated in Fig. 3. I_i is measured as proposed by Friedmann *et al.* [36]. The potential difference between the plasma

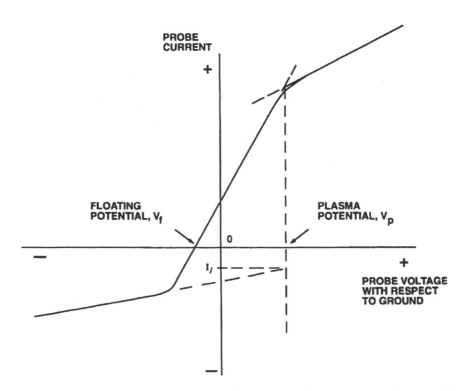

Figure 3. Representative Langmuir probe I–V characteristic curve.

and the web, ϕ_f, is given by $(V_p - V_f)$. The maximum energy of ions striking the web surface is, therefore, given by $q\phi_f$, where q is the ion's charge. A useful approximation applicable to cylindrical (and other) probe geometries states that the ion saturation current is proportional to $(1/2)A(n_+)(kT_e/M)^{1/2}$, where A is the area of the probe tip, (kT_e) is the electron temperature, and M is the mass of the ion [37]. T_e is inversely proportional to the slope of a plot of probe voltage vs. $\ln(I_e)$ in the region bounded by V_p and V_f, where I_e is the electron current. However, it is also true that T_e is proportional to ϕ_f [38]. Therefore, the total energy per unit time, or 'ion fluence', F_i, delivered to the web surface is proportional to $(\phi_f n_+)$ so that $F_i = \beta\phi_f I_i(\phi_f)^{-1/2}$, where β is a constant.

Optical emission from the plasma was monitored using a Jarrell-Ash model 1233 spectrometer and a Princeton Applied Research (PAR) model 1460 optical multi-channel analyzer configured with a PAR model 1456 detector. Relative oxygen atom number densities were determined by monitoring the emission spectral intensity from neutral atomic oxygen, OI at 845 nm. Argon was added in low concentrations for use as an actinometer [39, 40]. The emission line for neutral argon atoms, ArI at 750.4 nm, was monitored for this purpose.

Web temperature during plasma treatment was measured using chromel-alumel thermocouples (0.051 mm diameter) attached to the web by applying a bead of polyamic acid and curing at points in a line normal to the direction of web travel. These thermocouples were transported with the web through the plasma.

2.4. Surface analytical techniques

To obtain samples for contact angle measurement and XPS analysis, no metallization was performed. Advancing DI water contact angles on treated and untreated polyimide films were measured with a Ramé-Hart, Inc., model A-100 goniometer with optical protractor, using a sessile drop technique and a drop volume of 2 μl. Measurements were made immediately following removal of samples from the vacuum chamber. Contact angles were recorded within 30 s from initial application of the drop. Five measurements were made per sample. Values given in this report are the means of these measurements.

XPS analysis on microwave-treated samples was performed in a Perkin-Elmer Physical Electronics (PHI) model 560 instrument with a double-pass cylindrical mirror analyzer using $Mg\,K_\alpha$ rays for excitation. Survey and high-resolution spectra were collected with pass energies of 100 and 20 eV, respectively. Analysis of DC-glow treated samples was performed in a PHI-5500 Multiprobe spectrometer equipped with a hemispherical analyzer using monochromatized $Al\,K_\alpha$ rays for excitation. Survey and high-resolution spectra were collected with pass energies of 158 and 5.85 eV, respectively. For both instruments, binding energies were referenced to the hydrocarbon peak at 284.6 eV. Untreated samples were cleaned by successive rinses in acetone and methanol followed by drying in vacuum (at approximately 10^{-6} Torr). A comparison was made between the expected atomic ratios for carbon, oxygen, and nitrogen and the corresponding measured atomic concentrations for each of the untreated polymeric films, as reported in a previous paper [33]. High-resolution XPS spectra in the C1s region were used to determine the contributions due to different chemical environments, and to follow them as a function of plasma treatment.

2.5. Adhesion

The sputtered web was cut into 25.4 cm × 38 cm panels and copper plated to achieve a final copper film thickness of 7.6 μm. Rectangular masking decals, 1.0 mm wide, were placed on the panels for etching the copper in $FeCl_3$. $KMnO_4$ was used to remove the chromium layer. Final etched-line widths were on the order of 0.75 mm. An Instron model A1026 tensile tester was used to measure the force required to remove the line from the polyimide substrate, using a 90° peel test. Degradation of adhesion was measured on samples exposed to conditions of elevated temperature (85°C) and humidity (80% RH) in an environmental chamber for time periods of 100, 250, 750, and 1000 hours.

3. RESULTS AND DISCUSSION

Practical adhesion of metal films to Kapton-H and Upilex-S, as determined by 90° peel tests, is shown for the untreated polymers, with treatment downstream from an oxygen microwave plasma, and with treatment in a DC-glow discharge in Fig. 4. Microwave plasma treatment reduced practical adhesion levels for both Kapton-H and Upilex-S while DC-glow treatment produced adhesion about triple the value measured for untreated films. Analysis of polymer surfaces and characterization of the plasma environment are useful in understanding the dependence of resulting adhesion on the plasma system configuration.

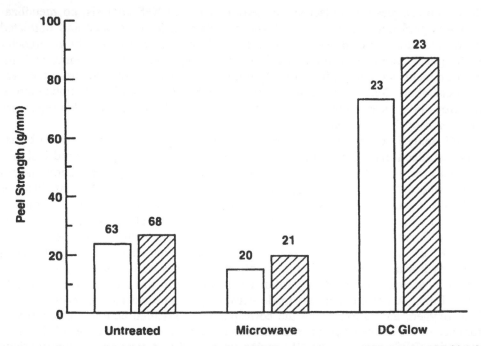

Figure 4. Peel strengths for Cu/Cr lines on films of Kapton-H (open bars) and Upilex-S (striped bars). Shown are results for the untreated polymers, and films treated using the two plasma configurations. Advancing DI water contact angles are shown above the bars.

3.1. Downstream microwave plasma treatment

High resolution XPS spectra of the C1s region for an untreated Kapton-H film and films treated at web speeds of 25, 100 and 500 feet per hour are shown in the inset of Fig. 5. The four curves are normalized to the highest peak in each spectrum. The concentration of carbonyl groups (peak ca. 288 eV) is inversely proportional to web speed.

DI water contact angles are shown for Kapton-H films treated at various web speeds in Fig. 5. The contact angle for untreated Kapton-H films was 63°. The trend with treatment time for contact angle change is typical of that observed in numerous previous investigations, with a rapid change for short treatment times (speeds greater than roughly 200 feet per hour) and a more gradual change for longer treatment (slower speeds).

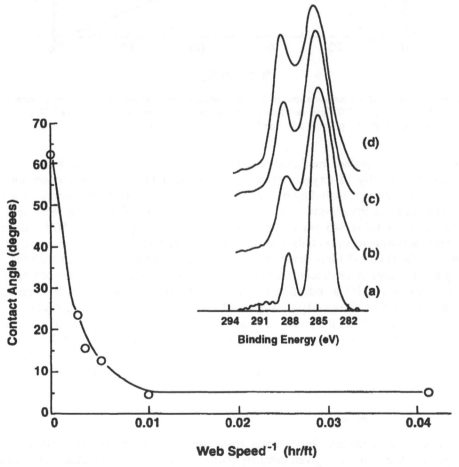

Figure 5. Advancing DI water contact angles on Kapton-H films treated at various web speeds downstream from an oxygen microwave plasma. The inset shows high resolution XPS spectra of the C1s region for an untreated Kapton-H film (a) and films treated at web speeds of 500 (b), 100 (c) and 25 (d) feet per hour.

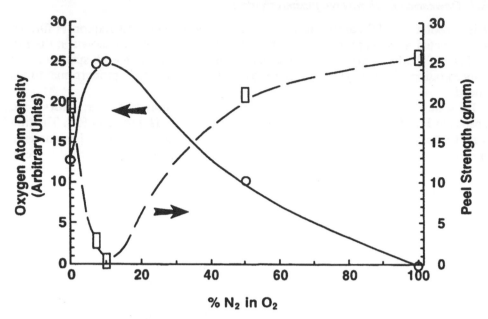

Figure 6. Intensity of the emission spectral line for neutral atomic oxygen, OI at 845 nm, corrected by actinometry using ArI at 750.4 nm, as a function of nitrogen concentration in the gas feed. Also shown is the resulting chromium–polymer adhesion for Upilex-S films treated using these gas mixtures.

The degree of surface modification depends not only on treatment time, but on the number density of oxygen atoms in the plasma [33]. As mentioned above, several gas additives, including nitrogen, can be used to increase the atomic oxygen concentrations in the plasma. Figure 6 plots the intensity of emission from neutral atomic oxygen, OI at 845 nm, corrected by actinometry using ArI at 750.4 nm, as a function of nitrogen concentration in the gas feed. Also shown is the metal–polymer adhesion for Upilex-S films treated using these gas mixtures. Consistent with Fig. 4, the greater degree of treatment resulting from enhancement of O atom concentrations in the plasma yielded a reduction in metal–polymer practical adhesion. The role played by reactive nitrogen species was not investigated in the present study.

These results suggest the presence of a weak boundary layer induced by treatment downstream from the oxygen microwave plasma. Formation of the weak boundary layer may be due to an excessive amount of chain scission (relative to crosslinking). Such behaviour has been observed for polystyrene treated downstream from oxygen microwave plasmas [29].

3.2. DC-glow treatment

As opposed to treatment downstream from the microwave plasmas, increases in oxygen atom density at a given pressure (constant web speed) resulted in enhanced adhesion for treatment using a DC-glow discharge. Figure 7 demonstrates that for a given chamber pressure, greater values of OI/ArI (achieved by increasing the DC power) resulted in enhanced peel strengths. However, for a given value of OI/ArI (at different pressures), widely varying values of adhesion are achieved. Adhesion

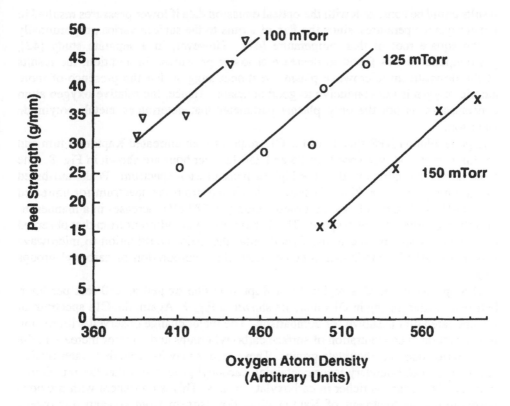

Figure 7. Line adhesion (for chromium on Kapton-H) as a function of oxygen atom density at various pressures and power levels used for DC-glow treatment. At a given pressure, greater power results in higher oxygen atom density.

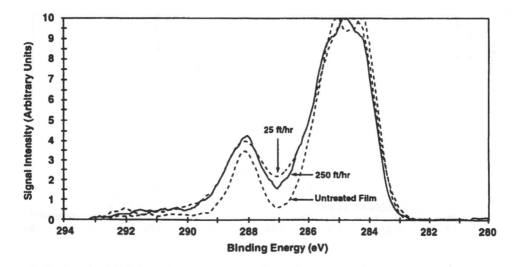

Figure 8. High resolution XPS spectra of the C1s region for an untreated Kapton-H film and films treated at web speeds of 25 and 250 feet per hour.

results could be consistent with the optical emission data if lower pressures resulted in greater gas temperatures, since the flux of atoms to the surface varies proportionally to the square root of that temperature [41]. However, in a separate study [42], gas temperature was found to decrease at lower pressures. In addition, the results of the downstream microwave plasma treatment suggest that the presence of more atomic oxygen is not conducive to good adhesion. Hence, the relative oxygen atom concentration is not the only plasma parameter that determines metal–polyimide adhesion.

High resolution XPS spectra of the C1s region for an untreated Kapton-H film and for films treated at web speeds of 25 and 250 feet per hour are shown in Fig. 8. The three curves are normalized to the highest peak in each spectrum. Peaks attributed to carbonyl groups (ca. 288 eV) show little change from the spectrum for untreated Kapton-H, while the carboxylate-group region (ca. 287 eV) increases in a manner inversely proportional to web speed. This behavior is quite different from that observed for the microwave treated films. Specifically, the major contribution to microwave plasma-treated film modification comes from the incorporation of carbonyl groups (Fig. 5).

XPS spectra in the C1s region for a Kapton-H film treated at 250 feet per hour, before and after rinsing in DI water, are shown in Fig. 9. Again, the C1s spectrum of the untreated film is shown for comparison. The major change observed with rinsing is a reduction in concentration of surface carbonyl groups and a minor increase in the contribution due to carboxylate groups. These results may indicate that water rinsing removes a weak boundary layer containing carbonyl groups and that the remaining, more stable, surface is richer in carboxylate groups. This is consistent with previous observations for treatment of Kapton films downstream from oxygen microwave plasmas [33]. In that study, the change in surface chemical composition with water

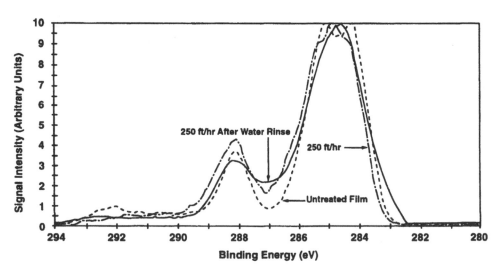

Figure 9. High resolution XPS spectra of the C1s region for a Kapton-H film treated at 250 feet per hour, before and after rinsing in DI water. The C1s spectrum of the untreated film is shown for comparison.

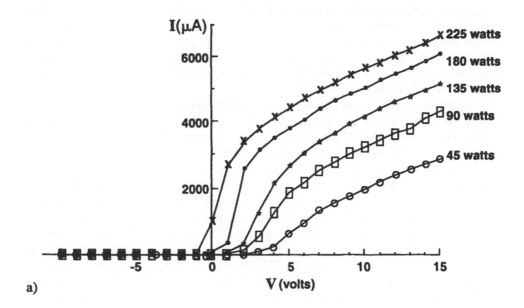

a)

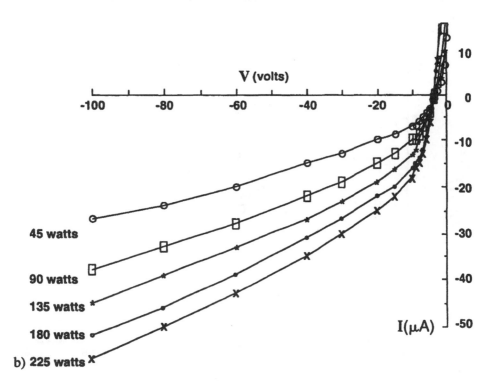

b) 225 watts

Figure 10. Langmuir probe I–V curves obtained at various powers at a pressure of 125 mTorr. In (b) the current axis has been enlarged and the voltage axis compressed in the ion current region.

rinsing was accompanied by an increase in DI water contact angle to 48°, regardless of the treatment duration and plasma parameters. XPS analysis of the rinsed Kapton surface indicated removal of carbonyl groups. In the present study, Kapton-H films treated using DC glow at low web speeds, such that steady state surface composition (and contact angle) were achieved, exhibited only a slight increase in contact angle, from 12° to 25°, upon water rinsing.

A similar tendency of ion bombardment toward formation of carboxylate groups on Kapton surfaces was demonstrated by Matienzo *et al.* [43]. Fourier transform infrared spectroscopy and XPS analysis revealed that reactions of high energy hellium ion beams (2 MeV) with Kapton-H films resulted in the formation of carboxylate groups by imide ring bond breaking reactions and a reduction in imide carbonyls.

Degree of ion bombardment during treatment of the web in the DC glow was quantified using a Langmuir probe to measure ϕ_f and $I_i(\phi_f)^{-1/2}$ (proportional to n_+). Langmuir probe I–V curves obtained at various powers at a pressure of 125 mTorr are shown in Fig. 10. Since ion currents are small compared to electron currents, an enlarged view of the curve in the ion current region is shown in Fig. 10b. Similar trends were observed for pressures of 75, 100 and 150 mTorr. To verify the linear relationship between ϕ_f and T_e, a plot of $\kappa I_i(\phi_f)^{-1/2}$ vs. $\alpha I_i(T_e)^{-1/2}$ is shown in Fig. 11. Each of these expressions is proportional to n_+. Here, α and κ are proportionality constants and T_e was calculated from the slope of the probe voltage vs. $\ln(I_e)$ curve.

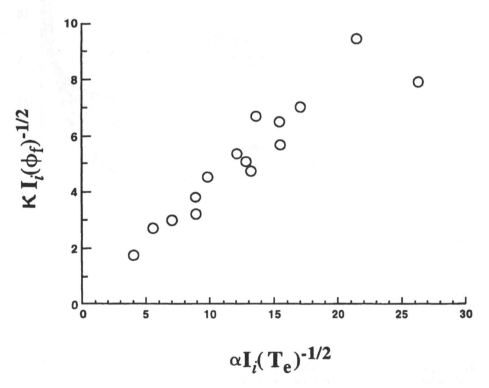

Figure 11. $\kappa I_i(\phi_f)^{-1/2}$ vs. $\alpha I_i(T_e)^{-1/2}$.

Peel strengths obtained at various values of pressure and DC power level are shown as a function of relative ion fluence, $\beta\phi_f I_i(\phi_f)^{-1/2}$, in Fig. 12a. In general, greater adhesion was obtained as the ion fluence to the surface was increased. Some dependence on pressure is observed. Specifically, for a given ion fluence, adhesion is greater at lower pressures. This pressure dependence can be the result of several factors. First, the surface flux of ions from the plasma into the sheath boundary (ions per unit area per unit time) is given by $1/4(n_+ v_+)$, where v_+ is the average velocity of ions in the plasma [41]. This velocity is proportional to $T_i^{1/2}$, where T_i is the ion temperature (much lower than T_e for the plasmas used in this study). The approximations used in the present probe analysis do not include a dependence on T_i. Greater ion temperatures (in the plasma) at lower pressures could enhance the flux of ions to the web, resulting in better adhesion. Second, although the probe analysis assumes a collisionless sheath (the equations for the cylindrical probe in the presence of collisions are somewhat unwieldy), at the pressures used in the present investigation, ions may lose some energy by way of collisions in traversing the space charge sheath. Therefore, the energy distribution for ions striking the web, very narrow at low pressures, may be spread toward lower energies at higher pressures such that the mean energy of ions striking the web surface will be somewhat less than ϕ_f. Third, the concentration of atomic oxygen in the plasma is greater at higher pressures (Fig. 7). As proposed by Burger and Gerenser [30], and as observed for the microwave plasma treatment, overtreatment in oxygen plasmas can degrade adhesion. Hence, enhanced oxygen concentrations at higher pressures could possibly reduce adhesion in this regime. In other words, although the presence of oxygenated polymer functional groups can promote adhesion, too much oxygen can be detrimental to adhesion, especially in the absence of energetic ion bombardment.

The first two phenomena can reduce the ion fluence to the sample by decreasing flux and average energy, respectively, of ions arriving at the web surface. The exact relationship between pressure, p, and ion surface flux at the sheath boundary or ion energy at the web are not known; a p^{-1} dependence will be assumed here. The corrected ion fluence to the web is therefore given by $\beta\phi_f I_i(\phi_f)^{-1/2}p^{-1}$. Adhesion is plotted as a function of this fluence in Fig. 12b.

These data suggest that adhesion is closely linked with ion dose, consistent with the work of Pappas *et al.* [12] using beams of oxygen ions at 200 eV to modify surfaces of PMDA-ODA and BPDA-PDA polyimides. Improved adhesion was observed for both polymers after ion beam pretreatment with a dose dependence observed for the latter material.

It is possible that the presence of ion bombardment inhibits the formation of a weak boundary layer. This may be due to an enhanced degree of crosslinking with ion bombardment, as suggested for polystyrene by Tead *et al.* [29]. Photon irradiation can also serve to modify polymers by virtue of crosslinking and chain scission. Webs treated in the DC glow are subjected to photon irradiation as well as ion bombardment. Such photon irradiation was absent in the treatment downstream from the microwave plasmas. The contribution of photons for promoting adhesion to films treated in the DC configuration of this study is not known. However, the correlation between ion fluence and adhesion strongly suggests that a sufficient degree of ion bombardment present at the surface of samples placed in the plasma imparts polymer-surface properties conducive to good adhesion.

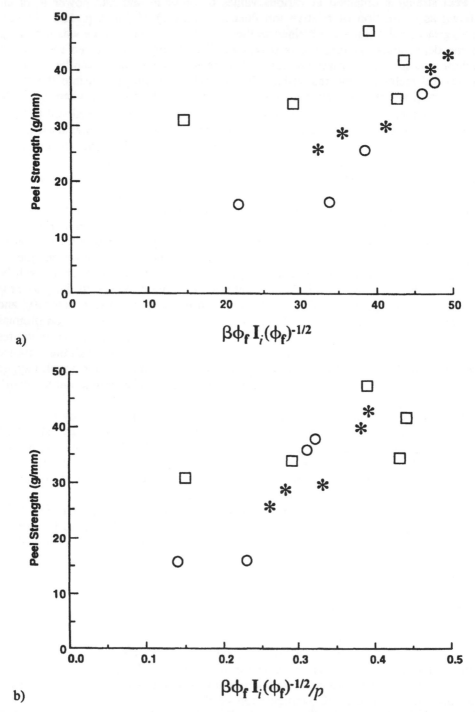

Figure 12. Peel strengths obtained at various values of pressure and DC power level plotted as a function of $\beta\phi_f I_i(\phi_f)^{-1/2}$, (a) and $\beta\phi_f I_i(\phi_f)^{-1/2}p^{-1}$ (b), where β is a proportionality constant. Data are shown for 100 mTorr (□), 125 mTorr (∗), and 150 mTorr (o).

Traces of Fe and Cr (from the stainless steel electrode) were observed on samples treated at slow (less than 50 feet per hour) web speeds, but were not detected at greater speeds. Since good adhesion is obtained at these higher web speeds, it is not likely that this contamination contributes to adhesion enhancement from DC-glow treatment.

3.3. Web heating in the plasma

A correlation was also observed between web temperature during DC-glow treatment and ion fluence, $\beta\phi_f I_i(\phi_f)^{-1/2}p^{-1}$ (Fig. 13). Hence, it is not surprising that changes in web temperature with variation in power and pressure are consistent with trends observed for metal to Kapton adhesion. Figure 14 shows the relationship between adhesion and web temperature during DC-glow treatment at several values of chamber pressure. At a given pressure, greater power results in higher temperatures. Although a general trend exists between web temperature and the resulting adhesion at any given pressure, high temperatures are not necessarily required to induce good adhesion. To verify this, an oxygen plasma was generated on a cathode located opposite the metallizer's chilled drum such that DC-glow treatment was performed with the web in contact with the chilled drum. A power density of 0.47 W/cm^2 was maintained. This is about twice the highest level used for experiments in the

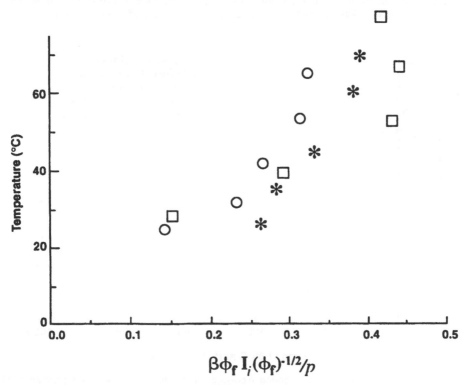

Figure 13. Maximum web temperature during DC-glow treatment as a function of $\beta\phi_f I_i(\phi_f)^{-1/2}p^{-1}$. Web speed was held constant. Symbols are the same as for Fig. 12.

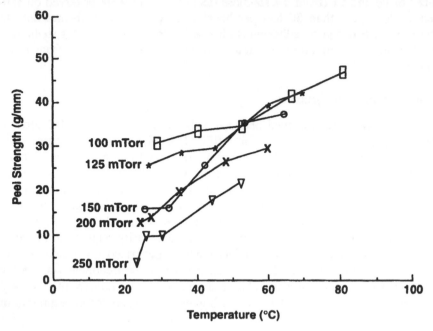

Figure 14. Line adhesion (of chromium on Kapton-H) vs. maximum temperature attained during DC-glow treatment at various pressures. Five power levels were examined at each pressure.

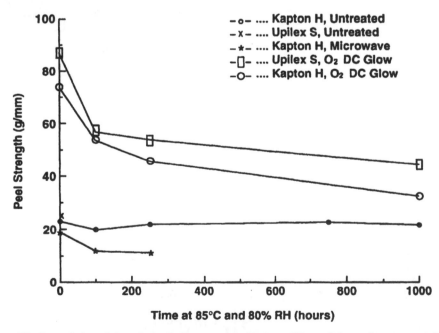

Figure 15. Degradation of chromium adhesion to polyimide at conditions of elevated temperature (85 °C) and humidity (80% RH) as a function of time. Type of polyimide and treatment conditions are shown in the legend.

'off-drum' configuration. Adhesion values of 66 g/mm were obtained with a maximum temperature of only 25°C. Similar power densities and pressures in the off-drum configuration yielded temperatures well over 120°C, with similar adhesion.

Excessive web temperatures induced during DC-glow treatment may have a negative impact on other finished web properties such as dimensional stability and heat creasing. Characterization of material heat creasing during metal deposition is highly subjective due to the inconsistent nature of crease formation. Heat creasing is related to the interaction of many processing factors and inherent web properties. However, under a constant set of deposition conditions with the raw material lot kept constant, trends in film creasing were observed relative to plasma processing conditions. Crease formation was observed to follow temperature excursions in plasma treatment. Higher powers and lower pressures induce more creasing during deposition, unless temperature is controlled by some means of heat sinking (i.e. on a chilled drum). As a consequence, an optimized process may require sacrificing some adhesion to reduce heat creasing.

3.4. Adhesion degradation

Degradation of adhesion at conditions of elevated temperature (85°C) and humidity (80% RH) is shown as a function of time in Fig. 15. Treatments were performed under an optimized set of operating parameters. In all cases, adhesion of treated films degrades after 1000 h to a value about half that measured at time zero. Adhesion for the untreated film, although poor initially, does not degrade appreciably with time. Although not investigated in this study, it is expected that DC-glow parameters (ion dose) could have an influence on this degradation behavior [12].

4. CONCLUSIONS

The most effective surface treatments for promoting metal–polyimide adhesion impart high concentrations of favorable functional groups without degrading the physical and mechanical properties of the polymer's surface region. The suitability of these functional groups depends on the nature (in an acid-base sense) of the deposited material. For deposition of chromium films onto polyimides, carbonyls and carboxylates are favorable. Plasma treatment is an effective means of incorporating these groups. However, concurrent with attachment of the functional groups, chain scission and crosslinking occur during plasma treatment. Crosslinking can serve to improve the cohesive strength of the surface region while excessive chain scission may result in a weak boundary layer and poor practical adhesion. Plasma processing configurations that promote ion bombardment tend to be more prone to crosslinking than those with little or no ion bombardment. Hence, the former processes have been shown to result in improved adhesion, while the latter can lead to degradation of adhesion. Hence, the presence of oxygen and ion bombardment are both important components in promoting metal/polyimide adhesion. Although it is difficult to separate their relative contributions, it appears that oxygen alone, without ion bombardment, is not sufficient to improve this adhesion. These trends have been demonstrated for highly aromatic like polyimides, but might not apply to other polymer structures, for example polyolefins.

Acknowledgements

The authors are grateful to K. Walinskas for assistance in design and construction of the DC cathode, and to W. E. Mlynko for useful discussions.

REFERENCES

1. W. M. Edward, US Pat. #3 179 6419 (1965).
2. L. J. Matienzo, unpublished results (1989).
3. R. D. Weale, L. J. Matienzo, K. J. Blackwell, C. Johnson and S. Deliman, in: *Proc. 36th Annual Tech. Conf. Society of Vacuum Coaters*, p. 242. Dallas, TX, April 25–30 (1993).
4. V. Fronz, in: *Proc. 32rd Annual Tech. Conf. Society of Vacuum Coaters*, p. 161. St. Louis, MO, April 24–28 (1989).
5. J. W. Dini, in: *Proc. Third Internat. Conf. Surface Modification Technol.* Neuchatel, Switzerland, August 28–September 1 (1989).
6. R. H. Hansen, J. V. Pascale, T. De Benedictis and P. M. Rentzepis, *J. Polym. Sci. A* **3**, 2205 (1965).
7. C. A. L. Westerdahl, J. R. Hall, E. C. Schramm and D. W. Levi, *J. Colloid Interface Sci.* **47**, 610 (1974).
8. S. I. Ivanov, D. M. Svirachev, V. P. Pechenyakova, Ch. V. Petrov and E. D. Dobreva, in: *Proc. 8th Internat. Symp. Plasma Chem.*, K. Akashi and A. Kinbara (Eds), p. 1827. International Union of Pure and Applied Chemistry, Tokyo (1987).
9. H. Yanazawa, Y. Suzuki, M. Suzuki, H. Obayashi and N. Hashimoto, in: *Proc. 1983 Dry Process Symp.*, p. 47. The Institute of Electrical Engineers of Japan, pp. 47. Kokutetsu-Rodo Kaikan, Japan (1983).
10. M. Kogoma and G. Turban, *Plasma Chem. Plasma Process* **6**, 349 (1986).
11. M. Kogoma, H. Kasai, K. Takahashi, T. Moriwaki and S. Okazaki, *J. Phys. D: Appl. Phys.* **20**, 147 (1987).
12. D. L. Pappas, J. J. Cuomo and K. G. Sachdev, *J. Vac. Sci. Technol.* **A9** (5), 2704 (1991).
13. F. D. Egitto and L. J. Matienzo, *Polym. Degrad. Stabil.* **30**, 293 (1990).
14. G. A. Takacs, V. Vukanovic, D. Tracy, J. X. Chen, F. D. Egitto, L. J. Matienzo and F. Emmi, *Polym. Degrad. Stabil.* **40**, 73 (1993).
15. E. M. Liston, in: *Proc. 9th Internat. Symp. on Plasma Chem.*, R. d'Agostino (Ed.), International Union of Pure and Applied Chemistry, Pugnochiuso, p. L7. Italy (1989).
16. F. Kokai, H. Saito and T. Fujioka, *J. Appl. Phys.* **66**, 3252 (1989).
17. J. Heitz, E. Arenholz, T. Kefer, D. Bauerle, H. Hibst and A. Hagemeyer, *Appl. Phys.* **A55**, 391 (1992).
18. A. D. Katnani, A. R. Knoll and M. A. Mycek, *J. Adhesion Sci. Technol.* **3** (6), 441 (1989).
19. K. J. Blackwell, F. D. Egitto and A. R. Knoll, in: *Proc. 35th Annual Tech. Conf. Society of Vacuum Coaters*, p. 279. Baltimore, MD (1992).
20. S. Tanigawa, K. Nakamae and T. Matsumoto, *Kobunshi Ronbunshu* **47** (1), 41 (1990).
21. A. Callegari, B. Furman, T. Graham, H. Clearfield, W. Price and S. Purushothaman, *Microelectron. Eng. (Netherlands)* **19**, 575 (1992).
22. K. W. Paik and A. L. Ruoff, *J. Adhesion Sci. Technol.* **4** (6), 465 (1990).
23. F. M. Fowkes, *J. Adhesion Sci. Technol.* **1** (1), 7 (1987).
24. K. L. Mittal and H. R. Anderson, Jr (Eds.), *Acid-Base Interactions: Relevance to Adhesion Science and Technology*. VSP, Zeist, The Netherlands (1991).
25. J. S. Arlow, D. F. Mitchell and M. J. Graham, *J. Vac. Sci. Technol.* **A5**, 572 (1987).
26. J. A. Thornton, in: *Deposition Technologies for Films and Coatings*, R. F. Bunshah (Ed.), p. 173. Noyes Publications, New Jersey (1982).
27. S. R. Cain and L. J. Matienzo, *J. Adhesion Sci. Technol.* **2** (5), 395 (1988).
28. J. M. Burkstrand, *Phys. Rev. B* **20** (12), 4853 (1979).
29. S. F. Tead, W. E. Vanderlinde, G. Marra, A. L. Ruoff, E. J. Kramer and F. D. Egitto, *J. Appl. Phys.* **68** (6), 2972 (1990).
30. R. W. Burger and L. J. Gerenser, in: *Metallized Plastics 3: Fundamental and Applied Aspects*, K. L. Mittal (Ed.), p. 179. Plenum Press, New York (1992).
31. J. W. Chin and J. P. Wightman, *SAMPE Q.* **23** (2), 2 (1992).

32. J. W. Chin, private communication, (1993).
33. L. J. Matienzo and F. D. Egitto, *Polym. Degrad. Stabil.* **35**, 181 (1992).
34. V. Vukanovic, G. A. Takacs, E. A. Matuszak, F. D. Egitto, F. Emmi and R. S. Horwath, *J. Vac. Sci. Technol.* **B6** (1), 66 (1988).
35. J. M. Cook and B. W. Benson, *J. Electrochem. Soc.* **130**, 2459 (1983).
36. J. B. Friedmann, C. Ritter, S. Bisgaard and J. L. Shohet, *J. Vac. Sci. Technol.* **A11** (4), 1145 (1993).
37. F. F. Chen, in: *Plasma Diagnostic Techniques*, R. H. Huddlestone and S. L. Leonard (Eds), p. 150. Academic Press, New York (1965).
38. E. Eser, R. E. Ogilvie and K. A. Taylor, *Thin Solid Films* **68**, 381 (1980).
39. J. W. Coburn and M. Chen, *J. Appl. Phys.* **51**, 3134 (1980).
40. R. E. Walkup, K. L. Saenger and G. S. Selwyn, *J. Chem. Phys.* **84** (5), 2668 (1986).
41. J. F. O'Hanlon, *A User's Guide to Vacuum Technology*, p. 10. John Wiley, New York, NY (1980).
42. K. J. Blackwell, F. D. Egitto and A. R. Knoll, in: *Proc. 33rd Annual Tech. Conf. Society of Vacuum Coaters*, p. 194. New Orleans, LA, April 29–May 4 (1990).
43. L. J. Matienzo, F. Emmi, D. C. VanHart and T. P. Gall, *J. Vac. Sci. Technol.* **A7**, 1784 (1989).

Plasma Surface Modification of Polymers, pp. 255–273
M. Strobel, C. Lyons and K. L. Mittal (Eds)
© VSP 1994

Enhancement of the sticking coefficient of Mg on polypropylene by *in situ* ECR-RF Ar and N₂ plasma treatments

M. COLLAUD,* S. NOWAK, O. M. KÜTTEL and L. SCHLAPBACH

Physics Department, University of Fribourg, Pérolles, 1700 Fribourg, Switzerland

Revised version received 8 October 1993

Abstract—A study of the sticking coefficient of Mg vapour on *in situ* Ar and N₂ plasma-treated polypropylene (PP) is presented. After exposure of the pretreated sample to a determined amount of Mg vapour, X-ray photoelectron spectroscopy (XPS) allows measurement of the adhered Mg amount at the polymer surface and the chemical nature of the interface. The sticking coefficient on an as-received sample is zero and is increased to several tenths depending on the pretreatment conditions, namely the nature and the pressure of the neutral gas, the treatment time, and the applied RF-bias. Relations between the plasma parameters, the XPS measured surface state before the metallization, and the sticking coefficient are investigated.

Keywords: Adhesion; metallization; polymer; sticking coefficient; plasma treatment; XPS; polypropylene.

1. INTRODUCTION

Polymers, particularly polyolefins because of their low cost and their ecologically favourable composition (carbon and hydrogen), are finding increasing use in a number of technological applications. However, their intrinsic adhesion ability is often not sufficient to ensure good adhesion with adhesives or metals, whereas more and more industrial applications require functional coatings. Therefore, a surface pretreatment is required [1]. Among a large number of methods, low-pressure plasma treatments have recently attracted great interest and use as surface pretreatments [2].

Plasmas are composed of electrons and a variety of ions, radicals, and neutral and excited species. They can be characterized by their electron and ion densities, ion flux and energy, and the ultraviolet (UV and VUV) radiation intensities. These 'internal' plasma parameters are determined by the 'external' conditions (ignition mode, pressure, geometry of the plasma chamber, gas composition, etc.). The main advantage of our low-pressure microwave-RF plasma chamber is that the electron cyclotron resonance (ECR) plasma allows us to obtain high densities of charged and excited species at very low pressures (< 0.1 Pa), whereas the RF-power permits us to control the energy of the ions striking the sample surface.

According to the broad literature on this subject [3–6], the efficiency of a plasma treatment to improve adhesion is generally accepted, but not understood. Increasing

---

*To whom correspondence should be addressed.

efforts have been made to understand the microscopic effects of a plasma on a polymer surface [7, 8].

Adhesion can be achieved by mechanical, physical, chemical, or electrical interactions. Plasma treatments can improve the adhesion in adhesive-polymer or metal-polymer systems by increasing one or several of these interactions. Cleaning by ablation of low-molecular-weight species, dehydrogenation and carbonization, chain scission and crosslinking, incorporation of radicals and reactive species at the surface, and structural modifications are all known effects of the plasma on a polymer surface. A number of necessary, but not always sufficient, conditions must be fulfilled to achieve good adhesion. Good wetting of the surface, a high sticking coefficient, homogeneous growth, and strong interactions at the interface are needed to ensure good macroscopic adhesion. This macroscopic adhesion depends not only on the characteristics of the interface, but also on the whole interphase region, namely on the successive layers joining the bulk phase of the substrate to the bulk phase of the overcoat. Therefore, good pretreatments have to ensure suitable surface conditions for adhesion without involving degradation of the near interface layer.

We have studied the variation of the sticking coefficient and the bonding of Mg on isotactic polypropylene as a function of the parameters of an *in situ* plasma treatment. Some interface properties (coverage ratio, chemical interactions, contamination layers, formation of stresses in the film, etc.) are closely related to the sticking coefficient. However, a good sticking coefficient is not a sufficient condition to ensure good interaction at the interface, and therefore to improve the adhesion. Magnesium is not of industrial interest, but the great variation in the thermally evaporated Mg sticking coefficient on hydrocarbon polymers with different pretreatments makes it a good sensor for interfacial investigation.

In our dual frequency ECR plasma chamber, the main adaptable parameters are the nature of the gas, the neutral gas pressure, the treatment time, and the RF-bias. We use XPS analysis to characterize the compositions of the surface and of the interface, and the chemical bonding of the elements. A noble gas (Ar) and a reactive gas (N_2) were used. The Ar plasma allows us to study the effects of the treatment without reactive species, and the N_2 plasma treatment has been found to lead to a better adhesion than Ar or O_2 plasmas [3, 9]. We try to relate the change in the sticking coefficient of the Mg to the plasma parameters (flux, density, energy, etc.) and to the state of the polymer surface before the metallization.

2. EXPERIMENTAL

The substrate investigated was Propaten HF 24 (ICI) isotactic polypropylene (PP, monomer: $CH_2=CH-CH_3$). The base material was mixed and extruded with 0.1% ®Irganox 1330 (Ciba-Geigy), 0.1% ®Irgafos 168 (Ciba-Geigy), 0.05% Ca-stearate, and 0.03% DHT-4A (Kyowa Chemical Industries, $Mg_6Al_2(OH)_{16}CO_3 \cdot 4\ H_2O$, a radical scavenger) at 260°C. After cooling, the PP pellets were pressed onto plates of 0.8 mm thickness in a heated mould at 230°C.

The PP surfaces were treated in a low-pressure plasma created by an ECR process with rare-earth permanent magnets at 2.45 GHz microwave frequency. The high-vacuum plasma chamber was pumped by a Balzers two-stage 180 l/s turbomolecular pump. The base pressure in the chamber was 5×10^{-6} Pa and the treatments were carried out in the pressure range between 2×10^{-2} and 10 Pa. The samples were treated in two different plasma gases, namely argon and nitrogen. The sample can be capacitively coupled to a 13.56 MHz RF-generator, so that a negative DC-bias (V_{RF}) develops on the sample. A Langmuir probe and an ion energy analyser were used to characterize the plasma. Both methods allow the plasma potential V_p and the ion flux to be determined. During the treatment, the floating potential V_f or the biased potential V_{RF} of the sample can be measured. The floating potential has always been measured to be between 0 and -4 V. The maximum bias value depends on the adjustable RF-power and the pressure; it is of the order of -200 to -300 V for RF-power of 20–23 W. The maximum energy of the ions is given by $(V_p - V_f)$ and $(V_p - V_{RF})$ for a floating and a biased sample, respectively. Details of the plasma chamber and of the plasma diagnostics and characteristics have been discussed elsewhere [10]. The plasma chamber was directly connected to the surface spectrometer, which enables *in situ* metallization and surface analysis to be performed, i.e. without any atmospheric contact. The transfer tube was kept at a pressure of 10^{-5} Pa.

The metal was thermally evaporated in the preparation chamber of the spectrometer at a constant rate of ~ 3 monolayers (ML) per minute. Although not of practical relevance for polymer metallization, magnesium was chosen for its great sensitivity to the surface state of the PP. Moreover, photoelectron spectroscopy analysis of Mg is well documented [11–14]. Mg plasmons are a sensitive probe for the chemical state of Mg and of the metal-polymer interface, and the MgKLL Auger spectra allow one to determine without doubt whether the Mg is in a metallic state. The film thickness was measured by a quartz crystal microbalance (QCM). For this study, the samples were exposed to a defined quantity of magnesium (13 ML). The effective amount of metal detected on the surface depends on the pretreatment. Therefore, the Mg concentration measured depends on the probability of Mg adsorption on the surface and is a measure of the sticking coefficient of Mg on the PP samples.

X-ray photoelectron spectroscopy (XPS) was performed using a VG ESCALAB 5 spectrometer at a base pressure of about 10^{-8} Pa, with non-monochromatized Si K_α (1740.0 eV) radiation. The spectra were taken with 20 eV pass energy in the constant analyser mode. The X-ray source was operated at 200 W (10 kV, 20 mA). The surface composition was deduced from the integral of the $1s$ core levels of C, O, N, and Mg after a Shirley background subtraction [15]. The sum of all these integrals corrected for the cross-section represents 100% of the signal detected by XPS. The sample charge was corrected by setting the C $1s$ peak at 285.0 eV [16] or, depending on the thickness of Mg and its chemical state, by setting the metallic state of Mg $1s$ as 1303.3 [11].

The PP samples exposed to X-rays suffer some deterioration. This effect induces a rise of the pressure in the analysis chamber due to desorption of water vapour and hydrogen, and subsequent oxidation of the Mg overlayer [17]. Therefore, the measurement time was kept as short as possible (< 25 min, or < 40 min if the valence band is measured). However, if the Mg concentration on an Ar plasma-treated PP surface is too small ($< 5\%$ of Mg), it is never found in a metallic state.

Repetition of the experiments yielded qualitatively reproducible results, but with significant scatter in the results. In fact, even if the treatment conditions are similar, other experimental factors (cleanliness of the plasma chamber window, amount of reflected microwave power, orientation of the Mg source) produce some variations between the sets of measurements.

3. RESULTS

3.1. Interfacial chemistry

The as-received polypropylene has a zero sticking coefficient towards Mg vapour. A previous study showed that the plasma pretreatments, and particularly the N_2 plasma treatments, improve the sticking coefficient of various metals on the PP surface [9]. Figure 1 shows the C $1s$, N $1s$, O $1s$, and Mg $1s$ peaks for 30 s Ar and N_2 plasma-treated and metallized PP samples. The pressure was 0.03 Pa for the Ar plasma treatment and 0.06 Pa for the N_2 plasma treatment. The peaks for the N_2 plasma-treated surface have been multiplied so that the Mg $1s$ peak has a maximum intensity as high as the Mg $1s$ peak after the Ar plasma treatment. However, the measured Mg atomic concentration is $\sim 75\%$ for both treatments. The C $1s$ peak spectra have been represented after the plasma pretreatment and after the metallization. For each treatment, the C $1s$ peaks have been normalized to allow a comparison between the non-metallized and the metallized sample surfaces.

Almost no oxygen was incorporated, proving that the *in situ* pretreatment and metallization prevent contamination effectively. In the case of the Ar plasma treatment, no nitrogen was found at the surface either. A small amount of nitrogen was incorporated if the sample was treated with the N_2 plasma (17% measured before the metallization and 4% after).

The elastic peak and the first surface and bulk plasmons are represented for the Mg $1s$ spectra. Up to four plasmons can be seen depending on the thickness of the Mg layer. The losses induced by the bulk and the surface plasmons are 10.7 eV and $h\omega_p/\sqrt{2} = 7.55$ eV, respectively [13, 14]. The metallic elastic peak is found at 1303.3 eV. It has been shown in a previous paper [9] that Mg$-$C bonds are formed at the interface between a low-energy Ar ion-bombarded PP and a Mg overlayer. Figure 1 shows that, in the case of an Ar plasma pretreatment, no evidence for Mg$-$C bonds can be found on the high-energy side of the Mg $1s$ peak. The superposition of the Mg $1s$ peaks aftert both plasma pretreatments and metallization suggests that if nitrogen is present at the interface, Mg$-$N bonds are formed. The shift between metallic Mg and Mg$-$N is ~ 1.8 eV, as we will see later. The intensity of the surface plasmon with respect to the bulk plasmon is smaller after the N_2 plasma treatment than after the Ar plasma treatment. It can be explained by a smaller amount of metallic Mg at the interface due to the formation of Mg$-$N bonds. The nitrogen peak is too wide due to the presence of various bonds (C$-$N, C$=$N, Mg$-$N, etc.) and too small in intensity to be of use for a detailed analysis of the Mg$-$N bonds.

After an Ar plasma pretreatment and metallization, no Mg$-$C bonds are visible on the low-energy side of the C $1s$ peak. The fact that the width of the carbon peak does not change with the metallization is also proof of the weak interactions with the Mg overlayer. Secondly, losses are detected on the high-energy side of the C$-$H peak and their energy shifts correspond to the first surface and bulk plasmons of the Mg. These plasmons are excited by the C $1s$ photoelectrons undergoing inelastic scattering as they pass through the Mg overlayer. Such losses in the C $1s$ peak have also been seen after Al deposition on PET [7]. However, the plasmon intensity of the C $1s$ peak does not represent more than 20% of the elastic peak, while the first plasmons of the Mg $1s$ peak represent $\sim 60\%$ of the elastic peak. This observation was taken as an indication that an island growth mode of the

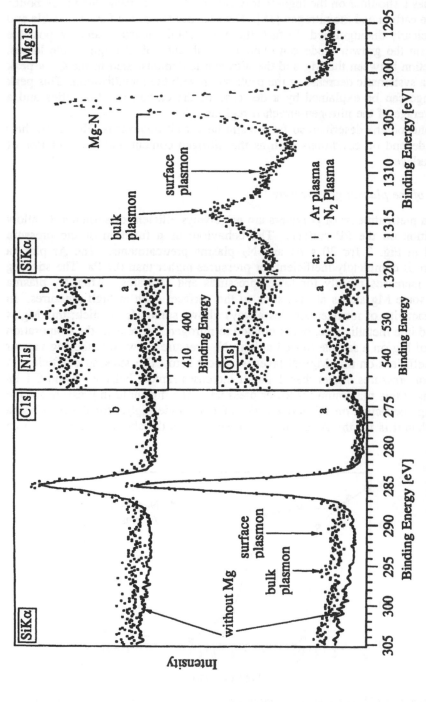

Figure 1. C 1*s*, N 1*s*, O 1*s* and Mg 1*s* core level spectra after a) a 30 s Ar plasma treatment at 0.03 Pa and an exposure equivalent to 13 ML of Mg and b) a 30 s N_2 plasma treatment at 0.06 Pa and an exposure to a dose equivalent to 13 ML of Mg. The normalized C 1*s* peaks before the metallization are also represented. All the peaks have been normalized, so that the maximal intensities of the Mg elastic peaks are the same for both treatments. For both treatments, the measured Mg percentage is ~ 75% of the total measured intensities.

Mg vapour takes place after Ar ion bombardment. The island growth mode has also been found after N_2 plasma treatment [9]. The C $1s$ peak after N_2 plasma treatment has a shoulder on the high-energy side of the C—H peak due to the bonds between the carbon and the nitrogen [18]. The plasmas losses are far less visible and cannot be clearly distinguished. In fact, the non-metallic interface and the possible differences in the growth mode can explain the absence of distinguishable losses. The interaction between the Mg and the nitrogen is demonstrated in the C $1s$ peak spectra by a systematic decrease of the peak width with the metallization. This peak width change can be explained by a decrease of the carbon functionalities and a greater coverage of the nitrogen-enriched regions.

All the phenomena described so far are similar for all treatments. However, their intensities depend on conditions such as the nitrogen content and the thickness of the Mg overlayer.

3.2. Effect of the plasma gas pressure

The plasma pretreatments of polymers are not always efficient or sufficient to allow Mg deposition on the PP surface. The behaviour as a function of the pressure is reported in Fig. 2 for 30 s Ar and N_2 plasma pretreatments. The Ar plasma pretreatment is completely inefficient for pressures higher than 0.1 Pa. The sticking coefficient increases rapidly for lower pressures and saturates. After N_2 plasma treatment, some Mg atoms always stick to the surface, even at high pressures. In fact, the reactivity of nitrogen species always allows some Mg to stick; this Mg is never found in a metallic sate, but is bonded to nitrogen. Moreover, the evaporation of 50 ML of Mg on a sample treated at high N_2 pressure shows that the Mg vapour sticking coefficient on this Mg—N layer is almost zero. The sticking coefficient is very low for pressures higher than 1 Pa. It increases rapidly between 1 and 0.1 Pa and remains at this maximum for lower pressures. The threshold in pressure for the Mg sticking coefficient on PP is one order of magnitude higher for the N_2 plasma treatment than it is for the Ar plasma treatment. These thresholds correspond to an

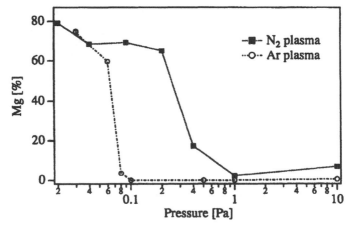

Figure 2. Percentage of Mg as a function of the pressure for 30 s of Ar and N_2 plasma treatments at a floating potential.

ion energy of ~ 15 eV for the Ar plasma treatment and ~ 9 eV for the N_2 plasma treatment. For pressures greater than 0.1 Pa, flux measurements under the same discharge conditions showed a higher flux of ions impinging on the substrate in the case of an Ar plasma than in the case of a N_2 plasma [10, 19]. Therefore, the ion energy and flux are both greater for an Ar plasma than they are for a N_2 plasma, and cannot explain the different limits of the sticking coefficient as a function of the pressure for the two gases. However, the maximum concentration of Mg found after evaporation of a dose equivalent to 13 ML is the same for both plasmas, namely 70–80%.

3.3. Effect of the treatment time

In Fig. 3, the Mg 1s percentage measured on the PP samples is represented for various pressures as a function of the treatment time, which corresponds to the ion dose received by the sample. For the Ar plasma treatment (Fig. 3a), we can see

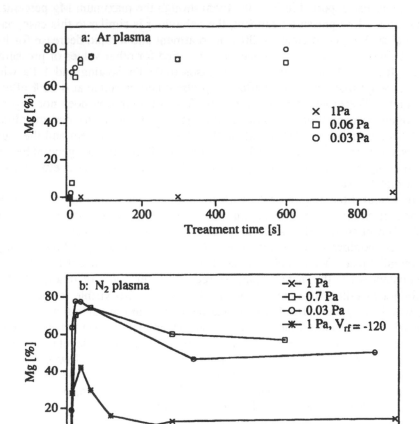

Figure 3. Percentage of Mg as a function of the treatment time for various pressures for (a) Ar plasma treatment and (b) N_2 plasma treatment. All the treatments but one were performed at a floating potential.

that at high pressure (1 Pa) the sticking coefficient of Mg on the PP surface remains zero whatever the treatment time. For pressures lower than the threshold pressure of some tenths Pa, the Mg sticking coefficient saturates after 60 s of treatment time.

In Fig. 3b, the behaviour after N_2 plasma treatment is represented. At 1 Pa and for higher pressures at a floating potential, the sticking coefficient of Mg increases slowly with the treatment time. However, the Mg is always entirely bonded to nitrogen and therefore it is in a non-metallic state. At this pressure, a previous study showed an increase of the nitrogen percentage with the treatment time up to a maximum of ~45% after 15 min [18]. Therefore, at 1 Pa, the Mg concentration increases with the nitrogen concentration.

At 0.03 Pa, the sticking coefficient of Mg reaches an optimum for 30 s of N_2 plasma treatment. For longer treatment times, the percentage of Mg sticking at the surface decreases. A similar dependence is observed for pressures in-between. The sticking coefficient as a function of the treatment time for samples treated at 1 Pa but with an RF-bias is also reported in Fig. 3b. Even though the maximum Mg percentage is half that after a low-pressure treatment, the behaviour is similar to this case, namely an optimal sticking coefficient for 30 s of treatment and a rapid decrease for longer treatment times. This dependence has been verified for other values of pressure and RF-bias. It is worth noting that in both cases (0.03 Pa floating and 1 Pa with an RF-bias), the nitrogen percentage after the pretreatment is stable at ~20% after 15 s of treatment. Therefore, in these cases, the Mg concentration does not depend on the nitrogen concentration. Moreover, for these two cases, the non-metallic part (Mg−N) remains more or less constant after 15–30 s of treatment and the metallic part shows a sharp decrease after 30 s of treatment. Therefore, the general behaviour of the sticking coefficient is closely connected with the sticking coefficient of the Mg in a metallic state.

Figure 4 shows the Mg $1s$ peaks with the first plasmons and the first Mg Auger peaks and plasmons for various N_2 plasma treatment times at 1 Pa with RF-bias of −80 V. The fit of the Mg $1s$ elastic peak is done with a constant width of 2 eV. The Mg−N shoulder shows a shift of ~1.8 eV towards the higher binding energies of the metallic peak. We can clearly see that after 30 s of treatment the Mg−N part stays constant while the metallic part decreases. After 300 s of treatment, the Mg is no longer metallic. The amount of metallic Mg is also visible in the plasmon's shape, which disappears after 120 s of treatment. The non-metallic state of the Mg is also clearly visible in the Auger peaks. In fact, when metallic Mg makes ionic bonds with more electronegative species (O, N, H), the $MgKL_2L_3(^1D)$ level of the metallic Mg shows a sharp decrease in intensity and a peak appears on its low kinetic energy side. For the Mg−N, the shift corresponds to the one between the $MgKL_2L_3(^1D)$ and the $MgKL_2L_2(^1S)$ levels, namely ~5.5 eV [13, 20]. Here, too, the plasmas losses (~565 eV) disappear with the metallic state of Mg.

3.4. Effect of RF-bias

The application of an RF-bias on the sample during the plasma pretreatment improves the sticking coefficient of Mg, as already discussed. In Fig. 5, a systematic study of the dependence of the sticking coefficient on the RF-potential is presented for 30 s of Ar plasma treatment. For the lowest pressure (0.03 Pa), the sticking coefficient

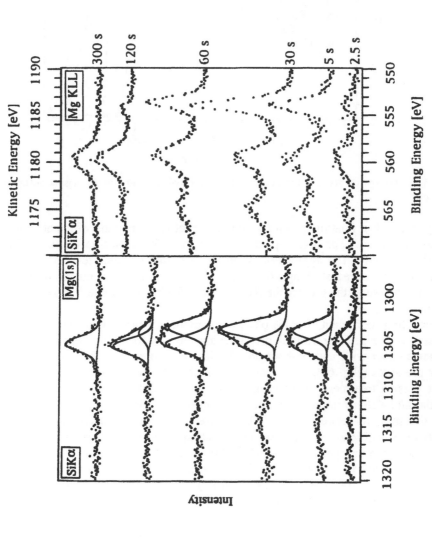

Figure 4. Mg 1*s* peak as a function of the treatment time at a pressure of 1 Pa and an RF-potential of −80 V. The peak positions have been corrected for charging by setting the C−H peak at 285.0 eV.

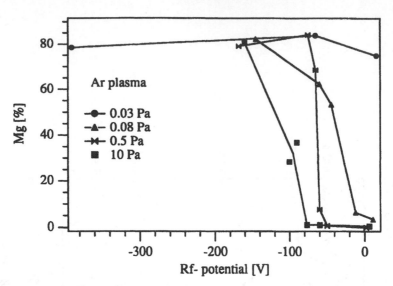

Figure 5. Percentage of Mg as a function of the RF-potential after 30 s of Ar plasma treatments at various pressures.

is very high at a floating potential and the treatment remains efficient when an RF-bias is applied. For higher pressures, the microwave plasma treatments alone become inefficient. The application of an RF-potential of some tens of volts gives a maximum measured sticking coefficient of Mg on PP. As a matter of fact, the greater the pressure, the greater the RF-potential needed to allow a non-zero sticking coefficient.

The sticking coefficient after N_2 plasma treatment as a function of the RF-potential is presented in Fig. 6 for various pressures and 30 s of treatment. Here, too, all the treatments, both with and without an RF-bias, are good at 0.03 Pa (Fig. 6a). At higher pressures (Fig. 6c), the dependence of the Mg percentage is also similar to the Ar plasma treatment, namely an almost zero sticking coefficient at a floating potential and a sharp improvement if an RF-bias is applied.

For pressures in-between (Fig. 6b), a more complicated behaviour is observed. The sticking coefficient is non-zero at a floating potential. However, when a bias is applied to the sample, the amount of Mg staying on the sample decreases. At 0.09 Pa, no more Mg atoms stick on the PP surface if the RF-potential is higher than −100 V. For the other pressures (e.g. 0.04 Pa), the sticking coefficient decreases at potentials between −50 and −150 V and increases for higher RF-potentials.

Figure 7 shows the Mg percentage versus the treatment time at 0.08 Pa for various RF-potentials. For all values of the RF-bias, the sticking coefficient shows a maximum and decreases sharply for longer treatment times. These maxima (circled points in Fig. 7) depend on the RF-bias, as shown in the inset. The lower the Mg sticking coefficient, the shorter the optimal treatment time. Figure 6 presents the RF-bias dependences for 30 s of treatment. These results are also not optimized as a function of the treatment time, i.e. as a function of the ion dose. However, Fig. 7 shows that a correct setting of the ion dose can increase the Mg percentage by a factor of 2 or 3, but in any case it does not allow the maximum sticking coefficient

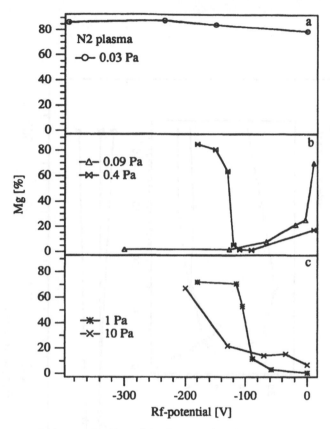

Figure 6. Percentage of Mg as a function of the RF-potential after 30 s of N_2 plasma treatments at (a) 0.03 Pa, (b) 0.09 and 0.4 Pa, and (c) 1 and 10 Pa.

to be reached. Therefore the general behaviour of the RF-bias dependence after N_2 plasma treatment is not influenced by the treatment time.

As has been reported in a previous paper [18], the nitrogen incorporated during the plasma treatment at a floating potential is far greater at higher pressures (45% at 1 Pa) than at low pressures (20% at 0.03 Pa). However, when a bias is applied, the nitrogen content drops to 10–15% for all pressures. Therefore, even if the nitrogen interacts with the Mg overlayer, the sticking coefficient is not dependent on the nitrogen percentage. As for the behaviour as a function of the treatment time, only the metallic part varies with the treatments, while the Mg–N shoulder stays almost constant. Therefore, the sticking coefficient depends on the ability of the Mg to form a metallic layer.

The metallic layer is clearly visible to the eye for both plasma treatments when there is 70–80% of Mg at the surface. It is worth noting that due to electric field inhomogeneities at the edge of the sample, the Mg coverage differs from the centre to the edge of the sample (diameter of 10 mm). The sticking coefficient at the edge corresponds systematically to a lower RF-bias than the applied one. Therefore, the variations seen when an RF-bias is applied depend on the intensity of the electric field at the sample surface.

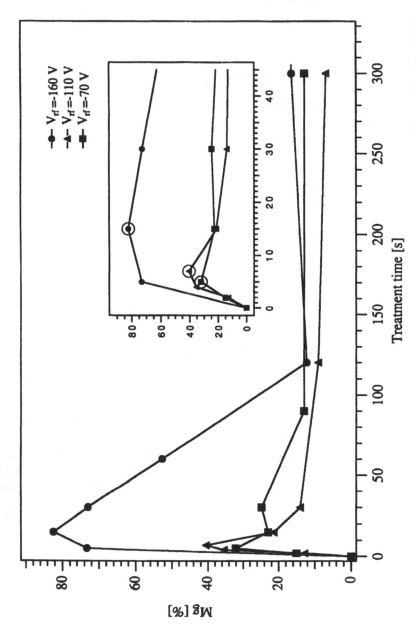

Figure 7. Percentage of Mg as a function of the treatment time for a N_2 plasma-treated sample at 0.08 Pa and various RF-potentials. The inset is an enlargement of the first 50 s of treatment.

3.5. Charge effect during XPS measurements

As PP is an insulator, the sample charges itself when photoelectrons are emitted. This charging results in a shift of the measured core level peaks towards higher binding energies. The C 1s measured binding energy after the pretreatment, i.e. the charge of the sample, decreases with increasing treatment time, increasing bias, and decreasing pressure. The electronic properties of the surface also depend on the ion dose and energy [18]. The Mg percentage as a function of the measured C 1s position before the metallization is reported in Fig. 8a for all the treatments with various treatment times, pressures, and RF-biases for the Ar plasma treatment. It is clear that the charge depends not only on the state of the surface, but also on some parameters of the

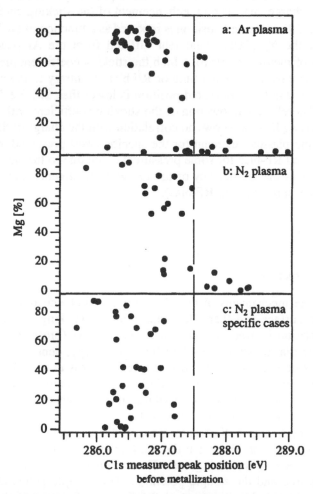

Figure 8. Percentage of Mg as a function of the C 1s measured peak position before the metallization for (a) all the Ar plasma treatments with various treatment times and RF-biases; (b) all the N₂ plasma treatments showing the same behaviour as the Ar plasma treatment, i.e. an increase of the sticking coefficient with increasing ion dose, flux, and energy; and (c) the N₂ plasma treatments showing a decrease of the sticking coefficient with increasing treatment time or RF-bias, namely with increasing ion dose, flux, or energy.

spectrometer (X-ray intensity, secondary electrons, sample position, etc.). Therefore, small variations between the sets of measurements are normal. However, we can see that, in principle, if the treatment leads to a C 1s measured position remaining higher than ~ 287.5 eV (288.3–289 eV for an as-received sample), the sticking coefficient of Mg is almost zero. When the Mg percentage is 70–80%, the C 1s charge before the metallization is found to be between 286 and 287.5 eV.

Figure 8b shows the C 1s measured position as a function of the Mg percentage for the N_2 plasma treatments, which show a similar behaviour with pressure, treatment time, and RF-bias to that of the Ar plasma treatment. For all these cases, we also found a low sticking coefficient for high C 1s measured positions after the pretreatment, and a C 1s position lower than 287.5 eV if the sticking coefficient was good or excellent. Therefore, the same internal plasma parameters induce, first, a lowering of the charge and, second, enhancement of the sticking coefficient.

In Fig. 8c, the C 1s measured position is reported as a function of the Mg percentage for cases when the N_2 plasma treatment differed from the Ar plasma treatment, namely the sets of measurements for which the sticking coefficient presented a sharp decrease with increasing treatment time or RF-bias. Contrary to the cases presented in Figs 8a and 8b, the C 1s measured position is lower than 287.5 eV, whatever the Mg coverage. Therefore, the lowering of the sticking coefficient with long treatment times and increasing RF-bias shows no correlation with the charge of the sample. The inefficiency of the plasma treatment for the specific cases presented in Fig. 8c (long treatment times and RF-bias for some pressures) is not due to the same phenomenon at the PP surface as the inefficiency of the too weak treatments (too short treatment times or high pressures without RF-bias).

4. DISCUSSION

4.1. *Plasma parameters*

To understand the mechanisms of the plasma-induced chemical and physical modifications at the surface and thereafter improving the sticking coefficient, we have to know the relation between the adjustable parameters (pressure, treatment time, RF-bias) and the plasma parameters (ion flux and energy, densities). Moreover, the VUV radiation can also play an important role [2]. Cros *et al.* [21] found that UV irradiation of polyphenylquinoxaline (PPQ) in a gaseous environment led to the appearance of new functionalities but no adhesion improvement. However, we have not done an estimation of this parameter so far.

The effect of the treatment time is to increase the received dose. Langmuir probe and ion flux measurements show that the ion flux and the plasma potential decrease with increasing pressure for both gases. The ion energy is given by the difference between the plasma and the sample potential. This sample potential is floating in the microwave plasma. Measurement of the flux as a function of the RF-bias in a comparable plasma chamber [22] gives non-significant variations for a pressure of 53 Pa and bias values between 0 and −200 V. Therefore, the main effect of the application of an RF-bias is to increase the ion energy. Finally, the total effect of the plasma treatment can be due to the synergy of some of these elements [23, 24].

4.2. Argon plasma treatment

The PP surface after Ar plasma treatment reacts only a little chemically with the Mg overlayer. New chemical species cannot be detected in either the Mg or the carbon peak. The carbon peak width does not change with the metallization. Moreover, the Auger parameter (Mg $1s$ binding energy – MgKLL binding energy) after Ar plasma treatment is 749.2±0.1 eV, whatever the treatment conditions. The Auger parameter of pure Mg has been measured to be 749.4 eV. This proves that almost no chemical interaction takes place at this interface. If there is oxygen contamination, the Auger parameter decreases by 0.4–0.5 eV, which means that a chemical interaction takes place. Due to the fact that no chemical interactions were demonstrated at the interface and that a correlation exists between the sticking coefficient and the charging of the PP under the X-rays, a modification of the electrical properties of the PP surface could be the reason for the improved sticking coefficient.

In order to attain an optimal pretreatment, a minimum treatment time is required. The calculated doses for the 60 s treatment for 0.2 and 0.03 Pa are 6×10^{16} and 1×10^{17} ions/cm^2, respectively. Measurements of the fluxes have been made with a metallic support (Langmuir probe or ion energy analyser). Compared with the real fluxes on an insulator, the difference can be of an order of magnitude. However, the dose always corresponds to more than a monolayer. A previous experiment has shown that for Ar ion bombardment, a minimum dose of 2×10^{14} ions/cm^2 is necessary to ensure a non-zero sticking coefficient; this dose does not depend on the ion energy between 700 and 2700 V [9]. The optimal treatment time for the Ar plasma treatment corresponds to a higher dose. Moreover, the sticking coefficient after treatments at 1 Pa is almost zero, whatever the treatment time; therefore, the efficiency of the plasma does not depend only on the dose. There is also a limit in pressure.

As discussed before, the main effect of the RF-bias is to increase the ion energy. The behaviours as a function of the pressure (Fig. 2) and of the RF-bias (Fig. 5) clearly correspond to a threshold in ion energy. The higher the pressure, the lower the plasma potential. As a consequence, the threshold in RF-potential to allow a non-zero sticking coefficient has to be greater at higher pressures. The threshold in energy cannot, however, be quantitatively estimated from the data, because the potential measurements (V_p, V_f, and V_{RF}) are done on conducting materials (Langmuir probe, sample holder). The actual values at the surface of an insulator can be considerably different. Qualitatively, the sticking coefficient after Ar plasma treatment is optimal when the ion flux and energy are maximal.

4.3. N$_2$ plasma treatment

From the Mg elastic and Auger peaks and from the C $1s$ peaks width, it is clear that Mg forms chemical bonds with nitrogen. Therefore, whatever the treatment, the presence of nitrogen in the surface sample ensures that the sticking coefficient of Mg on a N$_2$ plasma-treated sample is never zero. In cases where the nitrogen content is high (long treatment times at high pressures induce 45% nitrogen incorporation), the concentration of Mg can never exceed some per cents, even after evaporation of 50 ML. The sticking coefficient of Mg vapour on Mg–N also seems to be low. It has already been suggested that the sticking coefficient of Mg on Mg hydride is

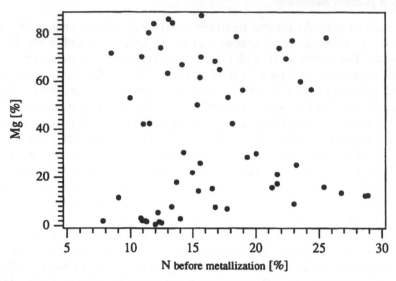

Figure 9. Percentage of Mg as a function of the nitrogen content before the metallization for all the N_2 plasma-treated samples.

almost zero [25]. Therefore, a good sticking coefficient is attained only if a Mg metallic layer can grow on the polymer. Perhaps nitrogen has a specific role in the adhesion phenomenon (multiplication of the nucleation sites for the growing film, stronger chemical interaction, etc.), but the content of nitrogen is, in any case, not a determinant parameter. To illustrate this, the Mg percentage is represented in Fig. 9 as a function of the nitrogen content before the metallization. It is clear from this graph that no correlation exists between only these two measured parameters. This assumption was also shown to be valid by André *et al.* [6].

The dependence of the sticking coefficient on the pressure and on the time and the RF-bias at high pressures (> 1 Pa) after nitrogen plasma treatment is qualitatively similar to that for the Ar plasma: higher ion energy and flux induce a higher maximum sticking coefficient.

At high pressures, the nitrogen concentration reaches its maximal percentage after a longer treatment time than at low pressures. For 0.04 Pa, 20% of nitrogen is incorporated in 15–30 s. At this pressure, longer treatments lead to a similar nitrogen concentration. Treatments with an RF-bias also induce a nitrogen incorporation of 20% after 15–30 s of treatment, whatever the pressure. Depth measurements have shown that the nitrogen is first incorporated deeper into the polymer and afterwards in the surface [18]. The maximum of the sticking coefficient corresponds approximately to the shortest treatment time where the nitrogen concentration is maximal. For longer treatments, the metallic part of the Mg decreases sharply and the sticking coefficient becomes very low. Strobel *et al.* [26] have shown that air corona induces a low-molecular-weight oxidized material at the surface of PP; this layer is soluble in a polar liquid. In the same way, Burger and Gerenser [27] explained the sharp decrease in adhesion of PE after O_2 plasma overtreatments (glow discharge at 0.01–0.06 Pa) by the fact that long treatment times (> 60 s) produced more chain scissioning of the polymer, a large amount of low-molecular-weight species, and, therefore, a new weak boundary layer. However, Hall *et al.* [28] found that the bond strength (PSI) towards epoxy resin of various polymers, including HDPE, LDPE, PP, PET, and PS, did not decrease with long H_2 and O_2 plasma treatment times (up to 30–50 min). In our case, the Ar or N_2 plasma-treated PP surfaces led to quiet different behaviours towards the sticking coefficient with overtreatment. The sticking coefficient itself should depend on the physical and chemical state of the surface, but is not directly influenced by a weak boundary layer. However, the behaviour of overtreated polymers towards both the sticking coefficient and the adhesion should depend not only on the substrate and the treatment time, but also on the treatment conditions, the neutral gas of the plasma, and the overlayer material.

The extinction of the sticking coefficient as a function of the RF-bias for certain pressures (0.09–0.4 Pa) cannot be explained by the nitrogen concentration before the metallization. Taking into account all the parameters that we can investigate with XPS (chemical composition, carbon and nitrogen chemical state, valence band, peak position, etc.), no difference between samples treated at a high RF-bias at different pressures can be detected on the surface before the metallization.

4.4. Estimation of the sticking coefficient

By evaporating 13 ML of Mg and by measuring the amount of Mg sticking on the surface of a PP sample after N_2 plasma treatment, a PP sample after Ar ion bombardment, and a copper sample, we tried to estimate the sticking coefficient on PP. We assumed that the film grows layer by layer and forms a homogeneous overlayer. With this approximation, the ratio of the intensities of the substrate (s) and the film (f) is given by

$$\frac{I_f}{I_s} = \frac{n_f s_f}{n_s s_s} \cdot \left(\frac{E_f}{E_s}\right)^{0.35} e^{d/\lambda_s} \cdot \left(1 - e^{d/\lambda_f}\right),$$

where I_f is the intensity of the film, I_s is the intensity of the substrate, n is the surface density, σ is the photoionization cross-section, E is the kinetic energy, d is the film thickness, λ is the mean free path, and the power 0.35 is given by the spectrometer specifications. The carbon mean free path has been taken as the substrate mean free path. The density of polypropylene is 0.9 g/cm^3. With this model, the thickness d of the Mg layer on the N_2 plasma-treated PP is a third of that on copper. Considering that Mg on metals has a sticking coefficient of 1 and taking into account the errors in this estimation (island growth mode of the Mg on PP, mean free path of electrons in PP, variation between the bulk and the surface densities, etc.), the plasma-treated samples have a sticking coefficient of the order of 0.3 ± 0.2. Comparison of the C 1s intensities before and after the metallization gives a coverage ratio of 45% at least, taking into account that the 45% of covered surface leads to a C 1s peak intensity shielded by the Mg overlayer (the thickness is given by the Mg overlayer on the Cu sample). The Ar ion-bombarded samples have a lower sticking coefficient towards the Mg vapour. Therefore the plasma pretreatments seem to be more efficient than Ar ion bombardment and the Ar and N_2 plasmas lead to the same maximum sticking coefficient.

5. CONCLUSION

Both Ar and N_2 plasma treatments enhance the sticking coefficient of Mg on the PP surface, but some limit conditions are required. The Ar plasma treatment needs 60 s to be at maximum efficiency and it has an efficiency threshold in pressure. The treatment has to be performed at pressures lower than 0.1 Pa to allow a non-zero sticking coefficient. At higher pressures, the application of an RF-bias increases the sticking coefficient from zero to the maximum measured on PP. This proves that there is a threshold in ion energy for the sticking coefficient. The incorporation of nitrogen during N_2 plasma treatment induces the formation of Mg–N bonds at the interface. Therefore, whenever the treatment is not efficient, the nitrogen acts as a reactive species and some Mg is always found at the surface, but in a non-metallic state. The behaviour as a function of pressure is similar to the Ar plasma case, namely the occurrence of a threshold at 1 Pa. For low pressures (0.04 Pa) or with an RF-bias, the sticking coefficient has a maximum for 5–30 s, and decreases for longer treatment times. There is no correlation between the nitrogen concentration at the surface before the metallization and the sticking coefficient. In short, treatment times between 10 and 60 s, low pressures (< 0.1 Pa), and the application of an RF-bias are all generally favourable factors for improving the sticking coefficient.

Acknowledgements

We thank H. P. Haerri (Ciba-Geigy, Fribourg, Switzerland) for providing us with the polymer samples. This research project was supported by the Swiss National Science Foundation, NFP 24, and Ciba-Geigy.

REFERENCES

1. D. M. Brewis and D. Briggs, *Polymer* **22**, 7–16 (1980).
2. E. M. Liston, *J. Adhesion* **30**, 199–217 (1989).
3. L. J. Gerenser, *J. Vac. Sci. Technol.* **A6**, 2897–2903 (1988).
4. L. J. Gerenser, *J. Vac. Sci. Technol.* **A8**, 3682–3691 (1990).
5. P. Bödö and J.-E. Sundgren, *Thin Solid Films* **136**, 147–159 (1986).
6. V. André, F. Arefi, J. Amouroux, Y. De Puydt, P. Bertrand, G. Lorang and M. Delamar, *Thin Solid Films* **181**, 451–460 (1989).
7. M. Bou, J. M. Martin and Th. LeMogue, *Appl. Surface Sci.* **47**, 149–161 (1991).
8. S. Akhter, X. L. Zhou and J. M. White, *Appl. Surface Sci.* **37**, 201–216 (1989).
9. S. Nowak, M. Collaud, G. Dietler, P. Gröning and L. Schlapbach, *J. Vac. Sci. Technol.* **A11**, 481–489 (1993).
10. S. Nowak, P. Gröning, O. M. Küttel, M. Collaud and G. Dietler, *J. Vac. Sci. Technol.* **A10**, 3419–3425 (1992).
11. X. D. Peng, D. S. Edwards and M. A. Barteau, *Surface Sci.* **195**, 103–114 (1988).
12. J. C. Fuggle, *Surface Sci.* **69**, 581–608 (1977).
13. L. Ley, F. R. McFeely, S. P. Kowalczyk, J. G. Jenkin and D. A. Shirley, *Phys. Rev.* **B11**, 600–612 (1975).
14. P. Steiner, H. Höchst and S. Hüfner, *Z. Phys.* **B30**, 129–143 (1978).
15. D. Briggs and M. D. Seah (Eds), *Practical Surface Analysis*. John Wiley, New York (1983).
16. D. T. Clark and A. Dilks, *J. Polym. Sci. Polym. Chem. Ed.* **17**, 957 (1979).
17. S. Nowak, H. P. Haerri, L. Schlapbach and J. Vogt, *Surface Interface Anal.* **16**, 418–423 (1990).
18. M. Collaud, S. Nowak, O. M. Küttel, P. Gröning and L. Schlapbach, *Appl. Surface Sci.* **72**, 19–29 (1993).
19. P. Gröning, S. Nowak and L. Schlapbach, *Appl. Surface Sci.* **64**, 265–273 (1993).
20. X. D. Peng, D. S. Edwards and M. A. Barteau, *Surface Sci.* **185**, 227–248 (1987).

21. A. Cros, H. Dallaporta, S. Lazare, F. Templier, J. Nechstchein, J. Palleau, H. Hiraoka and J. Torres, in: *Metallized Plastics 3: Fundamental and Applied Aspects*, K. L. Mittal (Ed.), pp. 201–213. Plenum Press, New York (1992).

22. O. M. Küttel, J. E. Klemberg-Sapieha, L. Martinu and M. R. Wertheimer, *Thin Solid Films* **193/194**, 155–163 (1990).

23. S. Veprek, *Plasma Chem. Plasma Process.* **9**, 29s–54s (1989).

24. P. Gomes de Lima, J. Lopez, B. Despax and C. Mayoux, *Rev. Phys. Appl.* **24**, 331–335 (1989).

25. A. Fischer, A. Krozer and L. Schlapbach, *Surface Sci.* **269/270**, 737–742 (1992).

26. M. Strobel, Ch. Dunatov, J. M. Strobel, C. S. Lyons, S. J. Perron and M. C. Morgen, *J. Adhesion Sci. Technol.* **3**, 321–335 (1989).

27. R. W. Burger and L. J. Gerenser, in: *Metallized Plastics 3: Fundamental and Applied Aspects*, K. L. Mittal (Ed.), pp. 179–193. Plenum Press, New York (1992).

28. J. R. Hall, C. A. L. Westerdahl, M. J. Bodnar and D. W. Levi, *J. Appl. Polym. Sci.* **16**, 1465–1477 (1972).

Plasma Surface Modification of Polymers, pp. 275–290
M. Strobel, C. Lyons and K. L. Mittal (Eds)
© VSP 1994

Improved adhesion between plasma-treated polyimide film and evaporated copper

N. INAGAKI,* S. TASAKA and K. HIBI

Laboratory of Polymer Chemistry, Faculty of Engineering, Shizuoka University,
3-5-1 Johoku, Hamamatsu, 432 Japan

Revised version received 3 January 1994

Abstract—The surface of polyimide film, Kapton® H, was modified with Ar, N_2, NO, NO_2, O_2, CO, and CO_2 plasmas to improve its adhesion to evaporated copper. The plasma treatments led to bond scission of the imide groups in the Kapton film to form carboxyl and secondary amide groups, and, as a result, the surface of the Kapton film changed from hydrophobic to hydrophilic. The Ar-, NO-, and NO_2-plasma treatments enhanced the adhesion between the Kapton film and copper, but the O_2-, CO-, and CO_2-plasma treatments did not. The roughness profile determined with an atomic force microscope showed that the plasma treatment removed a surface layer of the Kapton film, and the surface contained needle-shaped protuberances. The ATR IR spectra for the copper side torn-off from the Kapton film/copper joints showed the formation of coordinate bonds between carboxyl groups and copper atoms at the interface of the Kapton film and evaporated copper. The improved adhesion may be due to the formation of coordinate bonds between carboxyl groups and copper atoms, and the mechanical interlocking by penetration of the copper layer into the deep valleys between the protuberances.

Keywords: Polyimide film; Kapton; plasma treatment; adhesion; contact angle; surface energy; XPS; ATR IR spectra; roughness profile; coordination; anchor effect; copper.

1. INTRODUCTION

Poly[(N, N′-oxydiphenylene) pyromellitimide], polyimide, which is sold under the trade name Kapton® H by E. I. du Pont De Nemours & Co., is a super-performance polymeric material with high-temperature resistance, good mechanical properties, flame resistance, dimensional stability, and a low dielectric constant. Kapton is frequently used as an insulating and pattern-delineating material in microelectronics technology. In such applications, the adhesion between the polyimide film and the wiring copper metal becomes a topic of theoretical as well as practical interest.

Some surface modifications of polyimide films for the enhancement of adhesion between the polyimide film and copper metal have been actively studied by many investigators. The surface-modification techniques are complex formation of chromium or titanium with the carbonyl groups in polyimide [1–6], sputtering of the polyimide surface [7, 8], ion-beam etching [9], photo-reactions with UV or excimer-laser irradiation [10, 11] and implantation of $^{28}Si^+$ and $^{84}Kr^+$ ions [12]. The generally accepted mechanisms or theories of adhesion are (1) the mechanical interlocking

---

*To whom correspondence should be addressed.

theory, which is frequently called the 'anchor effect'; (2) theories based on surface energy, wetting, and adsorption; (3) the diffusion theory; (4) the electronic or electrostatic theory; (5) chemical bonding; and (6) the week boundary layer mechanism [13].

Chou and Tang [1] investigated the interfacial reactions occurring during metallization of the polyimide film surface. From the viewpoint of thermodynamics involving the oxidation reactions of metals at the metal/polyimide interface, the weak adhesion of copper and the strong adhesion of chromium were discussed. They concluded that the adhesion was due to the formation of metal-oxygen-carbon complexes at the metal/polyimide interface, and that a high concentration of oxidized metals at the metal/polyimide interface led to strong adhesion. On the other hand, Fowkes [14] emphasized that the formation of strong acid-base interactions at the metal/polymer interfaces was a good way to obtain strong adhesion between metals and polymers. These investigations point out the importance of special functionalities that facilitate interactions with metal atoms at the metal/polyimide interface.

In a chemical sense, plasma treatment is a radical-substitution reaction of the C—H bonds in polymers. Hydrogen abstraction by the collision of electrons, ions, or radicals leads to carbon radicals in the polymer chains, and then the carbon radicals react with simple radicals such as oxygen and nitrogen in the plasma to yield oxygen and nitrogen functionalities in the polymer chains. Therefore, hydrogen abstraction from polymer chains is the origin of the plasma treatment reactions, and facile hydrogen abstraction should lead to a successful plasma treatment. From the viewpoint of hydrogen abstraction, aromatic polymers are not favorable substrates for plasma treatment because hydrogen abstraction from aromatic carbons is not as easy as that from aliphatic carbons. In this study, we focus our interest on the functionalities that are generated at the polyimide surface by plasma treatment, and discuss how the plasma treatment improves the adhesion between the polyimide film and copper metal.

2. EXPERIMENTAL

2.1. Material

The polyimide film used in this study was poly[(N, N′-oxydiphenylene) pyromellitimide] (Kapton® H), kindly provided by Toray Du Pont Co., and was 508 mm wide and 25 μm thick. The low-density polyethylene film obtained from Idemitsu Petrochemical Co., Japan was 180 mm wide and 250 μm thick, and its density was 0.92 g/cm^3. These films were washed with acetone in an ultrasonic washer prior to the plasma treatment experiments.

2.2. Plasma treatment

The Kapton films, cut in a sample size of 25 mm wide × 120 mm long or 10 mm wide × 120 mm long and mounted on a sample frame, were plasma-treated using a home-made reactor which was a capacitively coupled system at a frequency of 20 kHz. It consisted of a bell-jar (400 mm diameter, 470 mm height) with a gas inlet, a pair of parallel flat-plate electrodes (150 × 150 mm) with a gap between the electrodes of 100 mm, a substrate stage which lay midway (separation of 50 mm from

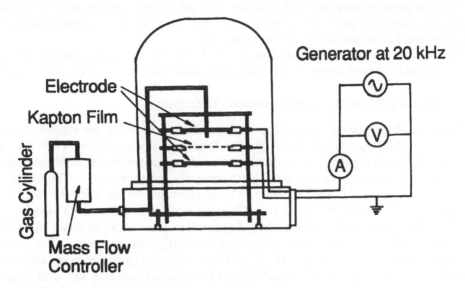

Figure 1. Schematic diagram of the plasma treatment chamber.

the upper electrode) between the electrodes, a pressure gauge, and a vacuum system consisting of a combination of a diffusion (550 l/s) and a rotary pump (320 l/min). The details of the reaction system have been reported elsewhere [15]. A schematic diagram of the reaction chamber is shown in Fig. 1.

The experimental procedures for the plasma treatment were essentially the same as those reported in [15]. The reaction chamber was evacuated to approximately 0.13 Pa, and then the given gas, Ar, N_2, O_2, CO, CO_2, NO, or NO_2, whose flow rate was adjusted to 10 cm^3 (STP)/min with a mass flow controller (piezo-valve type, STEC Co. Japan, model SEC-421 MKII), was introduced into the reaction chamber. The plasma was controlled by three parameters: a discharge current of 150 mA at a 20 kHz frequency, a system pressure of 13.3 Pa, and a treatment time of 3 min, because the kinetic energy of the electrons (electron temperature) and the concentration of electrons in the plasma are closely related to the discharge current and the pressure rather than to the electric power. The electric potential between the electrodes was about 700 V at the initiation of the discharge, but after initiation it decreased and reached a constant value of about 600 V 1 min after initiation. Therefore, the apparent input electric power is 90–105 W.

2.3. Contact angle of water and surface energy

The advancing contact angles in air of water, glycerol, formamide, diiodomethane, and tricresyl phosphate on the plasma-treated Kapton film surfaces were measured at 20°C using an Erma contact-angle meter with a goniometer, model G-1. Water for the test liquid was distilled twice in the presence of potassium permanganate. Glycerol, formamide, diiodomethane, and tricresyl phosphate were of analytical grade and were used as test liquids without further purification. A drop (1 μl) of these liquids pushed out from a microsyringe was transferred to the plasma-treated Kapton surface, and the height (h) and the width (w) of the drop on the Kapton surface were

measured using a telescope with a scale. The advancing contact angle (θ) of the
drop was estimated from the equation $\theta/2 = \tan^{-1}(2h/w)$. The reported advancing
contact angle is an average of ten measurements.

The surface energy of the plasma-treated Kapton film surface was estimated from
the advancing contact angles of the five liquids according to Kaelble's method [16].

2.4. Infrared (IR) and X-ray photoelectron (XPS) spectra

IR spectra of the plasma-treated Kapton films were recorded in the attenuated total
reflection (ATR) mode on a Horiba Fourier Transform Spectrometer FT300 with an
MCT detector. A crystal of germanium was used as a prism for the ATR technique,
and the incident angle of infrared light against the prism was 45°. The sampling
depth in the range of 2000 to 400 cm^{-1} is estimated to be 0.66–3.3 μm. The
spectral resolution was 1.0 cm^{-1} and 500 scans were recorded on each sample.

The XPS spectra of the surface of the Kapton films were obtained on an Ulvac-Phi
Spectrometer 5300 using a non-monochromatic AlK$_\alpha$ photon source. The anode
voltage was 15 kV, the wattage 400 W, and the background pressure in the analytical
chamber 1×10^{-7} Pa. The take-off angle of photoelectrons was 45° with respect
to the sample surface. The spectra were referenced with respect to the 285.0 eV
C_{1s} level observed for hydrocarbon. The C_{1s} spectra were deconvoluted by fitting
Gaussian functions to an experimental curve using a non-linear, least-squares curve-
fitting program supplied by Ulvac-Phi. A flood gun was used to eliminate the
charge effect. The sensitivity factors (S) for the core levels were $S(C_{1s}) = 1.00$,
$S(N_{1s}) = 1.61$, and $S(O_{1s}) = 2.40$.

2.5. Peel strength of joints between the Kapton film and evaporated copper metal

Kapton films (sample size 10 mm wide $\times$ 120 mm long) were treated with Ar, N_2,
O_2, CO, CO_2, NO, and NO_2 plasmas, and then removed from the plasma reactor.
The plasma-treated films were transferred to a vacuum evaporation reactor, and on
the plasma-treated Kapton film surface, copper metal (99.9% purity, purchased from
Nilaco Co., Japan) of 200 nm thickness was deposited by the vacuum evaporation
technique operated at a pressure of 1.3×10^{-3} Pa. The time to air exposure during the
transfer of the treated Kapton film from the plasma reactor to the vacuum evaporation
reactor was less than 5 min. Afterwards, the copper-deposited Kapton films were
adhered to sandblasted aluminum plates using an epoxy adhesive (Japan Ablestik Co.,
Japan). The epoxy adhesive was cured under a pressure of 5 kg/cm^2 at a temperature
of 95°C for 70 min. Therefore, the joint is the sandwich structure of plasma-treated
Kapton film/Cu film (200 nm thick)/epoxy adhesive/sandblasted aluminum plate,
and such joints were used as specimens for the peel-strength measurement.

The 180° peel strength of the joints between the Kapton film and the copper metal
was evaluated at a peel rate of 50 mm/min with an Instron. The peel strength was
determined from an average of ten specimens.

2.6. Surface morphology of the Kapton film

Kapton films treated with Ar and O_2 plasmas operated at a discharge current of
150 mA, at a 20 kHz frequency and at a system pressure of 13.3 Pa for 3 min were

used to examine the surface morphology. The surface morphology of the plasma-treated Kapton films was scanned with a Topo Metrix atomic force microscope (AFM) model TMX-2000 with a probe tip of 1 μm length, an aspect ratio (the ratio of the square of the tip length and the horizontal cross-sectional area) of 10:1, and a cusp of 10 nm (the curvature radius). The resolution in the $x - y$ direction was 0.1 nm, and in the z direction, 0.05 nm.

The roughness profile determined with the AFM was analyzed using a computer, and three parameters, R_a, Z_{max}, and Z_{av}, which are frequently used in the science of topology, were estimated for the mathematical description of the roughness profile. R_a is an arithmetic mean of departures of the roughness profile from the mean line. When the roughness profile is written as a function of $Z(x)$, as in Fig. 2, R_a is defined by the following equation:

$$R_a = \frac{1}{L} \int_0^L |Z(x)| \, dx, \tag{1}$$

where L is the sampling length and $Z(x)$ is the roughness profile as a function of the position in the x direction. Therefore, R_a represents the cross-section of the roughness profile per unit length, expressed in nm^2/nm. Z_{max} (in nm) is the maximum height of the roughness profile, and Z_{av} (in nm) is the mean height of the roughness profile. The five highest peaks (P_1, P_2, P_3, P_4, and P_5) and the five deepest valleys (V_1, V_2, V_3, V_4, and V_5) in the roughness profile (see Fig. 2) are chosen to define Z_{av} by the equation

$$Z_{av} = \frac{1}{5} \sum_{i=1}^{5} P_i - \frac{1}{5} \sum_{i=1}^{5} V_i. \tag{2}$$

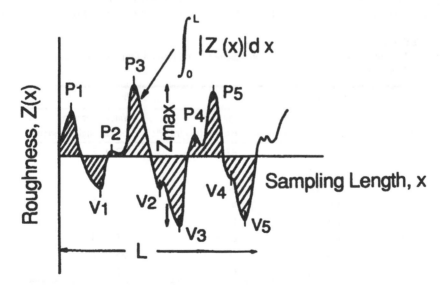

Figure 2. Definition of the parameters Z_{max} and R_a.

3. RESULTS AND DISCUSSION

3.1. Hydrophilic surface modification of Kapton films by plasma treatment

Although we have already reported [15] that Ar-, N_2-, O_2-, CO-, CO_2-, NO-, and NO_2-plasma treatments have the capability to make the surfaces of Kapton films hydrophilic, the surface modification with the plasmas is briefly described here. Typical results for the plasma treatments are shown in Table 1. The untreated Kapton film surface is hydrophobic and the advancing contact angle of water is 74.7°. The surface properties of the Kapton films are changed from hydrophobic to hydrophilic by the plasma treatment, and, as a result, the advancing contact angle of water is reduced from 74.7° to 7.6–11.6°. Regarding changes in the water advancing contact angle caused by the plasma treatment, Kapton film is distinguished from polyolefin films such as polyethylene. The plasma-treated Kapton films, irrespective of the gas used, show almost the same water contact angle. However, the contact angles of the plasma-treated polyethylene films show a strong dependence on the type of plasma gas [17]. As shown in Table 1, when polyethylene films were exposed to O_2, CO, CO_2, NO, and NO_2 plasmas under the same conditions as those used for the Kapton treatment, the water advancing contact angles were 35°, 16°, 8°, 25°, and 37°, respectively [17]. It is reasonable that the type of plasma gas has a strong effect on the efficiency of the hydrophilic surface modification because hydrophilic groups generated at the polymer surface by the plasma treatment are closely related to the active species that are generated in the plasma. Kapton films, when treated with an Ar, N_2, O_2, CO, CO_2, NO, or NO_2 plasma, show almost the same advancing contact angle of water of 7.6–11.6°. The lack of a dependence of the contact angle of water on the type of plasma treatment is a characteristic of the plasma-treated Kapton films.

Generally, there are two factors that lead to contact angle changes: a chemical factor and a physical factor. The chemical factor relates to changes in the chemical composition of the polymer surface caused by the plasma treatment. The physical factor involves changes in the surface geometry such as surface roughness. Increases

Table 1.

Advancing contact angles of water on plasma-treated Kapton films and plasma-treated polyethylene (PE) films

	Kapton film		PE film
Plasma	Water advancing contact angle (degrees)	Surface energy $(mJ/m^2)^a$	Water advancing contact angle (degrees)
None	74.7 ± 0.3	37 ± 3	102 ± 2
Ar	11.4 ± 0.9	67 ± 3	–
N_2	11.6 ± 0.5	65 ± 3	–
O_2	9.6 ± 0.5	67 ± 3	35 ± 1
CO	10.5 ± 0.7	67 ± 3	16 ± 0.8
CO_2	8.4 ± 0.7	67 ± 3	8 ± 0.8
NO	7.6 ± 0.9	69 ± 3	25 ± 1
NO_2	8.0 ± 0.8	69 ± 3	37 ± 1

^{a} Calculated according to the method described in [16].

in the surface roughness lead to changes in the contact angle of water. The magnitude of the change depends on whether the intrinsic, microscopic contact angle on the surface is greater or less than 90°. If the rough surface has an intrinsic, microscopic angle of less than 90°, a drop of water will penetrate into valleys between protuberances at the surface. Consequently, an increase in the surface roughness contributes to a decrease in the measured contact angle of water. However, if the surface has an intrinsic, microscopic contact angle greater than 90°, the increase of the surface roughness will make the observed contact angle of water larger. In addition to these two factors, the degradation products formed by the plasma treatment can also affect the contact angle of water. When a polymer surface is exposed to a plasma, degradation reactions occur at the polymer surface to produce degradation products with low molecular weights. A large portion of the degradation products is evaporated from the polymer surface because of the very low pressures in the reaction zone (about 10 Pa). If some of the degradation products still remain at the polymer surface after the plasma treatment process and these degradation products are soluble in water, then the degradation products will dissolve in any water drop that makes contact with the polymer surface during the contact angle measurement and, therefore, the surface tension of the water may be lowered. Consequently, the polymer surface displays a lower contact angle of water than a polymer surface without soluble materials would show.

Based on these concepts, there are three possible factors involved in interpreting why the plasma-treated Kapton film surfaces display a lower contact angle of water than the untreated Kapton film surface. These three factors are surface chemistry, surface roughness, and the contamination of the surface by water-soluble degradation products. When water-soluble degradation products are responsible for decreasing the contact angle of water, the contact angle of water may not be a constant value but a time-dependent value because of the dissolution of the degradation products. We observed no change in the contact angle of water of any plasma-treated Kapton film surface over a period of a few minutes after trickling down a water drop at the film surface. Therefore, we suggest that the main contribution to the decrease of the contact angle is surface chemistry and that some modification reactions occurred at the Kapton film surface to alter the chemical composition of the surface. This conclusion does not exclude minor contributions from roughness and contamination factors. Furthermore, the chemical reactions occurring in the plasma treatment are quite different on the Kapton and polyethylene films, because the Kapton films show an independence of the contact angle of water on the plasma treatment conditions, while the contact angle is dependent on the plasma conditions for the polyethylene films. The chemical reactions will be discussed later.

The XPS analyses show that the plasma treatments lead to a large increase in the O/C atomic ratio but a smaller increase in the N/C atomic ratio (Table 2). The O/C atomic ratio increases from 0.20 for the untreated Kapton film to 0.38–0.46 for the plasma-treated Kapton films. On the other hand, the N/C atomic ratio is 0.07 for the untreated Kapton film and 0.05–0.08 for the treated samples. The emphasizes that the plasma treatment, irrespective of the type of plasma gas, creates oxygen functionalities rather than nitrogen functionalities at the Kapton surface. Typical C_{1s} spectra for the O_2- and Ar-plasma-treated Kapton films and untreated Kapton film are shown in Fig. 3. The C_{1s} spectrum for the untreated Kapton film was deconvoluted into three components, which appear at 285.4 (CH), 286.3 (C−O and C−N),

Table 2.
The elemental composition of plasma-treated Kapton films

Plasma	Elemental composition (XPS atomic ratio)	
	O/C	N/C
None	0.20	0.07
Ar	0.39	0.05
N_2	0.38	0.07
O_2	0.42	0.06
CO	0.38	0.04
CO_2	0.47	0.06
NO	0.43	0.07
NO_2	0.46	0.08

The error in the estimation for the atomic ratios is less than 0.01.

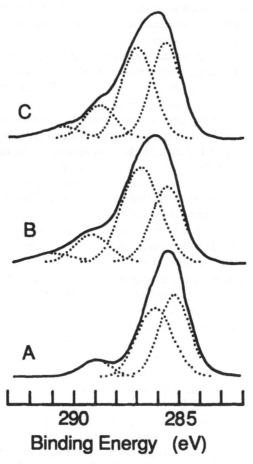

Figure 3. XPS (C_{1s}) spectra for an untreated Kapton film (A), an Ar-plasma-treated Kapton film (B), and an O_2-plasma-treated Kapton film (C).

and 289.0 eV (C=O). On the other hand, the O_2- and Ar-plasma-treated Kapton films possess four deconvoluted C_{1s} components at 285.3–285.7 (CH), 286.6–286.9 (C−O and C−N), 288.8–289.1 (C=O), and 290.0–290.7 eV (COOH and COOR). The new component appearing at 290.0–290.7 eV is a product generated by the plasma treatments and could be assigned to carboxyl and ester groups based on the binding energies reported in [18]. The new C_{1s} component is 5–8% of the total carbon in the plasma-treated Kapton films. Whether the new C_{1s} component is due to a carboxyl or an ester group could be decided from the IR data.

Although the C=O stretching vibration in both carboxyl and ester groups appears near 1720 cm^{-1}, when these groups are treated with sodium hydroxide, the carboxyl (COOH) groups will be neutralized and show an absorption shift to a low wavenumber (1560 cm^{-1}) [19] while the ester groups will show no shift. This neutralization can distinguish the carboxyl groups from the ester groups. Figure 4 shows the difference IR spectrum (spectrum A) between a NaOH-treated Kapton film and a non-NaOH-treated Kapton film, and also the difference IR spectrum (spectrum B) between an Ar-plasma-treated Kapton film and an untreated one. The plasma-treated Kapton films were immersed in 0.5 N NaOH aqueous solution at room temperature for 10 min. We confirmed that the imide groups of the Kapton film were never transformed to amide and carboxylate groups under the conditions of the NaOH treatment. We observe that a strong absorption peak appears at 1720 cm^{-1} in the difference

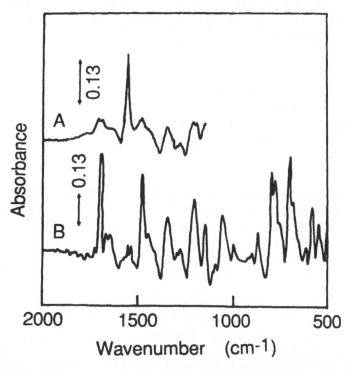

Figure 4. ATR IR difference spectra between an Ar-plasma-treated and NaOH-treated Kapton film and an Ar-plasma-treated but not NaOH-treated (A), and between an Ar-plasma-treated Kapton film and an untreated Kapton film (B). The arrow in the figure indicates the scale of apparent absorbance intensity in arbitrary, dimensionless units.

spectrum B. In the difference spectrum A, the absorption peak at 1720 cm^{-1} becomes weaker in intensity as a result of the neutralization with NaOH solution and a new absorption peak appears at 1560 cm^{-1}. These spectral changes suggest that the strong absorption peak at 1720 cm^{-1} in spectrum B should be assigned to carboxyl (COOH) groups and that the absorption peak at 1560 cm^{-1} in spectrum A should be assigned to carboxylate (COO$^-$) ion. Furthermore, spectrum B also shows characteristic absorption peaks at 1680 and 1655 cm^{-1} due to the CO—NHR-group, and at 1545 cm^{-1} due to N—H deformation in a secondary amide group [19]. Therefore, it is reasonable to state that carboxyl groups and secondary amide groups are formed by the plasma treatment. We did not observe the appearance or disappearance of IR absorption peaks, except the absorption peaks related to C=O and amide groups. In particular, the IR absorption peaks due to the C—H deformation vibrations in aromatic rings which were expected in the wavenumber range from 2000 to 1660 cm^{-1} are unchanged even after the plasma treatment (spectrum B). This indicates that no addition or substitution reaction occurred in the aromatic rings of the Kapton films.

A possible reaction process that could form the carboxyl and the secondary amide groups in the plasma-treated Kapton film is a combination of the cleavage of imide groups and the subsequent reaction with water:

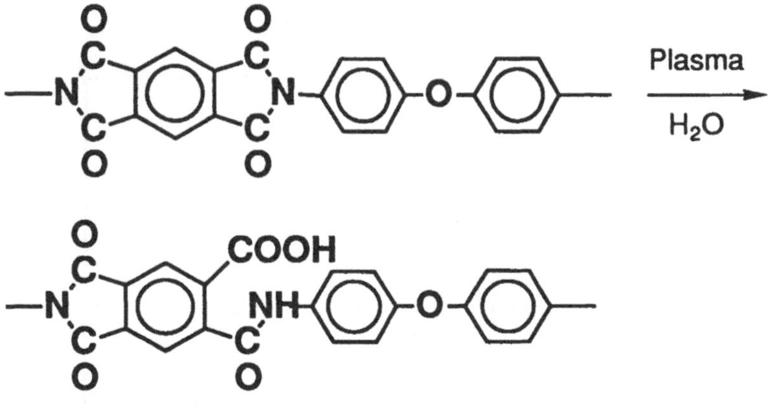

This suggests why all of the plasma-treated Kapton film surfaces show the same advancing contact angle of water. Generally, interactions between polymer surfaces and a plasma lead to the following possibilities: implantation reactions which form functional groups such as hydroxyl, carbonyl, carboxyl, and amino groups; etching reactions; creation of radicals in the middle of polymer chains or at the chain ends; and crosslinking reactions. These interactions, in a chemical sense, could be represented as (1) radical formation in the polymer chains, (2) degradation of polymers from the radical sites, (3) combination of the polymer radicals with simple radicals (implantation reactions), and (4) combination of two polymer radicals (crosslinking reactions). Therefore, the formation of radicals in the polymer chains is the origin of the chemical reactions occurring during the plasma treatment. The easy formation of radicals in the polymer chains would lead to a successful plasma treatment, especially for implantation reactions. Hydrogen atoms in Kapton are different in chemistry from those in aliphatic polymers such as polyethylene and polypropylene. Hydrogen atoms bonded to aromatic carbons are difficult to abstract by radicals, but hydrogen atoms bonded to aliphatic carbons are easily abstracted. The difficulty in hydrogen abstraction from the Kapton film would make the reaction course of the plasma interactions switch from hydrogen abstraction to the bond scission of imide groups. As a result, the implantation reactions scarcely occurred, and the imide groups in the Kapton film were predominantly broken down into carboxyl and secondary amide groups. This assumption is supported by the elemental composition of the plasma-treated Kapton film surface. When Kapton films are treated with N_2 and O_2 plasmas, as shown in Table 2, the two surfaces show the same elemental composition. The N/C and O/C atomic ratios for the N_2-plasma-treated Kapton film are 0.07 and 0.38, respectively, and those for the O_2-plasma-treated Kapton film are 0.06 and 0.42, respectively. The implantation of nitrogen atoms scarcely occurs in the N_2-plasma treatment, while oxygen atoms are minimally incorporated by the O_2-plasma treatment.

From these results, we conclude that the plasma treatment causes bond scission of imide rings in Kapton films and forms carboxyl and secondary amide groups. These new functional groups contribute to the decrease in the advancing contact angle of water.

3.2. Adhesion between the Kapton film and the evaporated copper layer

We consider that carboxyl groups, which are formed at the plasma-treated Kapton film surface, could contribute to the adhesion between the Kapton film and evaporated copper metal because of the possible formation of coordination bonds between the carboxyl groups and the copper metal. Table 3 shows the 180° peel strength of the adhesion joints between the plasma-treated Kapton films and the evaporated copper metal. The specimens for the adhesion measurement were the sandwich structure of the Kapton film/copper film/epoxy adhesive/aluminum sheet. In all the peel experiments, failure never occurred between the copper film and the epoxy adhesive or between the epoxy adhesive and the sandblasted aluminum sheet. We observed that the failure, in many cases, occurred cohesively in a layer of the Kapton films near the interface rather than at the interface between Kapton and copper because the characteristic absorption peaks of Kapton were observed in the ATR IR spectrum of the copper-film side torn-off from the joints. The spectrum will be discussed in a later

Table 3.

The peel strength of the joints consisting of plasma-treated Kapton film/evaporated Cu layer/epoxy adhesive/sandblasted aluminum plate

Modification of Kapton film	180° peel strength (N/10 mm)
Untreated	0.73 ± 0.06
Ar-plasma treatment	2.9 ± 0.09
N_2-plasma treatment	2.2 ± 0.09
NO-plasma treatment	2.2 ± 0.01
NO_2-plasma treatment	2.1 ± 0.08
CO-plasma treatment	1.2 ± 0.06
CO_2-plasma treatment	1.2 ± 0.01
O_2-plasma treatment	1.1 ± 0.16

section. We believe that the failure does not occur uniformly but in a mixed mode, i.e. cohesive failure and interfacial failure occur simultaneously. The 180° peel strength, as shown in Table 3, is enhanced by the plasma treatment. The peel strength for the joint of untreated Kapton film/Cu film (200 nm thick)/epoxy adhesive/sandblasted aluminum plate is 0.73 N/10 mm, while the peel strength for the plasma-treated Kapton film is 1.1–2.9 N/10 mm. The Ar-, N_2-, NO-, and NO_2-plasma treatments are effective in improving the adhesion between the Kapton and copper films. In particular, the Ar-plasma treatment causes a large increase in the peel strength from 0.73 to 2.9 N/10 mm. The CO-, CO_2-, and O_2-plasma treatments are not as effective in improving the peel strength.

The plasma treatment of Kapton films, as described in Section 3.1, is a sort of degradation reaction. A portion of the imide rings at the Kapton film surface is cleaved into carboxyl and amide groups by the plasma exposure. Furthermore, bond scission of the formed amide groups, especially in the case of heavy plasma exposure, will occur. This bond scission induces the formation of fragments of low molecular weight (a weak boundary layer) at the Kapton film surface. Therefore, we believe that mild plasma exposure rather than heavy plasma exposure enhances adhesion better. The surface morphology of the plasma-treated Kapton film surface gives us information on the degradation occurring at the Kapton film surface. The surfaces of the Ar- and O_2-plasma-treated Kapton films and the untreated film were inspected with an atomic force microscope (AFM). Figure 5 shows AFM pictures and Fig. 6 shows the profile of the surface roughness determined by AFM. The Ar-plasma-treated Kapton film gave good adhesion with the copper film (see Table 3). The O_2-plasma-treated Kapton films showed smaller advancing water contact angles than the Ar-plasma-treated films, but poor adhesion. The two plasma-treated Kapton films and even the untreated film have rough surfaces with needle-shaped protuberances. The AFM picture and the profile of surface roughness show changes in surface roughness caused by the plasma exposure. To evaluate the effects of the plasma treatment on the surface roughness, the three parameters R_a, Z_{max}, and Z_{av} for the two plasma-treated Kapton films and the untreated film are listed in Table 4. The surface roughness is related to the type of plasma gas. The Z_{max} and Z_{av} values for the Ar-plasma-treated Kapton film (22.89 and 10.08 nm) are much larger than those for the untreated Kapton film (13.37 and 7.39 nm), while the Z_{max} and Z_{av} values for the O_2-plasma-treated Kapton film (16.06 and 7.39 nm) are as low as those for the untreated film. However, the width of the protuberances (290–470 nm) for the O_2-plasma-treated Kapton film, as shown in Fig. 6, is greater than the width of the protuberances for the untreated films (180–300 nm wide), whereas the protuberance width (50–120 nm) for the Ar-plasma-treated Kapton film is narrower than that for the untreated films (180–300 nm). This means

Table 4.

The surface roughness of plasma-treated Kapton films

Plasma treatment	R_a (nm^2/nm)	Z_{max} (nm)	Z_{av} (nm)
Ar plasma	1.63	22.89	10.08
O_2 plasma	1.37	16.06	7.39
None	1.11	13.37	7.39

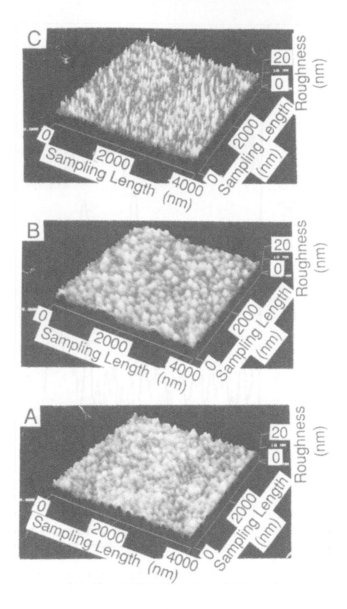

Figure 5. AFM photographs for a Kapton film surface (A), an O_2-plasma-treated Kapton film surface (B), and an Ar-plasma-treated Kapton film surface (C).

that the plasma treatment caused changes in the shape of the protuberances on the Kapton films: in the Ar-plasma treatment the protuberances became fine in width and increased in height, while in the O_2-plasma treatment the size of the protuberances increased. Therefore, there is little doubt that in both the Ar-plasma and the O_2-plasma-treatment some degradation occurs at the surface layer of the Kapton films. We believe that the degradation of the Kapton film in the O_2-plasma treatment may be more intense than it is in the Ar-plasma treatment because an increase

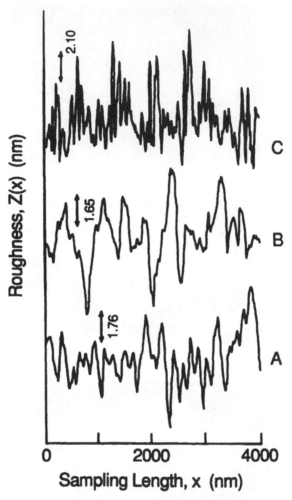

Figure 6. Surface roughness for a Kapton film (A), an O_2-plasma-treated Kapton film (B), and an Ar-plasma-treated Kapton film (C). The arrow in the figure indicates the dimension of roughness in nm.

in the width of a protuberance (occurring in the O_2-plasma treatment) is never possible without degradation of the whole film surface (protuberances + valleys), whereas the thinning of a protuberance (occurring in the Ar-plasma treatment) is possible by the degradation of only a restricted surface area (protuberances). As a result, the O_2-plasma-treated Kapton film surface contains a weak boundary layer while the Ar-plasma-treated film surface does not. From the above discussion, we could propose a possible contribution of surface topology to the adhesion improvement by the plasma treatment. This contribution may be (1) the anchor effect, which means mechanical interlocking by the penetration of the copper layer into the deep valleys of the rough Kapton surface which improves adhesion, and (2) the weak boundary layer (bond scission of imide rings) formed by the plasma treatment, which hampers adhesion. In addition to this contribution, the interaction (coordination) between carboxyl groups formed at the plasma-treated Kapton film surface and copper, as discussed in Section 3.3, can also contribute to adhesion. Which of the three factors is responsible for the adhesion between the plasma-treated Kapton film and the copper film cannot be concluded at the present time. A weak boundary layer may be the reason why the O_2-plasma-treated Kapton films showed a poorer adhesion, despite a smaller advancing water contact angle than that of the Ar-plasma-treated film.

3.3. ATR IR spectra of the failure surfaces

ATR IR spectra for the copper-film side torn-off from the Kapton/copper joints were recorded. Figure 7 shows the difference spectrum between the copper-film side torn-off from the joint of the Ar-plasma-treated Kapton film/copper film and the copper-film side torn-off from the untreated Kapton film/copper film joint. In the difference spectrum shown in Fig. 7, many absorption peaks appear in the ranges of 1738–1680 cm^{-1}, 1650–1615 cm^{-1}, 1558–1512 cm^{-1}, and at 1462, 1455, and 1395 cm^{-1}. The absorptions in the range of 1738–1689 cm^{-1} are related to the stretching vibrations of carbonyl groups [19]; the absorptions in the range of 1650–1615 cm^{-1} are related to the $C = O$ stretching vibrations in ionized carboxyl groups and $CO-NH$ groups [19]; and the absorptions in the range of 1558–1512 cm^{-1} are related to the $C = C$ skeletal in-plane vibrations of aromatic residues and the $N-H$ deformation in secondary amide groups [19].

Ionized carboxyl groups such as copper acetate show characteristic strong absorption peaks at 1617 and 1450 cm^{-1}, which are assigned to the anti-symmetrical and symmetrical vibrations of the COO^- structure, respectively [19]. So, the absorption peaks at 1615 and 1455 cm^{-1} that appear in the ATR IR difference spectrum for the copper-film side torn-off from the joint (Fig. 7) could be assigned to carboxyl groups coordinated with copper atoms. We conclude that coordination bonds between carboxyl groups and copper atoms are formed at the interface between the

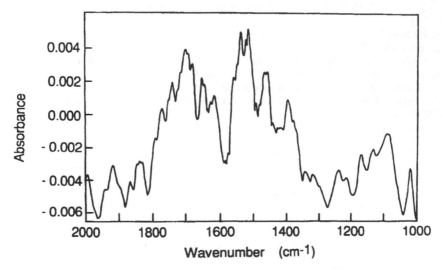

Figure 7. ATR IR difference spectrum between the copper-film side torn-off from the Ar-plasma-treated Kapton film/evaporated copper adhesion joint and the copper-film side torn-off from the untreated Kapton film/evaporated copper adhesion joint.

Ar-plasma-treated Kapton film substrate and the copper film. Such coordination bonds could contribute to the enhancement of the adhesion between the Kapton film and the evaporated copper layer.

4. CONCLUSION

The surface of polyimide film, Kapton® H, was modified with Ar, N_2, NO, NO_2, O_2, CO, and CO_2 plasmas and the adhesion between the plasma-treated Kapton film and the evaporated copper layer was investigated. The results are summarized as follows:

(1) The plasma treatments led to bond scission of imide groups in the Kapton film to form carboxyl and secondary amide groups, and, as a result, the Kapton film surface was converted from hydrophobic to hydrophilic.

(2) All the plasma treatments led to improved adhesion between the Kapton film and evaporated copper. The Ar-, NO-, and NO_2-plasma treatments were more effective than the O_2-, CO-, and CO_2-plasma treatments.

(3) The ATR IR spectra for the copper side torn-off from the Kapton film/copper joints showed the formation of coordination bonds between carboxyl groups and copper atoms at the interface of the plasma-treated Kapton film and the evaporated copper layer, which could lead to improved adhesion.

REFERENCES

1. N. J. Chou and C. H. Tang, *J. Vac. Sci. Technol.* **A2**, 751–755 (1984).
2. N. J. Chou, D. W. Dong, J. Kim and A. C. Liu, *J. Electrochem. Soc.* **131**, 2335–2340 (1984).
3. R. Haight, R. C. White, B. D. Silverman and P. S. Ho, *J. Vac. Sci. Technol.* **A6**, 2188–2199 (1988).
4. Y.-H. Kim, J. Kim, G. F. Walker, C. Feger and S. P. Kowalczyk, *J. Adhesion Sci. Technol.* **2**, 95–105 (1988).
5. F. Faupel, C. H. Yang, S. T. Chen and P. S. Ho, *J. Appl. Phys.* **65**, 1911–1917 (1989).
6. D. G. Kim, S. E. Molis, T. S. Oh, S. P. Kowalczyk and J. Kim, *J. Adhesion Sci. Technol.* **5**, 509–521 (1991).
7. N. J. Chou, J. Paraszczak, E. Babich, J. Heidenreich, Y. S. Chaug and R. D. Goldblatt, *J. Vac. Sci. Technol.* **A5**, 1321–1326 (1987).
8. T. S. Oh, S. P. Kowalczyk, D. J. Hunt and J. Kim, *J. Adhesion Sci. Technol.* **4**, 119–129 (1990).
9. K. W. Paik and A. Ruoff, *J. Adhesion Sci. Technol.* **4**, 465–474 (1990).
10. H. Hiraoka and S. Lazare, *Appl. Surface Sci.* **46**, 264–271 (1990).
11. G. Ulmer, B. Hasselberger, H.-G. Busmann and E. E. B. Campbell, *Appl. Surface Sci.* **46**, 272–278 (1990).
12. A. A. Galuska, *J. Vac. Sci. Technol.* **B8**, 470–481; **B8**, 482–487 (1990).
13. L. P. Buchwalter, *J. Adhesion Sci. Technol.* **4**, 687–721 (1990).
14. F. M. Fowkes, *J. Adhesion Sci. Technol.* **1**, 7–27 (1987).
15. N. Inagaki, S. Tasaka and K. Hibi, *J. Polym. Sci., Polym. Chem. Ed.* **30**, 1425 (1992).
16. D. H. Kaelble, *Physical Chemistry of Adhesion.* John Wiley, New York (1971).
17. N. Inagaki, S. Tasaka and K. Hibi, *J. Polym. Sci., Polym. Chem. Ed.* **30**, 1425–1431 (1992).
18. A. Dilks and D. T. Clark, *J. Polym. Sci., Polym. Chem. Ed.* **19**, 2847–2860 (1981).
19. L. J. Bellay, *The Infrared Spectra of Complex Molecules.* John Wiley, New York (1966).